A Short History of Science

ILLUSTRATED &

PUBLISHED BY

E-KİTAP PROJESİ & CHEAPEST BOOKS

www.cheapestboooks.com

Istanbul

CHEAPEST BOOKS

Your Books, Your Way...

978-625-6629-30-1

Isaac Newton

A SHORT HISTORY OF SCIENCE

BY
W. T. SEDGWICK AND H. W. TYLER
REVISED BY
H. W. TYLER AND
R. P. BIGELOW

The history of science is the real history of mankind.
—DU BOIS-REYMOND.

Those who try to understand the deeper meaning of science itself and its connections with other subjects of human thought and activity must know something of the story of its development.—DAMPIER.

Preface

The present book, while in part a revision of Sedgwick and Tyler's *Short History of Science* (Macmillan, 1917), is to a great extent new. Like the earlier work, it is the outgrowth of a course of lectures given for a number of years to undergraduates at the Massachusetts Institute of Technology.

Doctor Sedgwick having died some years after the publication of the first edition, the surviving author undertook the preparation of a revised edition and invited R. P. Bigelow, who had taken part in the lecture course, to share the editorial responsibility in coöperation with colleagues versed in other fields. The intervening years have heightened the appreciation of the enormous difficulty of the undertaking. On the other hand, the increased interest in the subject, as exemplified in college orientation courses, and the absence of a textbook of similar scope and aim have seemed to justify a second edition. The editors have deliberately abstained from any attempt to bring the history up to date in such matters as the new mathematical physics and the advances in the chemical and biological sciences characteristic of the twentieth century, since the available literature on these topics is abundant, and it may be doubted if they are not still too close to our generation for a just historical perspective.

In this edition, however, the attempt is made to trace briefly the history of the foundations upon which recent, as well as earlier, advances were based; to correlate the steps of progress with the spirit of the time; and to increase the emphasis on the evolution of scientific methods. It is hoped thus to give an impression of continuity in the history of science.

In comparison with the original volume, the space devoted here to mathematical science has been considerably reduced by the omission of proofs and relatively technical material, as well as of quotations. This has made possible more adequate treatment of certain other fields. At the same time it will be evident that no attempt at completeness is intended. The selection aims to be typical, and the large space devoted to mathematics and related subjects corresponds, not merely with the author's personal interests, but with the more extended development of these ancient themes. The extent to which quotations from various authors are included is due to the desire to bring some of the general literature of the subject to the attention of students having insufficient time for wide reading.

A serious difficulty arose from the diverse spelling of many proper names. Excepting Greek names, it was decided to follow, as a rule, the usage in Sarton's indispensable *Introduction to the History of Science*, so far as it goes. While the Greek form is much to be preferred, the Latin form of Greek names was retained in order to avoid a large number of changes in the text, and to avoid inconsistencies with quoted passages, in which the Latin form is used almost invariably.

The present edition has been prepared with the invaluable assistance of various persons competent in their respective fields. Substantial contributions have been made by Professors W. T. Hall and H. W. Shimer, both of the Institute of Technology. R. P. Bigelow in dealing with non-mathematical subjects has rewritten entirely or in part several chapters. Especial credit is due Professor Henry Crew for numerous amendments proposed after reading the first edition and to Professor Dirk J. Struik, who read the completed manuscript and made suggestions that add greatly to the value of the book. Some of the illustrations have been drawn by Mrs. Elizabeth Tyler Wolcott, and all have been arranged under her direction. In this connection, it is unfortunate that space prohibited greater use of the portraits of mathematicians and other historical material placed freely at her disposal by Professor David Eugene Smith.

Grateful acknowledgment is due to the following publishers for permission to quote from works listed under their names:

The Clarendon Press, Oxford — *Aristotle's Works in English Translation*, edited by W. D. Ross.

Harvard University Press, by permission of the President and Fellows of Harvard College — *Vitruvius, the Ten Books on Architecture*, translated by M. H. Morgan; three volumes of the Loeb Classical Library, viz.: *Aristophanes*, by B. B. Rogers, *Herodotus*, by A. D. Godley, and *Lucretius*, by W. H. D. Rouse.

Houghton Mifflin Company — *The Old Humanities and the New Science*, by Sir William Osler.

The Macmillan Company—*History of Science*, by Sir William Dampier; *Lectures on the History of Physiology*, by Sir Michael Foster; *Founders of Geology*, by Sir Archibald Geikie; *Short History of Greek Mathematics*, by James Gow; *The Copernicus of Antiquity*, by Sir Thomas L. Heath.

The Oxford University Press, London, Sir Humphrey Milford — *Science and Civilization*, edited by F. S. Marvin.

The University of California Press — *Sir Isaac Newton's Mathematical Principles of Natural Philosophy*, translation edited and revised by Florian Cajori.

The graceful Tyrolean quadrant that ornaments the title-page is from the collection of Professor Smith and the frontispiece is from the same source, obtained by the help of Mr. F. E. Brasch, formerly Secretary of the History of Science Society, with the permission of Ginn and Company and the McGraw-Hill Book Company. Thanks are due, also, to all whose permission or coöperation has allowed the use of illustrations as noted in each case in the text.

In the preparation of this work Doctor Tyler's connection with the Library of Congress has been especially advantageous. Doctor Bigelow has likewise made extensive use of the libraries of the Institute of Technology, Harvard University, the Boston Medical Library, Boston Society of Natural History, American Academy of Arts and Sciences, and the Marine Biological Laboratory at Woods Hole. The editors have been much indebted to the members of the staff in these libraries for many acts of kindness, and to many other friends without whose encouragement this work would not have been undertaken. May it contribute in some measure to a wider interest in a boundless field of knowledge, the record of intellectual conquests so much more significant and durable than those of either war or politics.

The last chapter was completed and the greater part of this preface had been written when a sudden heart failure abruptly brought the work of Doctor Tyler to an end. Thus left without the wise guidance of a cherished friend, the junior editor has had the unexpected and onerous task of preparing the book for publication. He, therefore, must accept responsibility for any errors or omissions that may appear on the printed page.

ROBERT PAYNE BIGELOW

Massachusetts Institute of Technology
September, 1939

Table of Contents

Illustrations

A Short History
of Science

Note: The map on the opposite page shows where civilization began in the Fertile Crescent and Nile Valley, and the location of some of the principal towns. Drawn by Elizabeth Tyler Wolcott.

Chapter I

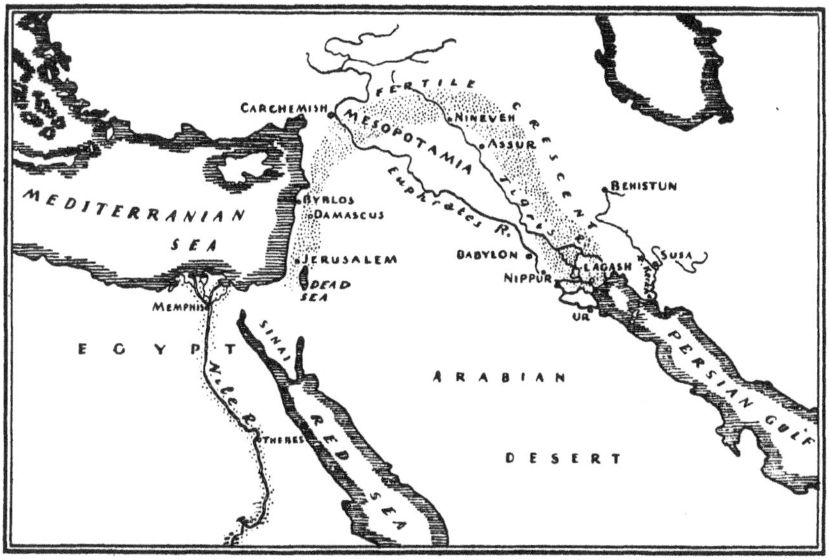

The Dawn of Civilization

We shall never get back to the beginning of things, never completely disperse the obscurity that enshrouds them. — JACQUES DE MORGAN, *Prehistoric Man*, p. 268.

In science, as in all other departments of human knowledge and inquiry, no thorough grasp of a subject can be gained, unless the history of its development is clearly appreciated.

If all history is only an amplification of biography, the history of science may be most instructively read in the life and work of the men by whom the realms of Nature have been successively won. — ARCHIBALD GEIKIE, *Founders of Geology*, pp. 1, 3.

THE ANTIQUITY AND ANCESTRY OF MAN

The history of human culture, in which the history of science is an important chapter, reveals at first a very slow growth with roots in the remote past. In his various biological aspects man shows evidence of descent from an ancestor related to the great apes. Many facts suggest a vast area in south central Asia north of the Himalayan mountains as the place where the human stem arose. The time when our ancestors became really human probably could not be stated definitely, even if all the circumstances were known, for the change must have been a very gradual one. However, it certainly was completed before the beginning of the Pleistocene. That geological epoch, following the Pliocene and preceding our own Recent Epoch, was distinguished by extraordinary cooling of the earth. Four times great ice sheets spread southward over lands of the northern hemisphere, and four times they retreated. During each of these Ice Ages, distinctive mammals appeared, some of gigantic proportions, and their skeletons, buried by dust storms or in the sediments of the swollen streams of the warm interglacial ages, enable geologists to recognize deposits laid down in any one age. On other evidence, geologists estimate the length of these ages in years; and the whole epoch is believed by American authorities to have lasted a million years ending about twenty-five thousand years ago.

Very early in the Pleistocene primitive men were living in widely separated localities, probably migrants escaping competition with more progressive races at home. The most primitive of these is the Trinil man (Pithecanthropus) of Java. He was very ape-like, but recent discoveries (1937) have shown anatomical features that are distinctively human. There is, however, no evidence of distinctively human behavior. It is different with Peking man (Sinanthropus), who inhabited caves in eastern China at about the same time. He had a larger brain, and he made tools and fire — activities as distinctively human as articulate speech. When he learned to kindle a fire from the sparks that flew as he chipped flints to make his crude implements, he made the first application of a physical principle to human needs. Perhaps earlier in time, but with more modern features, the Piltdown man (Eoanthropus) was established in southeastern England in the late Pliocene or earliest Pleistocene.

A somewhat later type, of Mid-Pleistocene age, the Neanderthal, pursuing the great beasts, overran Europe during the second interglacial period. Around their camp fires they made the first completely flaked flint implement, the hand-ax — a tool characteristic of the Old Stone Age, or Palæolithic. They, in turn, gave way during the last Ice Age, perhaps 150,000 years ago, to modern man, *Homo sapiens*, represented by the Brünn and Crô-Magnon races. The latter left in numerous cave dwellings implements of flint and bone, and drawings and sculptures, showing fine powers of observation and great manual dexterity.

With the milder climate following the last retreat of the ice came a long Middle Stone Age (Mesolithic), of several thousand years, when men made peculiar small flint implements. This was followed by the New Stone Age (Neolithic) with its polished stone axes and pottery. Somewhat later came the Metal Ages — (Copper, Bronze, and Iron). These stages of culture, while more or less world-wide, occurred at different times in different places in the three main branches of *Homo sapiens* — the Caucasian, the Mongolian, and the Negroid. The American Indians were Neolithic when discovered by Iron-Age Europeans.

PREHISTORIC MAN

> Man is a Tool-using Animal . . . Nowhere do you find him without Tools: without Tools he is nothing, with Tools he is all. — THOMAS CARLYLE, *Sartor Resartus.*

Turning far backward in imagination, we must think of the human race as barely superior to the beasts around it. The passing generations, absorbed in a desperate struggle for existence, had no history and left no permanent record. Uniquely equipped at some stage with the power of communicating with each other by intelligible speech, they gained the priceless advantage of transmitting the accumulated knowledge of one generation by oral *tradition* to its immediate descendants. Succeeding generations gradually began to leave enduring records, at first crude and fragmentary, in the form of tools, cairns, monuments, drawings, paintings, or carvings on ivory or rocks, on trees or the walls of caves. Out of this so-called "picture-writing" grew, though with extreme slowness, systems of writing in more or less conventional characters, hieroglyphics or crude alphabets, which it is the problem of the archeologist to decipher and translate.

We may think of the human race, or each of its divisions, as passing like the human individual through successive periods of infancy, childhood, and youth. As the child sees in most things that happen about him some human or living agency, so the primitive man imagines in the wind some hidden being shaking the tree or its leaves, a supernatural being or god roaring angrily in the thunder.

FIRST STEPS TOWARD CIVILIZATION

Civilization implies a fixed abode and means of recording and transmitting ideas and information. A fixed abode is adopted when men give up dependence on hunting as their main source of food and clothing, and begin to produce food by the cultivation of plants, notably barley and wheat. The earliest known evidence of such agriculture has been found in caves in Palestine (1929–33) containing deposits more than 60 feet deep with indications of periods of human habitation during an epoch of perhaps 100,000 years from the time of Neanderthal man to that of the modern Arab. In an upper layer of Mesolithic age many flint sickle blades and their bone handles were found, also a stone mortar and fragments of pestles, as well as small harpoon heads of bone, but no pottery. Besides being fishermen, these people were, at least, harvesters of grain.

THE "FERTILE CRESCENT"

In parts of the area extending from the delta of the Nile to the head of the Persian Gulf wheat and barley still grow wild. During the Late Stone Age this area was occupied by people who lived in dwellings of wattle or brick, made pottery, and used copper as well as stone implements, and cultivated wheat and barley. The southern extremities of the Fertile Crescent, Egypt, and southern Mesopotamia, with the advantage of a warm climate, abundant water for irrigation, and most fertile soil, were among the earliest seats of civilization. Each of them retained enduring monuments with inscriptions, but the art of reading them had been lost before the beginning of the Christian era.

CUNEIFORM TABLETS

About A.D. 1800 copies of cuneiform (wedge-shaped marks made by a metal tool in soft clay, afterwards baked) inscriptions from Persian palaces were brought to the attention of European scholars and in 1847 Sir Henry Rawlinson published a complete old Persian alphabet of 39 characters. The Emperor Darius (521–485 B.C.) had had a record of his victories cut into rock in each of the three languages of his empire (Behistun inscription). To him and to Rawlinson scholars of today owe their ability to read the records of Assyria and Babylonia.

In the ruins of ancient Nineveh, about A.D. 1850, there were found by Layard and his associates more than 20,000 tablets of baked clay with cuneiform inscriptions from the library of the Assyrian King Assurbanipal. In studying these Rawlinson found in a still more ancient language than that of the Assyrians, an account of the earlier civilization of Sumer and Akkad, an unsuspected vista of human history, reaching back towards the very dawn of civilization.

SUMER, AKKAD, AND ELAM

The Sumerians inhabited the region at the head of the Persian Gulf about the mouths of the Euphrates and the Tigris. To the north where these rivers converge was Akkad, inhabited by Semitic population long before the first dynasty of Babylon. To the east of Sumer was Elam, with Susa as its principal settlement. Here in 1891 French excavations revealed pottery of extraordinary thinness and beauty of design. Since the World War, Mesopotamia (Irak) has been more accessible to exploration and the results achieved by archeologists have cast a new light on the beginnings of civilization. The Sumerian social, religious, and legal ideas, the language, script, and art are said to have "dominated the whole valley of the Two Rivers for three millennia; long after the Sumerians had lost their national identity and their language was dead was the cultural edifice they had reared imposed upon and accepted by their conquerors and neighbors . . . a study of Asiatic prehistory must begin, where history begins, in Sumer" (CHILDE).

The Sumerians cultivated the date-palm and reaped grain with stone sickles; they used boats and copper-headed harpoons. Their pottery shows the use of the potter's wheel and

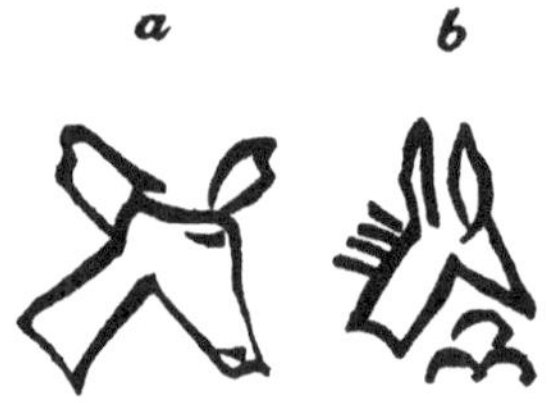

FIG. 1. — SUMERIAN PICTOGRAPHIC SCRIPT ON CLAY TABLETS OF JEMDET NASR, *c.* 4000 B.C. *a*, the word for " ass; " *b*, this sign combined with the sign for " mountain " is read " horse " (mountain ass). After Langdon, *Oxford Cuneiform Texts*, **7**, 19. Permission, The Clarendon Press, Oxford, publishers.

there is evidence of wheeled vehicles. They built houses and temples of clay, wrote on soft clay tablets, and adopted the use of seals for personal identification. Tablets found in their temples contain the earliest known picture-writing and a well-developed numerical notation. In the ruins of a nameless city burned before 3500 B.C. have been found clay tablets with writing pictographic in form, but with phonetic signs combined to form words, showing the first steps in the evolution of the cuneiform from the previous picture-writing.

THE DAWN OF HISTORY

History begins when written tradition can be verified by the study of contemporary monuments. Such a monument is a stone tablet from the foundation of a temple at al-Ubaid. The builder of the temple inscribed here the name of his king, the same to whom late Sumerian tradition attributed the founding of the first Dynasty of Ur. The time when this first historical dynasty ruled Sumer is believed from recent studies to have been from about 3150 to 3000 B.C. This early dynastic period was distinguished from the prehistoric by its exceptional wealth. Thus arose a civilization which, in spite of temporary political or military reverses, dominated Western Asia for more than a thousand years. Its influence is still felt in the daily life of all civilized people. Any account of other early civilizations in India and China lies outside the range of this book, but something must be said of Egypt and the Ægean.

PREHISTORIC EGYPT

History begins in Egypt with the first Pharaoh, called Menes, who united Upper and Lower Egypt in one kingdom. This event is attested by written documents confirmed by an inscription on the "Palermo Stone." The estimated date is about 3360 B.C. Egypt is the creation of one great river, the Nile, as Sumer and Babylonia were of the Euphrates and the Tigris. Rising in lofty snow-mountains far to the south, the Nile cuts a deep trench through the desert of northern Africa on its way to the Mediterranean. Boundaries were fixed by nature in the deserts east and west, the cateracts of the Nile to the south, and the Mediterranean at the north.

The period just before the time when the two kingdoms were united is called Predynastic. It stretches back toward a Mesolithic culture of unknown date. During the late Predynastic period flint was the chief material for implements and weapons, copper, gold, and silver were in use, and excellent pottery was produced. Of particular interest to archeologists are the remarkable parallels at this time between Egypt and Sumer.

Sumerian civilization, whatever its ultimate origin, had developed in its own country and on its own lines for so long that it could fairly by now be called endemic while that of Egypt was inspired and made possible by the introduction of foreign models and foreign blood. The character of the borrowings and the proximity of the superior culture leave no alternative source for the influence which affected Egypt at the close of its predynastic age; directly or indirectly that came from southern Mesopotamia. — C. L. WOOLLEY, *The Sumerians*, pp. 187–188.

HIEROGLYPHICS: THE ROSETTA STONE

In 1799 a French officer found near the mouth of the Nile a slab of basalt about four feet by three with inscriptions in Greek and two forms of unknown Egyptian characters. Copies of the inscriptions were distributed for study, while the stone itself found its way to the British Museum where it may still be seen. Thomas Young (1773–1830) showed the order in which the unknown characters — hieroglyphics — should be read, and recognized their syllabic and alphabetic significance. J. F. Champollion (1790–1832) finally succeeded in 1822 in publishing a full Egyptian alphabet with values of many syllabic signs.

THE ÆGEAN, OR MINOAN, CIVILIZATION

The center of this civilization was discovered in Crete by excavations of Sir Arthur Evans and others begun in 1900, the name Minoan being derived from Minos, in Greek tradition the greatest king of that island. In the ruins of the palace at Cnossus finally destroyed about 1200 B.C. have been found remains of a culture perhaps the finest flower of the Bronze age, the germs of the Greek civilization. The Minoans left numerous records, but in the absence of any bilingual inscription it has not yet been possible to decipher them. There is evidence, however, of trade with Egypt and the study of objects of Egyptian origin makes it possible to correlate the phases of Minoan civilization with known dates in Egyptian history.

. . . From whatever point of view one considers the basin of the Ægean, it accentuates with remarkable force the features which distinguish the Mediterranean in its entirety and thereby appoints itself the cradle of the aforesaid civilization. What strikes one elsewhere in great countries of the East is hugeness and uniformity; production, power, and even beauty are all matters of quantity. Here, the continual diversity of nature leaves no room anywhere for large agglomerations of plants, animals, or men. Here is the triumph of autonomy and individualism, the free blossoming of natural gifts, with no other restriction than the necessity of a harmonious organization.— G. GLOTZ, *Ægean Civilization*, p. 12.

HISTORICAL PERIODS IN THE ANCIENT EAST

There are three great periods in Egyptian history. The Old Kingdom or Pyramid Age, 3000–2475 B.C. — is a period of extraordinarily rapid progress in the control of mechanical power. The second, Middle Kingdom or Feudal Age, 2160–1788 B.C. — is a period of intellectual advance and of foreign commerce. Libraries of papyrus rolls were assembled and ships navigated the Ægean and Red Seas. The third period, the Empire or New Kingdom, 1580–1150 B.C. — was a time of great buildings and of foreign conquest.

Of the pyramids some 80 are more or less completely preserved. The largest of these royal tombs, that of Cheops from about 2600 B.C., is 493 feet in height and 775 feet square. The four sides correspond with the points of the compass and a shaft in the pyramid was directed to the pole star of that period, while another shaft was at such an angle that Sirius at its culmination shone on the sarcophagus within.

In the small fraction of Egyptian literature which has come down to us works of pure imagination are rare; art, science, and literature were subordinated to religious and utilitarian purposes. Authority, discipline, honesty are fundamental. The monumental works of architecture and engineering, pyramids, obelisks, statues, temples, dikes, and irrigation works afford ample evidence of technical competence and of skill in elementary mathematics. Metal work, pottery, enamel, and colored glass; the mummification and dissection of bodies tell a similar story in other fields. But speculation and investigation are missing.

The political power of the Sumerians so long dominating the Fertile Crescent was crushed by Amorites and Elamites, the former founding an empire with a capital at Babylon, in which however Sumerian culture and literature survived. The reign of Hammurabi (1948–05) was distinguished by a high degree of organization and prosperity. The next great period in the history of Western Asia is due to the Assyrians, a Semitic tribe east of the Tigris which had absorbed the elements of Sumerian culture and in 732 B.C. captured Damascus and reached the Mediterranean. Under Sargon and his successors the Assyrians conquered the entire Fertile Crescent destroying Babylon and establishing a new capital at Nineveh. The last great period of Semitic dominance was that of the Chaldeans with their capital at restored Babylon, until conquered by Cyrus of Persia 539 B.C.

WRITING AND THE ALPHABET

The invention of writing and of a convenient system of records on paper has had a greater influence in uplifting the human race than any other intellectual achievement in the career of man. — J. H. BREASTED, *Conquest of Civilization*, p. 53.

While Sumer was the cradle of writing we are indebted to Egypt for papyrus. Primitive forms of writing have of course appeared elsewhere, but it is in these two lands that we find the beginnings of our own alphabet. Writing was at first pictorial, then became linear, and from this there developed the cuneiform in Sumer, hieroglyphics in Egypt, letters in Phœnicia, Syria, and Greece.

The earliest Sumerian writing is pictorial with conventional signs representing ideas not words. Thus the picture of a human foot meant "foot," "go," "stand," or with additional strokes "hasten," "carry," "foundation." By about 3150 B.C. the evolution from pictograph into cuneiform was complete. The signs were largely syllabic, each representing a vowel with one or two consonants. The system was adopted by other peoples to their own languages and its use was continued in Babylonia, Assyria, and Chaldea. Even the oldest known Hebrew inscription is cuneiform.

In Egypt a linear script scratched upon slate indicates a probable relation with Sumer preceding the introduction of hieroglyphics. Egypt shows before and during early historic time considerable evidence of artistic and religious influences from Sumer; though differences in environment naturally led to gradually increasing divergence. In Egypt, for example, building stone was abundant while in Sumer the use of brick and clay tablets was general. Along the banks of the Nile grew the papyrus reed, from which cut into long thin strips the Egyptians could make a smooth, tough, yellowish paper. On this it was easy to write with a reed pen dipped in ink made of soot and water thickened with gum. About 3000 B.C. the Egyptian hieroglyphics had reached their final form, in which 24 consonants were included. For ordinary business the writ-

FIG. 2. — AN EXAMPLE OF HIERATIC WRITING. Papyrus of Hor-djeser-hou I-em-hotep, Ptolemaic period (332–30 B.C.). Courtesy of the Metropolitan Museum of Art.

ing was in a running hand called hieratic. The Semitic peoples at the eastern shore of the Mediterranean in contact with both Egypt and Mesopotamia used Egyptian papyrus and an alphabet of 22 signs, each representing a single consonant — the vowel sound for reading being indicated by the context. Their priceless gift of this simple method of writing was carried by the seafaring Phœnicians to the Ionian Greeks about 900 B.C. By the Greeks vowels were added and from their alphabet of 27 letters the Latin and thence our own have been derived.

REFERENCES FOR READING

BOAS, FRANZ, *Mind of Primitive Man*, Rev. Ed., 1938.
BREASTED, J. H., *Conquest of Civilization*, 1926.
CHILDE, V. G., *New Light on the Most Ancient East*, 1934.
COLE, F. C., *The Long Road from Savagery to Civilization*, 1933.
DEMORGAN, JACQUES, *Prehistoric Man*, 1925.
GLOTZ, G., *The Ægean Civilization*, 1925.

LUBBOCK, (SIR) JOHN (LORD AVEBURY), *Prehistoric Times*, Ed. 7, 1913.
MORET, ALEXANDER, *The Nile and Egyptian Civilization*, 1927.
OSBORN, H. F., *Men of the Old Stone Age*, Ed. 3, 1921.
PEAKE, H. J., *Early Steps in Human Progress*, 1933.
SCHMIDT, R. R., *The Dawn of the Human Mind*, 1936.
WOOLLEY, C. L., *The Sumerians*, 1928.

NOTE: The figure at the head of the opposite page represents an Egyptian shadow clock as described by Borchardt, *Geschichte der Zeitmessung*, Abb. 9. Redrawn by E. T. W.

Chapter II

Science Begins in the East

The oldest known treatise on surgery . . . written in Egypt nearly 5,000 years ago, discloses to us the thoughts of the earliest man who reveals a scientific attitude of mind. This treatise is therefore the earliest document in the history of science. — BREASTED, *Science*, **74**, 641, Dec. 25, 1931.

It is one of the lessons of the history of science that each age steps on the shoulders of the ages which have gone before. — SIR MICHAEL FOSTER, *Lectures*, 1901, p. 300.

BEGINNINGS OF SCIENCE

A history of science might be based on some more or less logical system of definitions and classification. Such systems and such points of view belong to relatively recent and mature periods. Science has grown without much self-consciousness as to how it is defined, without any great concern as to the distinction between pure and applied science, or as to the boundaries between the different sciences.

The periods at which primitive man of different races began to have conscious appreciation of the phenomena of nature, of number, magnitude, and geometric forms can never be known, nor the time at which their elementary notions began to be so classified and associated as to deserve the name of science. Very early in any civilization, however, there must obviously have been developed simple processes of counting and adding, of time and distance measurement, of the geometry and arithmetic involved in land measurement and in architectural design and construction.

PRIMITIVE COUNTING AND NUMBER SYSTEMS

There is no language without some numerals, though in extreme cases the range may be merely *one*, *two*, *many* (i.e., more

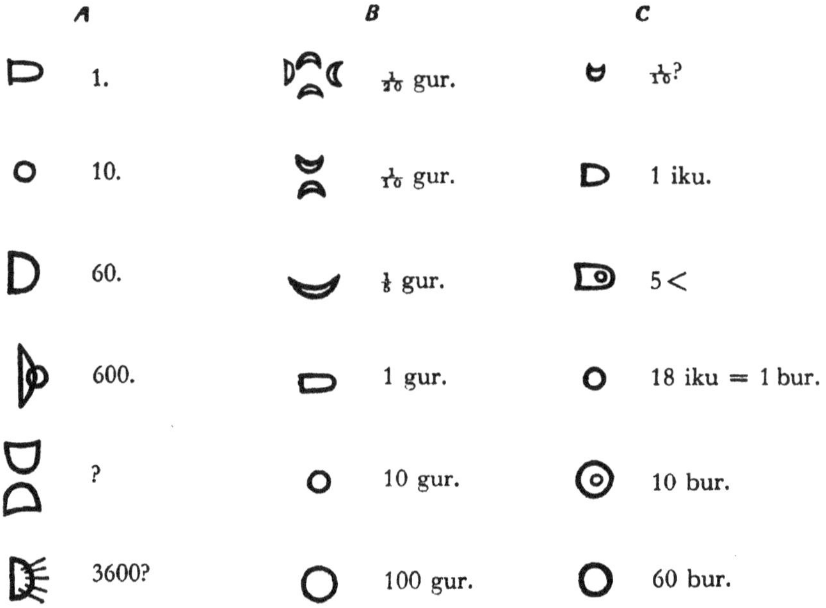

FIG. 3. — ARCHAIC SUMERIAN NUMERALS FROM JEMDET NASR, *c.* 4000 B.C. *A*, ordinary system; *B*, grain; *C*, land measure. After Langdon, *loc. cit.*, 63–67. Permission, The Clarendon Press, Oxford, publishers.

than two); and for most of us such a word as *million* is nearly equivalent to an innumerable multitude. The process of counting is naturally facilitated by the use of fingers and toes as counters, their number 10 being the well-known anatomical basis for our decimal number system. Thus the word digit is derived from the Latin word *digitus*, meaning finger. The mathematical advantage of 12 as a base conveniently divisible by many factors has often been pointed out, but the choice unfortunately had to be made long before its real significance could possibly be apprehended. Vestiges of the use of 5 and 20 are familiar; the former, for example, in the Roman numerals IV, VI, etc., the latter in such expressions as "three score and ten."

The earliest known numerals are those found in Sumer, Elam, and Egypt. Among the former are accounts of temple revenues using both sexagesimal and decimal systems.

From these archaic systems were developed the cuneiform numerals of Babylonia and Assyria. The earliest Egyptian numerals from about 3400 B.C. are straight marks for small numbers with a special sign for 10. Later signs were made for

$= 10 + 10 - 1 = 19$

$= 10 + 10 + 10 + 10 - 3 = 37$

FIG. 4. — CUNEIFORM NUMERALS. After Smith and Ginsburg, *Numbers and Numerals*, p. 16. Permission of Professor W. D. Reeve, Teachers College, Columbia University, and by courtesy of Professor David Eugene Smith.

100, 1,000, and 10,000. The system was purely decimal. It is difficult for us to realize the extent of our indebtedness to the comparatively recent so-called Arabic — or more properly Hindu — notation, in which numbers of whatever magnitude may be expressed by means of only ten symbols.

ANCIENT WEIGHTS AND MEASURES

Measurement, fundamental in science, had had its origin in trade and construction. The values of weights and measures in the Ancient East are known either from the actual instruments or from other sources, units of the same name differing considerably in value from place to place. The oldest known stone weights are from a Sumerian temple at Lagash (about 3000 B.C.), each inscribed, "1 mana, Dudu High-priest," — in our scale about one ounce. A later Assyrian scale included the shekel, the mana = 60 shekels (about 1.1 lb.), and the talent = 60 mana. The early Sumerian carpenters used a scale of digits equaling 0.65 inch. The Babylonian cubit (fore-arm) was 20.6 inches in our measure, and was divided into 30 digits (fingers). The higher units were sexagesimal, ending in a parasang, or league, of about 3.5 English miles.

The Egyptians used decimal systems of weights and measures. The largest unit of weight, for measuring wheat, was about two pounds. The cubit of the Pyramid Age, nearly the same length as the Babylonian cubit, was divided into hundredths. But apparently for the convenience of workmen, the scale was usually marked also approximately into 7 palms, a palm being 4 digits. The two systems were incommensurate, like our yard and meter.

PRIMITIVE ASTRONOMICAL NOTIONS

On the astronomical side the most obvious fact is the division of time into periods of light and darkness by the apparent motion of the sun about the earth. With closer attention it must soon have been observed that the relative length of day and night gradually changes, and that this change is attended by a wide range of remarkable phenomena. At the time of shortest days, vegetable and animal life (in the north temperate zone) is checked by severe cold. With the gradually lengthening days, however, snow and ice sooner or later disappear, vegetation is revived, birds return from the warmer south, all nature is quickened. In the symbolism of the beautiful old myth, the sleeping princess, our earth, is aroused by the kiss of the sun-prince. The longest days and those which succeed them are a period of excessive heat and of luxuriant vegetation, followed by harvests as the days shorten, towards the completion of the great annual cycle. In time, closer observers, noting the stars, discovered that corresponding with this great periodic change are gradual variations in the starry hemisphere visible at night, that, in other words, the sun's place among the stars is progressively changing, that it is in fact describing a path completed in a large number of days, which after many years of counting is found to be 365. It is also found that the midday height of the sun above the southern horizon shares in the annual cycle. The determination of the number of days in the year is a matter of very gradual approximation, possible only to men who have already attained some command of numbers and the habit of preserving records extending over a long series of years. For there is no well-marked beginning of the year as of the day. An erroneous determination of the number of days becomes apparent only after a number of years, increasing with the accuracy of the original approximation. For example, if the year is assumed to be exactly 365 days, that is, about six hours too short, the festivals and other dates will slip back about 24 days in a century, and thus lose their original correspondence with climatic conditions. A revision of the calendar will be necessary. (See pages 20–21, and 38.)

Still another natural period is introduced by the motion of the moon, which seems like the sun to have a daily motion about the earth, and also to describe a closed path among the stars in a period of about 29 days. Unlike the sun, however, the moon has during this period a remarkable change of apparent shape and luminosity from "new" to "full" and back again. The difficulty of expressing the precise length of the month and the year in days, causing the imperfection of early calendars, has on the other hand reacted to the advantage of mathematical astronomy by demanding more and more precision both in observation and in the computation based upon it.

THE PLANETS

Another celestial phenomenon, though less obvious than the foregoing, must have found wide recognition in prehistoric times. The stars vary widely in grouping and individual brilliancy, but in general their relative positions are sensibly constant. To this constancy, however, five exceptions are discovered in the wandering motion of the planets, Mercury, Venus, Mars, Jupiter, and Saturn, which like sun and moon have their several paths among the stars but with seemingly irregular motions. Corresponding to the seven bodies there was set up by prehistoric people an arbitrary division of time into weeks of seven days, "the most ancient monument of astronomical knowledge." The correspondence with the planets is still preserved in the names of the days of the week in several modern languages.[1] Our own names for them have come indirectly through the Saxons.

ANCIENT CHRONOLOGY; CALENDARS AND MEASUREMENT OF TIME[2]

The frequent and obvious cycles of the moon furnished the earliest measure of time, each cycle being a lunar month. In Sumer each month began with the new moon, which was carefully observed by the priests. There were twelve months in the year and when the accumulated shortage (about 11 days each year) became too great, an extra month was intercalated, a practice that continued until the reign of Darius (521–485 B.C.).

At the summer solstice the rising sun shining straight down the main street of Babylon, furnished a gauge for adjustment of the calendar to the solar year. In Lower Egypt the year commenced with the annual rise of the Nile, which begins rather suddenly about the middle of July, a date marked astronomically by the rising at sunrise of the Dog Star, Sirius, the brightest of the fixed stars. This is a striking event, nowhere more brilliant than in the vicinity of Cairo. In course of time it became evident that the return of Sirius was five days later than the completion of the year of twelve 30-day months, and the year was accordingly lengthened by the addition of five holidays. The residual shortage of about 6 hours resulted in a slow periodic change of the beginning of the year which would return to its original position in about 1,458 years. By means of

[1]

English	*French*	*Italian*	
Sunday	dimanche	domenica	(Sun)
Monday	lundi	lunedi	(Moon, Luna)
Tuesday	mardi	martedi	(Tiw, Mars)
Wednesday	mercredi	mercoledi	(Woden, Mercury)
Thursday	jeudi	giovedi	(Thor, Jupiter)
Friday	vendredi	venerdi	(Frig, Venus)
Saturday	samedi	sabato	(Saturn)

[2] See J. H. BREASTED, "The Beginnings of Time Measurement and the Origins of our Calendar," *The Scientific Monthly*, **41**, 289–304, Oct., 1935.

this "Sothic [1] cycle," says Breasted, "We may therefore conclude that the civil calendar of Egypt was introduced in 4236 B.C. . . . not only the earliest fixed date in history, but also the earliest date in the intellectual history of mankind." The year was divided into three seasons — the inundation, the season of cultivation, and the harvest season. The months, probably lunar at first, became 30 days, and this continued into historic time. An edict of 238 B.C. introduced the leap-year, but the innovation was afterwards forgotten.

The artificial subdivision of the day and night into hours was achieved in Egypt by means of some form of the sun-dial for the day and of the water-clock, or clepsydra, for the night — day and night having each twelve hours of varying lengths. The earliest time-piece known is an elaborate water-clock described by its maker Amenemhet on the wall of his tomb-chapel in the cemetery of Egyptian Thebes about 1550 B.C.

The Babylonians on the other hand by measuring the amount of water escaping slowly from a vessel kept full ascertained that the time from the first appearance of the upper edge of the Sun to the moment the whole disk was visible was $\frac{1}{720}$ of the entire time from one sunrise to the next. This period of day and night was then divided into 12 equal parts, one of which would thus equal twice the modern hour, and is the time required for the sun to travel 60 times its own diameter.

BABYLONIAN ASTROLOGY AND ASTRONOMY

Primitive people generally attribute life to anything that moves, including the heavenly bodies. The Sumerians personified the forces of nature and worshiped them as gods.

From the Sumerians the Semites of Babylonia inherited their gods and a religion of fear, a dread of calamity from the mysterious forces of nature. Their two chief gods were the Sun and the Moon. They went further and identified the other great gods with the five visible planets, known to us by their Roman names (p. 19). Movements of these heavenly bodies represented activity of the gods foreshadowing terrestrial

[1] From the Greek *Sothis*, an Egyptian name of Sirius.

events. The instinctive desire to foresee the future called for a priestly act of interpreting the signs in the sky, and Astrology began. Since it involved careful observations of the position not only of sun and moon but also of the five planets and the keeping of careful records extending over long periods, the foundations of the future science of astronomy were firmly laid. There is however no evidence of a systematic geometrical theory of the celestial motions or even of any recognition of the need of such a theory. Early observers were impressed with the possible analogy between certain phenomena in the heavens and those in human life. Thus sunrise and sunset, the phases of the moon, the appearances and disappearances of the planets and their relative positions were carefully studied. A continuous series of observations has been found, for example, of the appearances and disappearances of Venus from 1921 to 1901 B.C., with a table for computing the next disappearance for any given time of reappearance.

The celestial equator was divided into 360 degrees. Evidently suggested by the twelve full moons of the year and their successive positions among the stars, the "Twelve Signs of the Zodiac" were mapped out for the first time as equal divisions of the ecliptic, the starry path of the sun and the planets; and in each sign three stars were chosen as being representative of that sign. A text of the fifth century B.C. enumerates 71 recognized constellations in three groups, each ruled by a god. Particularly noteworthy is the discovery of "the Saros," a period of 6,585 days — 223 moons, a little more than 18 years — for the recurrence of solar eclipses. This must have been based on a long series of observations, but without taking account of the region of visibility of eclipses of the sun. A list of eclipses of the moon from 747 B.C. was known to Ptolemy in the second century A.D.

The periods of the planets in their orbits were determined with surprising exactness. As to accuracy of direct observations, it is said that in later Babylonian times angles were measured to within 6 minutes and time to less than a minute. Not long before 500 B.C. the great Babylonian astronomer

Nabu-ri-mannu, Naburianos in Greek, computed the length of the solar year as 365 days, 6 hours, 15 minutes, and 41 seconds, "a result which is only 26 minutes and 55 seconds too long. This is the earliest known close approximation to the length of the solar year." (BREASTED, *Sci. Mon.*, **41**, p. 295.)

In the following century the Chaldean, Kidinnu (Greek Kidenas), with records of observations for 360 years at his disposal, made a still more accurate determination of the length of the year, and even made the remarkable discovery of the very slow variation in the direction of the earth's axis, known to Greek and later astronomers as precession of the equinoxes.

BABYLONIAN MATHEMATICS

Our knowledge of Babylonian mathematical science has received a considerable increase in recent years mainly through the work of the Austrian scholar (now in Providence, R. I.), Otto Neugebauer, who has deciphered many of the original inscriptions preserved in museums in Istanbul and many other cities. About 200 such texts have been deciphered. In Babylonian arithmetic whole numbers were expressed in general by two fundamental characters, 1 = 𒁹, 10 = 𒌋. The numbers known to have been used ran into the hundred thousands, this naturally implying a highly developed command of the fundamental operations by means of which larger numbers are connected with smaller ones. From 1 to 59 numbers are expressed by repeated use of the signs for 1 and 10, e.g., 23 = 𒌋𒌋𒐈, the tens preceding the ones. But superimposed upon this decimal system is a sexagesimal system, expressed by the use of the same signs, but multiplied by a power of 60. In this position-system 𒁹𒐉, which we may write 1,4, may represent 64, but also 3,604 or even $1\frac{40}{60} = 1\frac{2}{3}$. In later texts (after *c.* 500 B.C.) a special symbol for non-existing inner sexagesimal places exists, a sign which we may represent by 0. Then 1,0,4 may mean 3,604, but also $60\frac{1}{15}$.

A table of squares of the natural numbers presents, for example, nothing novel for the first seven numbers, after which follow, however, the equivalent of

1,4 is the square of 8
1,21 is the square of 9
1,40 is the square of 10
2,1 is the square of 11.

Just as in our notation, for example, 325 means three times the square of ten plus twice ten, plus five, so this table must mean:

once sixty plus four
once sixty plus twenty-one, etc.

implying the representation of 60 by 1 in the second place.

Babylonian multiplication tables are very numerous. There are also tables for squares and square roots, cube roots, and tables for the solution of $x^2 + x^3 = c$, when x runs through the value from 1 to 60:

5,4,12 where base $x = 26$
5,40,12 where base $x = 27$,

which means that $5,4,12 = 5 \times 60^2 + 4 \times 60 + 12 = 18,252 = 26^2 + 26^3$; $5,40,12 = 20,412 = 27^2 + 27^3$. In a few places square roots occur from numbers which are not perfect squares. A great number of tables give reciprocal numbers, viz., pairs of numbers n and m such that their product, nm, is a power of 60. Some of these tables of later date go to 17 sexagesimal places. Such tables reduce division to multiplication.

In the field of algebra, Babylonians of 2000 B.C. had the equivalent of a formula for the solution of the quadratic equation, which until 1929 was attributed to the late Alexandrian Greeks, more than 2,000 years later. We also find systems of linear equations, and of quadratic equations and some problems leading to cubic equations, which were solved with the aid of the tables for $x^3 + x^2$.

In plane geometry there was a correct knowledge of area for rectangles, triangles, and trapezoids, of similar triangles, of the right triangle inscribed in the semicircle, and of the important theorem known long afterwards as Pythagorean.

The use of the circle quickly leads to the discovery that a chord equal to the radius subtends one-sixth of the four right angles at the center, and is thus one side of a regular inscribed hexagon, a figure found on Babylonian monuments. A failure to distinguish between the length of the arc and that of its chord led to the first crude approximation to the ratio of a circumference to its diameter, $\pi = 3$, which occurs in the Old Testament where King Solomon's molten sea is said to be "ten cubits from the one brim to the other; it was round all about, . . . and a line of thirty cubits did compass it round about." (I Kings vii, 23.)

In solid geometry, volumes were computed not only for right prisms and circular cylinders, but even for frustums of cones and of square pyramids.

Surveys were made with numbered plans, and a single map of the known world has come down to us. But maps in general were hardly more than mere itineraries for travellers with lists of places, and their distances from each other. Practical problems deal with the measurement of grain, workers digging a canal, etc. It is noteworthy that in all this study of Babylonian mathematics, as also in the case of the Egyptians, we have the advantage of dealing with the actual original material whether in pottery or papyrus. On the other hand the Greek manuscripts with which we shall deal in later chapters exist only in forms many centuries later than the originals.

EGYPTIAN ASTRONOMY

Under a cloudless sky, on the Nile as on the banks of the Euphrates, astronomy had its origin in a religious motive. It was not astrology in Egypt but Astrolatry, the adoration of the stars, that led the priests to watch for their risings that they might be worshiped.

Well before 3000 B.C. the year was divided into 36 decades of ten days each for grouping the constellations along the celestial equator — the oldest appearance of a circle of 360 divisions. The oldest surviving astronomical instrument is a forked stick and a plumb-line with a handle at the top,

Fig. 5, used by an Egyptian astronomer to determine the position of stars in the northern sky. Tables showing the time when important stars crossed the meridian have been preserved from about 1100 B.C., also records of more than 350 solar and more than 800 lunar eclipses before the Alexandrian period.

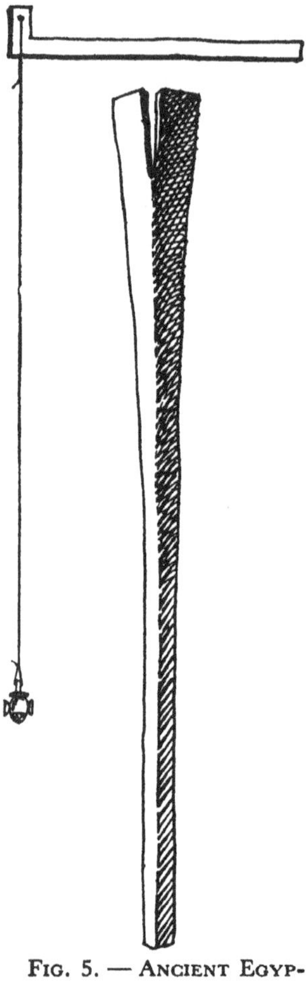

FIG. 5. — ANCIENT EGYPTIAN ASTRONOMICAL INSTRUMENT. After Borchardt, *loc. cit.* Taf. 16.

MATHEMATICAL SCIENCE IN EGYPT

The pyramids are monumental evidence of appreciation of geometric form as well as of a high development of engineering construction. In nearly all of them the slope of the lateral faces is 52° and their base lines are nearly uniform in direction.

Our most important source of information is the *Ahmes Papyrus*, or *Rhind*[1] *Papyrus*, dating from some time between 1700 and 2000 B.C. "Entrance into the knowledge of all existing things and all obscure secrets" are the opening words of this oldest known mathematical treatise. Rules follow for computing the capacity of barns and the area of fields, the text consisting, however, rather of actual examples than of rules. Reference is made to writings some 500 years older, presumably based in their turn on centuries of tradition.

In the computations fractions are used as well as whole numbers, but fractions other than $\frac{2}{3}$ are expressed in terms of fractions with unit numerators. The problem of decomposing other fractions into a limited number of such reciprocals is

[1] Written by a scribe named A'hmosé, discovered at Thebes, and purchased in 1858 by A. H. Rhind, and is now in the British Museum.

interestingly treated. It would appear that such decompositions, probably effected by special devices, were tabulated as records of experiments with essentially additive properties.

The problems discussed include a class equivalent to our algebraic equations of the first degree with one unknown quantity.

Example of dividing 700 loaves among four men in the proportion of the numbers $\frac{2}{3}$, $\frac{1}{2}$, $\frac{1}{3}$, and $\frac{1}{4}$. Let me know the share that each man receives.

Add $\frac{2}{3}$, $\frac{1}{2}$, $\frac{1}{3}$, and $\frac{1}{4}$; it makes $1 + \frac{1}{2} + \frac{1}{4}$. Get 1 by operating on $1 + \frac{1}{2} + \frac{1}{4}$; it makes $\frac{1}{2} + \frac{1}{14}$. Take $\frac{1}{2} + \frac{1}{14}$ of 700; it is 400. $\frac{2}{3}$, $\frac{1}{2}$, $\frac{1}{3}$, and $\frac{1}{4}$ of 400 will give the shares of the four men. — A. B. CHACE, *The Rhind Mathematical Papyrus*, Vol. I, p. 101.

Certain problems show an acquaintance with arithmetic and geometric progressions. Thus, a series of the numbers 7, 49, 343, 2,401, 16,807, accompanied by the words person, cat, mouse, barley, measure, has been interpreted to mean: 7 persons have each 7 cats, each cat catches 7 mice, each mouse eats 7 stalks of barley, each stalk can yield 7 measures of grain; what are the numbers of each?

Special symbols are used for addition, subtraction, and equality. The Egyptian seems never to have had a multiplication table. Multiplication by 13, for example, was accomplished by repeated doubling, and then by adding to the number itself, its products by 4 and by 8.

Herodotus reports from the fifth century B.C. that the Egyptians reckoned with stones, a practice independently developed in many lands, notably in the form of the abacus. This little computing machine of beads on wires was invented independently in different parts of the ancient world. In the Orient it is still widely employed.

The handbook also contains information in regard to weights and measures, and treats of the conversion from one denomination into another. On the geometrical side problems are given, depending on the use of formulas not derived in the text. They include computation of areas of fields bounded either by straight lines or circular arcs, including in the former case only isosceles triangles, rectangles, and trapezoids; also the volume of the frustrum of a square pyramid.

The classical problem of "squaring the circle" is attempted, the result being equivalent to the approximation $\pi = \frac{256}{81} = 3.16+$, as against the actual $3.14+$, an excellent result for the time.

EGYPTIAN LAND MEASUREMENT

Greek writers emphasize the methods of land measurement of the Egyptians consequent on the obliteration of boundaries by floods of the Nile. Herodotus relates that Sesostris had so divided the land among all Egyptians that each received a rectangle of the same size, and was taxed accordingly. Whoever lost any of his land by the action of the river must report to the king, who would then send an overseer to measure the loss, and make a proportionate abatement of the tax. Thus arose geometry (geometria = earth measurement).

Egyptian geometry was in the possession of the formula for the area of rectangle, triangle, trapezoid, and of the volume of elementary solids, including the volume of a frustum of a pyramid of altitude h and square bases of sides a and b: $V = \frac{h}{3}(a^2 + ab + b^2)$.[1] There are several approximation formulas in use. Both in Babylonian and Egyptian mathematics, however, we never find an abstract formula, we only get numerical data.

The main features of Egyptian mathematical science are then as follows: about 2000 B.C. a well-developed use of whole numbers and fractions, a method of solving equations of the first degree with one unknown quantity, several formulas for area and volume of elementary figures, both exact and approximate, an approximate method for finding the circumference of a circle of given radius.

ANCIENT CHEMICAL INDUSTRIES

The hieroglyphic name for Egypt is *quemi* or *chemi*, the black land, Chemistry thus owing its name to Egypt. The actual be-

[1] See R. C. Archibald, *Science*, Jan. 31, 1930.

ginnings of the arts related to chemistry are lost, and the earliest evidence of chemical processes is based upon things made by ancient peoples: metals, pottery, glass, cement, pigments, and dyed materials. From the 15th to the 12th century B.C. the ancient Egyptians evolved systems of magic, astrology, necromancy, together with those of theology and philosophy. Plato, Pythagoras, and Democritus are said to have sought instruction in Egypt; but most of our information has been obtained from Greek writings of later date. We know nothing of the theories that prevailed at the periods of the oldest objects found.

In prehistoric Egypt, considerable skill had been acquired in the use of metals, glass, and pottery; and in the practice of tanning and dyeing. Gold and silver were apparently the first metals known, since these metals occur free in nature and are easily extracted from their ores. Silver appears as a constituent of very ancient pieces of gold; but from 3000 to 1500 B.C. it seems to have been rarer and more valuable than gold.

While Egypt is by no means rich in metal ores, at the beginning of the historical period metal was in fairly common use in Egypt. One of the oldest pieces of gold jewelry dates from 3400 B.C. A piece of cast metal found in Sumer of about the same period was nearly pure copper. Bronze, an alloy of tin and copper, was used in prehistoric days, but the alloy was probably obtained by smelting ores containing both copper and tin rather than by adding tin to molten copper. Bronze, however, was not in common use until about 2500 B.C. Mercury is said to have been found in tombs of 1600–1500 B.C., and the amalgamation of gold with mercury was known at a very early date.

The Egyptians regarded lead as the mother of metals, probably because small quantities of silver are found with rich lead ores. The idea of the transmutation of metals is thus very ancient. Tin was called white lead by the early Romans, and mercury was, for a long time, regarded as liquid silver.

A few objects of iron appeared in Egypt and in Babylonia at very early times. But iron was not commonly used in Egypt or in Greece until about 1200 B.C., when men learned by heat treatment to convert iron into steel, thus obtaining for the first time an alloy superior to bronze.

The art of making glass and pottery was well developed in Egypt at the beginning of the historical period. From the analysis of very old glass and by deciphering ancient Egyptian formulas, it is clear that glass was made then as now, by fusing together quartz sand, limestone, and the ashes of plants. Assyrian blue glass contained copper and their red glass cuprous oxide. White glass contained oxide of tin and yellow glass lead antimonate. Lead oxide is found in glass from the Fourth Dynasty in Egypt. Glass was also known to the earliest civilizations of India and Asia Minor, as well as Egypt. This is also true of the arts of weaving and dyeing; a fine specimen of dyed mummy cloth dates from about 3000 B.C. or earlier. Dyestuffs such as indigo from India and the red safflower were certainly used as early as 1550 B.C. Among the pigments, besides preparations from the juice of plants, were red oxide of iron, yellow ochre, basic copper carbonates (blue azurite and green malachite), charcoal or bone-black, and a gray mixture of charcoal and limestone.

Leather was made even in predynastic times by curing the hides of animals with vegetable tannins; and ointments and pomades were used to protect the body from heat. Perfumes from flowers date back to at least 2500 B.C. Beer was made in Egypt and in Babylonia at an early date; at about 2900 B.C. mention is made of a special kind of beer made from red barley. Wine from the grape must have been used from the earliest dynasties.

Very old records mention natron (sodium carbonate and bicarbonate) and this was included together with sodium chloride and other inorganic substances under the general name *salt*. In the early days alum was used as a charm and later for fire-proofing.

The practical artisans of antiquity left no theories that are available to us, but we may assume that speculations concerning the fundamental nature of matter and concerning those transformations which we now regard as chemical reactions were not unknown. The extent of the practical knowledge possessed by the ancients commands our admiration in spite of long periods of retrogression.

ANCIENT BIOLOGY, MEDICINE, AND SURGERY

Their writings show that the ancient Egyptians and the peoples of the Fertile Crescent (p. 6) knew more than a mere farmer, herdsman, or hunter would know about living beings. Engravings on slate palettes and ivory knife handles of predynastic Egypt (p. 9) and on monuments of later times testify to keen powers of observation and great skill in portrayal of a large number of animal forms, native and exotic. In some cases the drawing is so good that the exact species can be identified.

The Egyptians were renowned in ancient times for their knowledge of plants and herbs, but only what is in a few medical texts has survived. Our knowledge of Babylonian and Assyrian botany is much greater. From the remains of the libraries of Nineveh, now preserved in the British Museum, have been selected 660 medical tablets and 120 plant-lists containing Sumerian names with Assyrian equivalents and synonyms. In the plant-lists there was an intelligent and methodical arrangement, generally in a fixed order beginning with the grasses. Of the date-palm, probably the chief source of Babylonian wealth, there remain entire texts devoted to the parts of the plant, the names of its varieties, and the technical terms relative to its cultivation. Two tablets prove that in Sumer as early as 2327 B.C. there was practical knowledge of the sexuality of this plant, in which male and female flowers grow on separate trees and must be brought together artificially to insure abundant fruit. This practice, of immemorial antiquity in Arabia, appears to be represented symbolically on the magnificent Assyrian sculptures to be seen in modern museums.

Some notions of the anatomy of man and animals were possessed by both Egyptians and Babylonians. The ancient Egyptian language held more than 100 anatomical terms for at least 36 different organs or parts of the body. External anatomy their surgeons learned from the study of wounds. General ideas of the appearance and connections of internal organs were afforded by the embalming of distinguished dead to be preserved for religious reasons as mummies. This process involved removal, cleansing, and separate preservation of organs from the abdominal and chest cavities.

Babylonian anatomy, like astronomy, came from a desire to know the will of the gods. A very good clay model of a sheep's liver with the parts labeled in cuneiform inscriptions for students in a Babylonian temple, shows careful study of this organ, believed to be in man the seat of the soul and to reflect in the sacrificed animal the mind of the god. The search for omens led also to careful observations at the birth of babies and young animals and gave rise to a considerable vocabulary of the external anatomy and abnormalities of the new-born.

Medicine defined as the attempt to prevent, cure, or alleviate disease, includes certain practices of magic. These were very similar in Egypt and Babylonia and arose from a belief, prevalent among primitive people, that, unlike the well-known effects of a blow or a bite, the mysteries of disease and death were supernatural — the acts of a demon who must be driven from the patient by prayers, incantations, and magic rites. A magic act might be the tying of seven knots in a string to imprison the demon or the giving of a repulsive medicine to the patient. Thus it came about that certain drugs, mostly vegetable, were recognized as effective. Magic became less important, and persons skilled in the knowledge and preparation of drugs were in request in cases of sickness. In both countries a large number of substances, about 550 in Assyria — vegetable, animal, and mineral — were used as drugs. In each prescription a considerable number of ingredients were offensive to smell or taste, like dung, asafoetida, and probably all had magic significance. But many prescriptions had a nucleus of materials that were wholesome and appropriate, e.g., poppy, castor-oil, onion, fig, hartshorn, sulphur, etc. For each disease in both countries there were many prescriptions. Some gave fairly well-localized symptoms and diagnosis.

Babylonian medicine is known chiefly from the translations of more than 200 of the 660 medical tablets remaining from the library of Assurbanipal (p. 7). A large proportion are purely medical, i.e., without magic. The Egyptians attributed great antiquity to their own medicine. It is known that Imhotep the architect of the Step Pyramid (2980 B.C.) was also physician to King Zoser. Their medicine is known to us from eight or more papyri (p. 13). The most truly medical of these is the *Ebers Papyrus*, believed to have been copied about 1500 B.C. from much older books. It contains a long series of prescriptions with some incantations, and two sections of special interest because they give a theory of disease based, not on demons, but on the supposed failure of vessels believed to convey certain fluids from the heart to various parts of the body.

Ancient surgery had this important difference from medicine: it offered rational treatment of injuries caused by visible agents. Little is known of Babylonian surgeons, beyond the regulations in the *Code of Hammurabi* (p. 11). This *Code*, the oldest known unification of the law, carved on a black diorite column, was compiled in part from known Sumerian laws and decisions.

Nearly all that is known of Egyptian surgery is in the *Edwin Smith Surgical Papyrus*. It is 15 feet long when unrolled, a copy made during the 17th century B.C. of a book believed to have been composed during the Pyramid Age (p. 11) and to have been edited with explanatory glosses not later than 2500 B.C. Forty-eight cases are discussed in order from the top of the head to the middle of the chest, where the copyist stopped. Each case has a title, examination, diagnosis, and treatment (unless incurable). Sixteen cases left without treatment show the surgeon's interest in the human body, apart from healing. In his examinations he gathered a considerable knowledge of anatomy, physiology, and pathology. The brain received its first name in literature and its coverings were recognized.

The paralyzing effects of injury to the brain and to the spine were described; but the function of the nerves remained unknown. The heart action was taken as an index of the patient's condition for the first time in medical history. The movements were observed by probing with the finger, and that the pulse was counted may be inferred from the mutilated first section of the papyrus; but there is no hint of the circulation of the blood. The muscles of the jaw are discussed in some detail. The authors of the text and glosses are the earliest known natural scientists, "who, confronting a world of objective phenomena, made and organized their observations and based inductive conclusions upon bodies of observed fact." (BREASTED, *Edwin Smith Surgical Papyrus*, p. 15.)

REFERENCES FOR READING

BREASTED, J. H., *Ancient Times* (Ed. 2, 1935), or *The Conquest of Civilization*, 1926.

BUDGE, (SIR) E. A. W., *The Divine Origin of the Craft of the Herbalist*, 1928.

CHACE, A. B., *The Rhind Mathematical Papyrus*, 1927–29.

DAVY, G., AND ALEXANDER MORET, *The Nile and Egyptian Civilization*, 1927.

DAWSON, W. R., *The Beginnings: Egypt and Assyria* (*Clio Medica*, Vol. I), 1930.

FARRINGTON, BENJAMIN, *Science in Antiquity* (Home Univ. Lib.), 1936.

PARTINGTON, J. R., *Origins and Development of Applied Chemistry*, 1935.

READ, JOHN, *Prelude to Chemistry*, 1937.

STILLMAN, J. M., *The Story of Early Chemistry*, 1924.

THOMPSON, R. C., *Dictionary of Assyrian Chemistry and Geology*, 1936.

ULLMAN, B. L., *Ancient Writing and its Influence*, 1932.

NOTE: The figure of Pythagoras on the facing page is on a Greek coin of Samos of the time of Trajan (A.D. 98–117), now in the British Museum; redrawn by Elizabeth Tyler Wolcott from an engraving in the collection of Professor David Eugene Smith.

Chapter III

The Earliest Greek Science

The Greeks are the most remarkable people that ever existed. . . . They were the beginners of nearly everything . . . of which the modern world makes its boast. — J. S. MILL, *Dissertations.* (*Legacy of Greece*, p. 251.)

It was not the *practice* of science which the Greeks invented, but the *scientific* idea, the conception that the world was knowable. — CHARLES SINGER. (*Science and Civilization*, p. 45.)

Aristotle observes that science was preceded by the arts. . . . those arts naturally came first which are directed to supplying the necessities of life, and next came those which look to its amenities. It was only when all such arts had been established that the sciences . . . were in turn discovered, and this happened first in the places where men began to have leisure. This is why the mathematical arts were founded in Egypt. — T. L. HEATH, *Hist. Greek Math.*, I, p. 8.

GEOGRAPHICAL BOUNDARIES

From the twilight of civilization and the first suggestions of science in Chaldea and Egypt, we pass to the brilliant dawn of science and civilization in Greece. Geographically we shall be concerned not merely with Greece itself, but with other Hellenic countries, especially the Ionian shores and islands of western Asia Minor, and, as time passes, with the Greek colonies in southern Italy, Sicily, and, after its conquest by Alexander the Great, northern Egypt.

Greece and its civilization seem immeasurably closer to us both in time and in spirit than do ancient Babylonia and Egypt. In these more remote civilizations Science had been cultivated chiefly as a tool, either for immediate practical applications or as a part of the professional lore of a conservative priesthood. In Greece, on the other hand, for the first time in the history of our race, human thought achieved a high degree of freedom, and real science became possible.

Mathematics as a science commenced when first some one, probably a Greek, proved propositions about *any* things or about *some* things, without specification of definite particular things. — A. N. WHITEHEAD, *Intro. to Math.*, p. 15.

INDEBTEDNESS OF GREECE TO BABYLONIA AND EGYPT

Greek civilization and Greek science owed much both to Egypt and Chaldea. Herodotus has been quoted already, and Theon of Smyrna (second century A.D.) says:

In the study of the planetary movements the Egyptians had employed constructive methods and drawing, while the Chaldeans preferred to compute, and to these two nations the Greek astronomers owed the beginnings of their knowledge of the subject.

Mathematics, considered as a science owes its origin to the *idealistic* needs of the Greek philosophers, and not as fable has it, to the *practical* demands of Egyptian economics. . . . Adam was no zoologist when he gave names to the beasts of the field, nor were the Egyptian surveyors mathematicians. — H. HANKEL. (Moritz, *Memo. Math.*, p. 211.)

The Babylonians and the Egyptians had done work upon astronomy before the Greeks, and had accumulated material to hand on to them; but it was the Greeks alone who were capable of development. It is not without due cause that we look up to the Ionian thinkers of the sixth century as our forerunners in the whole range of modern science. — J. L. HEIBERG, *Math. and Phys. Sci. in Classical Antiquity*, pp. 5–6.

THE GREEK POINT OF VIEW

It is not, however, so much the achievements of the Greeks in positive science which compel our attention and admiration as it is the remarkable spirit which they displayed toward man and the universe.

From the gods of the old Greek mythology we get an insight into the genius of the Greeks that nothing else can give. We see the picture of a race, false, boastful, and licentious perhaps, but with a sense of beauty, a confident joy in life and a warmth of affection that bespeak a gallant, vigorous, open-hearted, conquering people; . . . As the ages passed and the intellect mastered the emotions . . . man turned from a belief in capricious happenings dependent on the chance will of irresponsible gods, to a vision of the uniformity of nature under divine and universal law. — SIR WILLIAM DAMPIER, *History of Science*, pp. 12, 14.

In a striking essay, W. L. Butcher has portrayed with extraordinary clearness those characteristics of the Greeks which lifted them above all of their predecessors and above most, if not all, of those that have come after them:

The Greeks before any other people of antiquity possessed the love of knowledge for its own sake. To see things as they really are, to discern their meanings and adjust their relations, was with them an instinct and a passion. Their methods in science and philosophy might be very faulty, and their conclusions often absurd, but they had that fearlessness of intellect which is the first condition of seeing truly. . . . Greece, first smitten with the passion for truth, had the courage to put faith in reason, and in following its guidance to take no account of consequences. "Those," says Aristotle, "who would rightly judge the truth must be arbitrators and not litigants." "Let us follow the argument wheresoever it leads" may be taken not only as the motto of the Platonic philosophy, but as expressing one side of the Greek genius. — W. L. BUTCHER, *What We Owe to Greece*, pp. 1–2.

It was the privilege of the Greeks to discover the sovereign efficacy of reason. . . . And it was Ionia that gave birth to an idea, which was foreign to the East, but has become the starting-point of modern science, — the idea that Nature works by fixed laws. . . .

Again, in history, the Greeks were the first who combined science and art, reason and imagination. . . . The application of a clear and fearless intellect to every domain of life was then one of the services rendered by Greece to the world. . . . The Greek genius is the European genius in its first and brightest bloom. — W. L. BUTCHER, *Some Aspects of Greek Genius*, pp. 1–40.

SOURCES

The sources of our information as to the scientific ideas of the Greeks are exceedingly meagre, some of the most important treatises being known to us only by title or by detached quotations, or indirectly through Arabic translations. Among ancient sources of information in regard to Greek mathematical science the following may be mentioned:

The poems of Homer, composed after 900 B.C., and those of Hesiod a century later give us a general picture of Greek life and thought, and the historical work of Herodotus in the fifth century B.C. connects the Greeks with the neighboring peoples.

About 330 B.C., Eudemus of Rhodes, a disciple of Aristotle, wrote a history of geometry of which a summary has been preserved by Proclus in the commentary mentioned below.

About 70 B.C., Geminus of Rhodes wrote an Arrangement of Mathematics with historical data. This has been lost, but quotations are preserved in later authors.

About A.D. 140, Theon of Smyrna wrote *Mathematical Rules necessary for the Study of Plato.*

About A.D. 300, Pappus's *Collection* contains much information in regard to the previous development of geometry.

In the fifth century A.D., Proclus published a commentary on Euclid's *Elements* with valuable historical data.

THE CALENDAR

The Greek calendar was based at an early period on the lunar month, the year consisting, as in the Babylonian calendar, of twelve months of 30 days each. About 600 B.C. a correction was made by Solon, making every two years contain 13 months of 30 days and 12 of 29 days each, giving thus 369 days per year. In the following century a much closer approximation — 365¼ days — was attained by confining the thirteenth month to three years out of eight. This arrangement naturally failed, however, to meet the Greek desire that the months begin regularly at or near new moon, and Aristophanes makes the Moon complain:

CHORUS OF CLOUDS [1]

First, All hail to noble Athens,
  and her faithful true Allies;
Then, she said, your shameful conduct
  made her angry passions rise,
Treating her so ill who always
  aids you, not in words, but clearly;
Saves you, first of all, in torch light
  every month a drachma nearly,
So that each one says, if business
  calls him out from home by night,
"Buy no link, my boy, this evening,
  for the Moon will lend her light."
Other blessings too she sends you,
  yet you will not mark your days
As she bids you, but confuse them,
  jumbling them all sorts of ways,
And, she says, the Gods in chorus
  shower reproaches on her head,
When in bitter disappointment
  they go supperless to bed,
Not obtaining festal banquets
  duly on the festal day;
Ye are badgering in the law-courts
  when ye should arise and slay!

About 400 B.C., Meton the Athenian observed that 19 years consist of almost exactly 235 lunar months, and accordingly proposed a new calendar with 125 months of 30 days and 110 of 29 days, corresponding to an average year of 365 days, 6 hours and 19 minutes — only about 30 minutes too long. Of this "Meton's cycle" still preserves traces. On account of so much confusion in the official calendar the almanacs of the time designated the dates for agricultural operations by means of the constellations visible at the corresponding time.

[1] ARISTOPHANES, *The Clouds* (B. B. Rogers, *Aristophanes*, Vol. 1, pp. 323–324).

TIME MEASUREMENT

We have seen in Chapter II that in Babylonia and Egypt time measurement depended either on some form of sun-dial as a natural means, or on an apparatus analogous to the hour-glass as an artificial method. Anaximander (p. 47) is said to have seen the Babylonian sun-dial and to have introduced it into Greece during the first half of the sixth century. But we have no knowledge of the form of the instrument until a much later period.

The first Greek sun-dial of which a description is preserved belongs to the time of Alexander the Great, and consisted of a hollow hemisphere with its rim horizontal and a bead at the center to cast the shadow. Curves drawn on the concave interior divided the period from sunrise to sunset into twelve parts, these lengths being thus proportionate to the lengths of the daylight period.

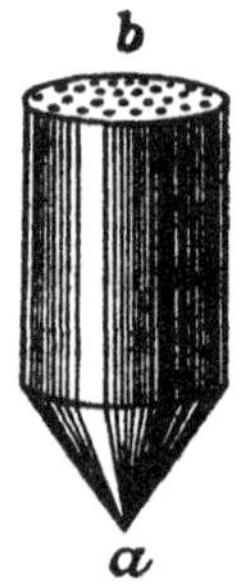

FIG. 6. — THE CLEPSYDRA AS DESCRIBED BY EMPEDOCLES. *a*, orifice; *b*, perforated top. Redrawn after Burnet, *Early Greek Philosophy*, p. 230.

The use of the clepsydra, or water-clock, in Greece dates from the fifth century B.C. It consisted there of a spherical bottle with a minute outlet for the gradual escape of water. Its use in regulating public speaking is illustrated by Demosthenes's demand when interrupted, "You there: stop the water."

For the sake of conformity with the sun-dial division of each day and each night into twelve equal parts, the rate of flow in the clepsydra required continual adjustment. Ingenious improvements were made in the mechanism in course of time, but in considering the work of the Greek astronomers, the impossibility of accurate time measurement must not be forgotten.

GREEK ARITHMETIC

In Greek arithmetic the earliest known numerals are merely the initials of the respective number words. This "Attic" system was used in Athens as late as about 95 B.C. Another system, perhaps almost equally ancient, employed all of the Ionian alphabet and added three archaic letters. The 27 characters formed three groups of nine consecutive letters representing units, tens, and hundreds, respectively. Thus the first letter $\alpha = 1$; the tenth $\iota = 10$; and the nineteenth $\rho = 100$. By this alphabetical notation all numbers from 1 to 999 could be expressed. The signs for thousands were the same as for units prefixed by a stroke, or inverted accent, e.g., $_{\prime}\beta = 2{,}000$; and the sign for myriads, or tens of thousands, was M with the number of myriads indicated above, $\overset{\gamma}{\mathrm{M}} = 30{,}000$. While certainly known by 450 B.C., this system was not in general use until much later, and was not official until the Hellenistic period. This use of letters for numbers was not confined to Greece, but appears to have originated there. The Greeks had no zero, and never discovered the immense advantage of a position-system, such as that by which we are able to express all numbers by only ten symbols. Fractions occur not infrequently. The change from the earlier notation to that with 27 characters was detrimental. There were not only more characters to memorize, but computation became materially more complicated. These disadvantages far more than offset the superior compactness, the sole merit of the new system. The special importance of such compactness for coins has led to the suggestion that they were the medium through which this notation was introduced.

In a simple numerical computation of late date the Greek alphabetic numerals and their modern equivalents are:

$\overline{\sigma\xi\epsilon}$			
$\overline{\sigma\xi\epsilon}$	265		
	265		
$\overset{\delta}{M}\ \overset{\alpha}{M},\beta\ ,\alpha$	40000,	12000,	1000
$\overset{\alpha}{M},\beta\ ,\gamma\overline{\chi}\ \overline{\tau}$	12000,	3600,	300
$,\alpha\ \tau\ \kappa\epsilon$	1000,	300,	25
$\overset{\zeta}{M}\ \overline{\sigma\kappa\epsilon}$	70225.		

An Example of Greek Computation — Gow, *Hist. Greek Math.*, p. 50.

Actually for such a computation, the Greeks used pebbles or some kind of abacus.

Division was an exceedingly laborious process of repeated subtraction.

Probably nothing in the modern world would have more astonished a Greek mathematician than to learn that, under the influence of compulsory education, the whole population of Western Europe, from the highest to the lowest, could perform the operation of division for the largest numbers. — Whitehead, *Introduction to Mathematics*, p. 59.

Approximate square roots were found by the later Greeks. The reckoning board, or abacus — known in so many different forms throughout the world — came into very early use, but actual evidence in regard to its form is meagre.

A sharp distinction was made between the art of calculation (*logistica*), and the science of numbers (*arithmetica*). The former was deemed unworthy the attention of philosophers, and to their attitude may be fairly attributed the fact that Greek mathematics was always weak on the analytical side, and seemed in a few centuries to reach the limit of its possible development.

GREEK GEOMETRY

It was in geometry that Greek mathematics chiefly developed, and for several fundamental reasons. The Greek mind had a strong predilection for formal logic, a keen æsthetic appreciation of beauty of form, and, on the other hand, with no adequate symbolism for arithmetic or algebra, a distinct disdain, at any rate among the educated, for the commercialized mathematics of computation. The history of Greek mathematics is therefore to a great extent the history of geometry. Formal geometry as distinguished from the solving of particular geometrical problems, had, indeed, no previous existence, and we have to do with the beginnings of elementary geometry as we now know it.

THE IONIAN PHILOSOPHERS

The sense of curiosity, the feeling of wonder, the spirit of inquiry — these are the common elements of philosophy and science.

In the childhood and youth of the race specialization has not begun, all knowledge lies invitingly open to the expanding mind. We have seen how much had been accumulated in Egypt and Babylonia of knowledge and skill in observing and recording the phenomena of the heavens, in irrigation, and in measurement of land. Much of the same general character was doubtless true of the Phœnicians, the Trojans, the Cretans, and other precursors of the Greeks. But nothing deserving the name of science has come down to us from the Ægean or Greek civilization before the time of Thales of Miletus, chief of the Ionian philosophers, and one of the "seven wise men of Greece."

THALES (*c. 624–? B.C.*)

The ancient and fragmentary register of Greek mathematicians, the *Summary of Proclus*, attributed to Eudemus, reads:

> Geometry is said by many to have been invented among the Egyptians, its origin being due to the measurement of plots of land. This was necessary there because of the rising of the Nile, which obliterated the boundaries appertaining to separate owners. Nor is it marvellous that the discovery of this and the other sciences should have arisen from such an occasion, since everything which moves in development will advance from the imperfect to the perfect. From mere sense-perception to calculation, and from this to reasoning, is

a natural transition. Just as among the Phœnicians, through commerce and exchange, an accurate knowledge of numbers was originated, so also among the Egyptians geometry was invented for the reason above stated.

Thales first went to Egypt, and thence introduced this study into Greece. He discovered much himself, and suggested to his successors the sources of much more; some questions he attacked in their general form, others empirically. — PROCLUS. (Gow, *Short Hist. Greek Math.*, pp. 134–135.)

The significance packed into this terse quotation may be emphasized. The mathematics of the Mesopotamians, the Egyptians, and the Phœnicians was essentially a tool to meet concrete needs and lacked abstract formulæ. Nevertheless, as we have seen, the Egyptians had developed a solid geometry, and the Babylonians had methods for solving quadratic and even some cubic equations. The Greek intellect, seizing upon the knowledge of these practical races, refined from it the germs of a new pure science, making the knowledge "more general" and "more comprehensible," and at the same time discovering much that was new. On the other hand, inclining in its zeal for pure science to the opposite extreme of disregard for the concrete applications, Greek science eventually reached its own limit of possible growth. In the long run scientific progress must depend on due recognition of the complementary importance of both pure and applied science.

Thales was of Phœnician descent, a native of Miletus, a city of Ionia, a flourishing Greek colony in what is now Asia Minor. As an engineer he was employed to construct an embankment for the river Halys. As a merchant he dealt in salt and oil, and, visiting Egypt, learned there something of the wisdom of the Egyptian priesthood. He occupied himself with the study of the stars as well as of geometry, and in particular,

announced to the inhabitants of Miletus that night would enter upon the day, the sun hide himself, the moon place herself in front, so that his light and radiance would be intercepted.

Herodotus relates that there was a war between the Lydians and the Medes, and after various turns of fortune,

They were still warring with equal success, when it chanced, at an encounter which happened in the sixth year, that during the battle the day was turned to night. Thales of Miletus had foretold this loss of daylight to the Ionians, fixing it within the year in which the change did indeed happen. So when the Lydians and Medes saw the day turned to night they ceased from fighting, and both were the more zealous to make peace. — A. D. Godley, *Herodotus*, I, p. 91.

This eclipse is supposed to have taken place in 585 B.C. The prediction of the year of an eclipse gained Thales a great reputation with his contemporaries, though his designation with six others as "wise men of Greece" appears to have had a primarily political significance. None of the other six at any rate had any scientific standing. He taught that the year has 365 days; that the equinoxes divide the year unequally; that the moon is illuminated by the sun. The geometrical attainments attributed to him include the following elementary theorems: the angles at the base of an isosceles triangle are equal; the circle is bisected by its diameter. He was acquainted with the inscription of the right triangle in the semicircle and the measurement of height by shadow, involving the principle of similar triangles. Plutarch relates that Niloxenus, conversing with Thales concerning King Amasis, says:

Although he also admires you on account of other things, he prizes above everything the measurement of the pyramids, in that you have without any trouble and without needing an instrument, merely placed your staff at the end of the shadow cast by the pyramid, showing from the two triangles formed by the contact of the solar rays that one shadow has the same relation to the other as the pyramid to the staff. — Plutarch, *Banquet of the Seven Wise Men*.

In connection with his shadow measurements it is interesting that his scholar Anaximander, introduced the sun-dial into Greece.

While our knowledge of Thales and his work is extremely meagre, the mathematical results above mentioned have considerable significance in connection with the comparison between Greece and Egypt. Most of the facts ascribed to Thales may well have been known to the Egyptians. For them these facts would have remained unrelated; for the Greeks they were the beginnings of an extraordinary development of the science of geometry.

To Thales the earth is a circular disk floating in an ocean of water. This water is the fundamental element of the whole. Ice, snow, and frost turn readily into water, even rocks wear away and disappear in it. Man himself seems capable of turning into it, while the waters of sea and land shrink into solid

FIG. 7. — THE WORLD AS DESCRIBED BY HECATAEUS OF MILETUS (p. 66). Redrawn by Elizabeth Tyler Wolcott after Breasted, *Ancient Times*, courtesy of Messrs. Ginn and Company.

residues. By evaporation of the water air is formed, its agitation causing earthquakes. The stars between their setting and rising pass behind the earth.

As to the quantity and form of this first principle, there is a difference of opinion; but Thales, the founder of this sort of philosophy, says that it is water (accordingly he declares that the earth rests on water), getting the idea, I suppose, because he saw that the nourishment of all beings is moist, and that warmth itself is generated from moisture and persists in it (for that from which all things spring is the first principle of them); and getting the idea also from the fact that the germs of all beings are of a moist nature, while water is the first principle of the nature of what is moist. . . .

Some say that the earth rests on water. We have ascertained that the oldest statement of this character is the one accredited to Thales the Milesian, to the effect that it rests on water, floating like a piece of wood or something else of that sort. . . . And Thales, according to what is related of him, seems to have regarded the soul as something endowed with the power of motion, if indeed he said that the loadstone has a soul because it moves iron. . . . Some say that soul is diffused throughout the whole universe; and it may have been this which led Thales to think that all things are full of gods. — ARISTOTLE. (Arthur Fairbanks, *First Philosophers of Greece*, p. 2.)

ANAXIMANDER (c. 611–545 B.C.)

A second philosopher of Miletus, Anaximander, made a different interpretation of nature, holding that the fundamental stuff, out of which all things are made, is something between air and water. He believed the earth to be balanced in the center of the world, because being in the center and having the same relation to all parts of the circumference, it ought not to tend to fall in one direction rather than in any other. This point of view illustrates the natural tendency of the Greek philosopher to emphasize geometrical symmetry.

Among those who say that the first principle is one and movable and infinite, is Anaximandros of Miletos, . . . pupil and successor of Thales. He said that the first principle and element of all things is infinite, and he was the first to apply this word to the first principle; and he says that it is neither water nor any other one of the things called elements, but the infinite is something of a different nature, from which came all the heavens and the worlds in them; and from what source things arise, to that they return of necessity when they are destroyed; . . . Evidently when he sees the four elements changing into one another, he does not deem it right to make any one of these the underlying substance, but something else besides them. — THEOPHRASTUS. (Fairbanks, p. 11.)

Anaximander, collecting data from the Ionian sailors frequenting Miletus, constructed a map of the earth, and speculated on the relative distances of the heavenly bodies.

ANAXIMENES (sixth century B.C.)

A third native of Miletus is Anaximenes. For him the stars are fixed upon the celestial vault, and pass behind the northern (highest) part of the earth on setting. Air, not water, is the first cause of all things, the others being formed by its compression or rarefaction. The heat of the sun is due to its rapid motion, but the stars are too remote to give out heat.

Most of the earlier students of the heavenly bodies believed that the sun did not go underneath the earth, but rather around the earth and this region, and that it disappeared from view and produced night because the earth was so high toward the north.

Anaximenes and Anaxagoras and Demokritos say that the breadth of the earth is the reason why it remains where it is.

Anaximenes says that the earth was wet, and when it dried it broke apart, and that earthquakes are due to the breaking and falling of hills. — ARISTOTLE. (Fairbanks, p. 18.)

The school of Thales and his successors in this Ionian outpost of Greek civilization was soon succeeded by developments of still greater importance in the more remote Italian colonies.

PYTHAGORAS AND HIS SCHOOL

The register of mathematicians (Proclus) proceeds: — "But next Pythagoras changed the study of geometry into the form of a liberal education, for he examined its principles to the bottom and investigated its theorems in an immaterial and intellectual manner. It was he who discovered the subject of irrational quantities and the composition of the cosmical figures." These few words, like those quoted of Thales, are full of meaning. Pythagoras, assembling the facts known to the Egyptians and Babylonians and the theorems of Thales, began the systematic organization of scientific philosophy and in particular of mathematics.

Pythagoras, about 532 B.C., founded in Crotona, a Greek city of southern Italy, a school which had much of the character of a fraternity or secret society, this with political tendencies ultimately arousing hostility which proved destructive to it. Beyond these undisputed facts his life and work are obscured by a great mass of tradition and myth, even the date of his birth being doubtful. A native of the island of Samos not far from Miletus, he appears to have been much affected by Egyptian influences due to a prolonged residence in that country. A visit to Babylon is alleged, but with doubtful authority.

The etiquette of the Pythagorean school required that all discoveries should be attributed to the "Master" and not revealed to outsiders. Pythagoras, himself, appears to have been interested in geometry, as witnessed by the so-called Pythagorean theorem, and in the theory of numbers, particularly in connection with music and geometry. He is said to have first introduced weights and measures among the Greeks.

The attribution of particular results or beliefs to individuals of this period is, however, very doubtful on account of the fact that Pythagoras left no writings whatever, that his school was essentially a secret society, and that in later centuries it became the custom to credit its founder with all sorts of knowledge which he could not possibly have possessed.

Pythagoras makes the classification, arithmetic (numbers absolute), music (numbers applied), geometry (magnitudes at rest), astronomy (magnitudes in motion), this fourfold division or "quadrivium" continuing in vogue for some two thousand years. The distinction between abstract and concrete arithmetic had been emphasized among the Greeks in comparatively early times. Arithmetic and geometry were distinguished on one side from mechanics, astronomy, optics, surveying, music, and computation on the other. The aim of Greek arithmetic "was entirely different from that of the ordinary calculator, and it was natural that the philosopher who sought in numbers to find the plan on which the Creator worked, should begin to regard with contempt the merchant who wanted only to know how many sardines, at 10 for an obol, he could buy for a talent."

PYTHAGOREAN ARITHMETIC

In pure arithmetic, or number theory as we should call it, the Pythagoreans enunciated such dicta as, for example, "Unity is the origin and beginning of all numbers but not itself a number." Prime and composite numbers were distinguished, and theorems of considerable algebraic complexity

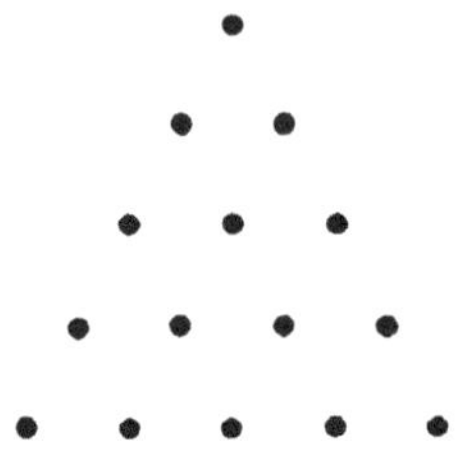

FIG. 8. — PYTHAGOREAN TRIANGULAR NUMBERS.

discovered. There is naturally no algebraic symbolism, but "unknown" and "given" quantities are employed in the modern sense. Odd and even numbers received special names, and besides the series of squares and cubes and the arithmetic and geometric progressions previously known, other series were derived from these, for example, the triangular numbers: 1, 3, 6, 10, 15, etc., by successive addition of the natural numbers. The reason for the name triangular will be clear if one counts the dots in the triangle formed by taking one, two, three or more rows beginning at the top of the figure.

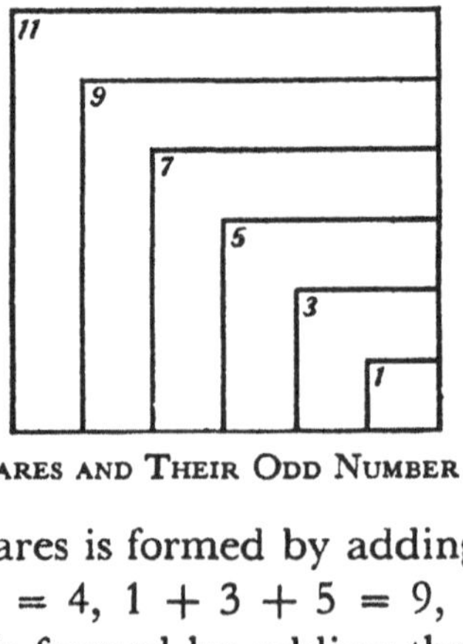

FIG. 9. — SQUARES AND THEIR ODD NUMBER DIFFERENCES.

The series of squares is formed by adding the odd numbers successively: $1 + 3 = 4$, $1 + 3 + 5 = 9$, etc. The series 2, 6, 12, 20, 30, etc., is formed by adding the even numbers, or again by multiplying adjacent natural numbers. If we construct a series of squares or parallelograms with a common angle and sides of length 1, 2, 3, 4, 5, etc., the figure which must be added to any one to produce the next larger was called by the Greeks a *gnomon*, the area of which would be represented by one of the series of odd numbers — an interesting and typical example of the Greek habit of combining geometry with number-theory. As products of two numbers were associated with areas — "square" or "oblong" — so products of three factors were interpreted as volumes.

PYTHAGOREAN GEOMETRY

In geometry the Pythagoreans formulated definitions of the fundamental elements, line, surface, angle, etc. They are

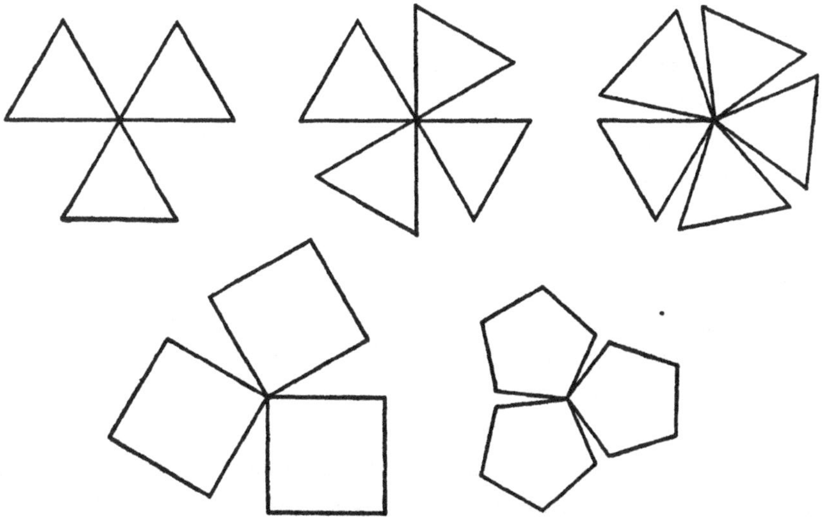

FIG. 10. — FORMATION OF COSMICAL BODIES FROM EQUILATERAL TRIANGLES, SQUARES, AND PENTAGONS.

credited with a number of theorems, implying a knowledge of methods of determining area and of the properties of parallel lines. They developed a fairly complete theory of the triangle, including the fundamental proof that the sum of the angles of a triangle is two right angles, by a method not very different from our own. The theory of the "cosmical bodies," mentioned in the register, is of special interest. Any solid angle must have at least three faces. If three equal equilateral triangles have a common vertex they will when cut or folded so that their edges are brought together, form a solid angle, and a fourth equal triangle will complete a regular tetrahedron. Similarly, if we start with four triangles, we may build up by folding with four others a regular octahedron, or starting with five, an icosahedron with 20 faces. Six triangles, however, will fill the angular space about a point, and thus not permit the formation of a regular polyhedron. Using squares instead of triangles, we obtain only the cube; using pentagons (angle 108°), the regular dodecahedron — 12 faces, 3 at each vertex. The Egyptians must have been familiar with the cube, the regular tetrahedron, and the octahedron. To these, with the icosahedron, the Pythagoreans associated the four elements — earth, air, fire, and water. Their discovery of an additional body, the regular dodecahedron, formed by 12 pentagons, made a break in the correspondence, and the need was met by the addition of the universe, or, according to others, the ether, as a fifth term in the cosmical series. This correspondence was not merely symbolical, but physical, the earth being supposed to consist of cubical particles, etc. We cannot infer that the impossibility of a sixth regular polyhedron was known. That only these five regular polyhedra are possible was in fact first actually proved by Euclid. There is a tradition that the Pythagorean discoverer of the dodecahedron was drowned at sea on account of the sacrilege of announcing his discovery publicly. A later commentator records a similar tradition that the discoverer of the irrational perished by shipwreck, since the inexpressible should remain forever concealed, and that he who touched and opened up this picture of life was transported to the place of creation and there washed in eternal floods.

The regular polygons naturally were studied, and in particular the decomposition of them into right triangles of 45° and 30°–60°. With the pentagon the attempt naturally failed, but the five-pointed star formed by drawing diagonals was a special emblem of the Pythagoreans. With the inscribed pentagon connects itself naturally the division of a line in extreme and mean ratio, or, as it was later characterized, the "golden section." This division, by which the square on the greater segment of a line is equivalent to the rectangle whose sides are the other segment and the whole line, occurs repeatedly in Greek architecture of the fifth century, with fine effect, and must have been systematically employed.

To Pythagoras must probably be ascribed a proof in the modern sense of the celebrated theorem of Babylonian origin which bears his name: — that the sum of the squares on the two sides of the right triangle is equivalent to the square of the hypotenuse — this forming the basis for the theory of the irrational mentioned by Proclus. If familiar with the triangle having sides of length 3, 4, and 5, it would be natural to investigate whether a similar relation could be verified for other right triangles. In the most familiar case of the isosceles right triangle it soon appears that the length of the equal sides being taken as 1, the length of the hypotenuse could be only approximately expressed. Though it can be so easily constructed, its length cannot indeed be exactly expressed by any whole number, or fraction; it is *irrational*.

If it is true as Whewell says, that the essence of the triumphs of science and its progress consists in that it enables us to consider evident and necessary, views which our ancestors held to be unintelligible and were unable to comprehend, then the extension of the number concept to include the irrational, and we will at once add, the imaginary, is the greatest forward step which pure mathematics has ever taken. — H. HANKEL. (Moritz, p. 281.)

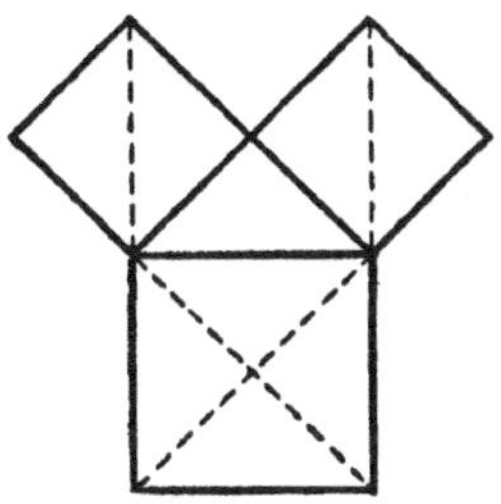

FIG. 11. — A SPECIAL CASE OF THE PYTHAGOREAN THEOREM.

In this special case the proof of the Pythagorean theorem is easily effected by a simple graphical construction, involving merely the drawing of diagonals of squares. The smaller triangles in the figure are evidently all equal. The larger square contains four of them, the smaller squares, two each. It seems possible that this was the Pythagorean method, but as to how the proof was accomplished in other cases we have no information, the simpler proof of Euclid having completely superseded the earlier. On the other hand, for the corresponding arithmetical problem of finding three whole numbers which can be the sides of a right triangle, Pythagoras is said to have given a correct solution, equivalent in our notation to

$$(2a + 1)^2 + (2a^2 + 2a)^2 = (2a^2 + 2a + 1)^2,$$

a denoting any positive integer. How this method was discovered remains a matter of conjecture.

We may recognize here the characteristic elements of the inductive method, first, observation of the particular fact that in a certain right triangle, with sides 3, 4, and 5, the sum of the squares on the two sides is equal to that on the hypotenuse; second, the formation of the hypothesis that this may be true also for right triangles in general; third, the verification of the hypothesis in other particular cases. Then follows the deductive confirmation of the hypothesis as a law for all right triangles.

PYTHAGOREAN ASTRONOMY AND PHYSICAL SCIENCE

It has been already noted that one of the most fundamental principles of the Pythagorean school was the significance attached to number in connection with all sorts of phenomena, the regular motions of the heavenly bodies, the musical tones, etc. There is a tradition that Pythagoras, walking one day, meditating on the means of measuring musical notes, happened to pass near a blacksmith's shop, and had his attention arrested by hearing the hammers as they struck the anvil produce sounds which had a musical relation to each other. It was observed that vibrating cords emitted tones dependent in a simple way on their length; for example, cords of lengths 2, 3, and 4 giving a tone, its fifth, and its octave, respectively.

Pythagoras and his school evidently understood that the shorter string gives the higher pitch; but as yet there was no comprehension of the fact that the tone emitted by a musical instrument is completely determined by the physical structure of the instrument. It was even supposed that each of the various heavenly bodies and the sphere of the fixed stars had a characteristic tone, these all uniting to produce the so-called "music of the spheres." The quantitative description of phenomena introduced by Pythagoras foreshadows, in a certain sense, the whole story of mathematical physics.

Pythagoras himself seems to have been the first among the Greeks to adopt the idea that the Earth was a sphere while, as we shall see in the next chapter, a later member of his school took the further great step of giving it an orbital motion. The common sense view at the time was naturally that the Earth was evidently flat and immovable.

From this period dates the long-lived theory that the universe consisted of the four elements, earth, air, fire, and water. Pythagoras identified the morning and evening stars and attributed the Moon's light to reflection. His universe was a sphere with a spherical earth at its center and endowed with a kind of life.

The activity of the Pythagorean school continued to be important until about 400 B.C.; that is, until the rise of the Athenian school under Plato and his successors. It had not only created the science of mathematics; it had developed, however vaguely and imperfectly, the idea of a world of physical phenomena governed by mathematical laws.

Dr. Allman says of Pythagoras:

> In establishing the existence of the regular solids he showed his deductive power; in investigating the elementary laws of sound he proved his capacity for induction; and in combining arithmetic with geometry . . . he gave an instance of his philosophic power.
>
> These services, though great, do not form, however, the chief title of this Sage to the gratitude of mankind. He resolved that the knowledge which he had acquired with so great labour, and the doctrine which he had taken such pains to elaborate, should not be lost; and . . . devoted himself to the formation of a society *d'élite*, which would

be fit for the reception and transmission of his science and philosophy; and thus became one of the chief benefactors of humanity, and earned the gratitude of countless generations. — G. J. ALLMAN, *Greek Geometry from Thales to Euclid*, p. 50.

GREEK CHEMISTRY

One branch, at least, of applied chemistry flourished in Athens. "The people were mostly craftsmen, merchants, and sailors, with pottery as the chief article of export, for Athens led the world in ceramics." Egypt is generally regarded as the mother country of the chemical arts, but it is from Greek and Roman literature that we learn most about early chemical theory, and it is not altogether clear how much of this originated in Egypt. The early Greek philosophers had great influence in developing chemical theory, but their social positions and the prevailing idea that all manual labor was degrading, kept them out of touch with first hand practical knowledge, "pure thought" being alone worthy of their philosophy. For this reason, progress in chemistry was retarded and did not keep pace with the development of mathematics and metaphysics along rational lines.

Mention has been made of attempts to discover a primordial element, believed to be water by Thales (p. 43) and air by Anaximenes (p. 48), who thought fire to be produced from air by rarefaction, and material substances by its condensation. It is interesting to note that the indeterminate primordial substance of Anaximander, the *apeiron*, is somewhat similar to the ancient Hindu concept of the *akasa*, or *ether;* and that the ideas of Pythagoras in regard to number and form as applied to substances have analogies in the most modern theories concerning the constitution of matter.

EARLY GREEK MEDICINE

We have seen that in the primitive man and in the great civilizations of Egypt and Babylonia, the physician evolved from the priest — in Greece he had a dual origin, philosophy and religion. — WILLIAM OSLER, *Evolution of Modern Medicine*, p. 37.

If we would understand the intellectual development of a people, we must take into account its religious problems and feelings. — G. SARTON, *Introduction to the History of Science.* I, p. 4.

The earliest recorded religion of the Greeks is set forth in the poems of Homer and of Hesiod. Whatever their origin, the poems were put into writing about 700 B.C. The religion they describe was very different from the fearful beliefs of the Sumerians and the Egyptians.

"It was a religion that in itself gave scope for all the joyousness of a sunny race, all its gift of humanity and beauty and sociability." — W. C. GREENE, *Achievement of Greece*, p. 226.

Apollo, the sun god, was the original god of healing. But the Greeks attributed their first knowledge of medicine to Asclepios (Latin, Aesculapius), who appears to have been a real chieftain and physician of Thessaly, and to have fought in the Trojan War. The Homeric poems reveal the existence at that time of a developed practice of medicine and wound surgery. Later Asclepios was deified, as was Imhotep in Egypt (p. 33), and temples were erected for his worship. According to the myth, he was the son of Apollo and had two daughters, Hygeia and Panacea, who attended the temples and cared for the sacred, harmless snakes, the symbol of the god. This serpent cult suggests a Minoan influence (Singer '28, fig. 2), which in turn had its origin in remote Sumer (p. 7).

A temple of healing was called an Asclepieion, and the site was usually selected with intelligence on a hill or mountain or in a forest near a source of pure water or a thermal or mineral spring. By the time of Alexander (356–323 B.C.) it is estimated that over 200 such temples existed. The famous ones had become great health resorts, as has been shown by excavations of Epidaurus, about 30 miles from Athens, on the island of Cos, and at Pergamum in Asia Minor. At each temple a staff of trained priests received selected cases, excluding anything unclean, and after prolonged preparation by baths, massage, and diet, followed by prayers and sacrifice, the patient was brought into the temple for the *incubation* sleep, during which in a dream, or, failing that, in the person of a priest, the god appeared and gave directions for the cure. Testimony to these temple cures and descriptions of them have been found in numerous votive offerings and in a great variety of documents. There is evidence that much of the medical lore of the priests, and of other physicians as well, had its origin, probably through the conquered Minoans, in Mesopotamia and in Egypt.

We may turn from the mysticism of the temples to an entirely different aspect of the Greek genius —

> The veil of Nature the Greek lifted and herein lies his value to us. . . . and for the first time in history, man had a clear vision of the world about him — "had gazed on Nature's naked loveliness" ("Adonais") unabashed and unaffrighted by the supernatural powers about him. — WILLIAM OSLER, *Evolution of Modern Medicine*, p. 36.

Like mathematics, medicine was regarded as a branch of philosophy, and it was the early philosophers — "particularly the Ionian Physiologists whether at home or as colonists in the south of Italy, in whose work the beginnings of scientific medicine may be found." (OSLER, *loc. cit.*)

There were Greek physicians before the temples. No mention is made of Asclepieia in the poems of Homer, but these poems reveal the existence of professional physicians, not priests, and of women having some knowledge of drugs. Anatomy is represented by 150 special words, and wounds were accurately described. But it is to the Ionian philosophers that we must look for ideas that formed much of the theoretical basis of Greek medicine. Their outlook upon Nature, and their spirit of investigation had a profound influence. To Anaximenes (p. 48) may be traced the doctrine of the *pneuma*, or breath of life; and Pythagoras (p. 49) had an extraordinary influence upon medicine through his theory of numbers and idea of the importance of critical days, which was favored by certain common diseases — malaria and typhus. Probably attached to his school at Crotona was Alcmaeon, the most famous of Greek physicians before Hippocrates, the first to recognize the brain as the seat of the mind and the origin of the nerves. In the fragments of his writings are found the earliest records of anatomical observations in Greece. By dissection of animals he began the construction of a positive basis for medical science. He described the optic nerve and the Eustachian tube. His use of the word "pore" for nerve is historically important. His study of the distribution of blood vessels led to anatomical investigations of others in the centuries to follow.

By the sixth century B.C. in Ionia there were in active operation two medical schools that were to become famous. The earliest was at Cnidos, a peninsula on the southwest coast of Asia Minor; and the other, a little later, was on the neighboring island of Cos. At about the same time Darius the Great, king of Persia from 521 to 485 B.C., founded a medical school in Egypt, the earliest known scientific foundation by royal mandate.

REFERENCES FOR READING

ALLMAN, G. J., *Greek Geometry from Thales to Euclid.* 1889. Chapters I, II.
CAJORI, FLORIAN, *A History of Mathematics.* Ed. 2, 1919. pp. 16–23.
DREYER, J. L. E., *Planetary Systems.* 1906. Chapters I, II.
FAIRBANKS, ARTHUR, *The First Philosophers of Greece.* London. 1898.
GOW, JAMES, *A Short History of Greek Mathematics.* 1884. Chapters III, IV, VI.
HEATH, T. L., *Aristarchus of Samos.* 1913.
——, *History of Greek Mathematics.* 1921.
The Legacy of Greece,[1] Ed. by R. W. Livingstone. 1922.
OSLER, (SIR) WILLIAM, *The Evolution of Modern Medicine.* 1921. pp. 35–58.
SARTON, GEORGE, *The History of Science and the New Humanism.* 1931.
——, *Introduction to the History of Science,* Vol. I, pp. 65–80.
Science and Civilization,[1] Ed. by F. S. Marvin. 1923.
SINGER, CHARLES, *Short History of Medicine.* 1928. pp. 1–13.
SMITH, D. E., *History of Mathematics.* 1923–25.

[1] The collective title will be used in references to books of essays.

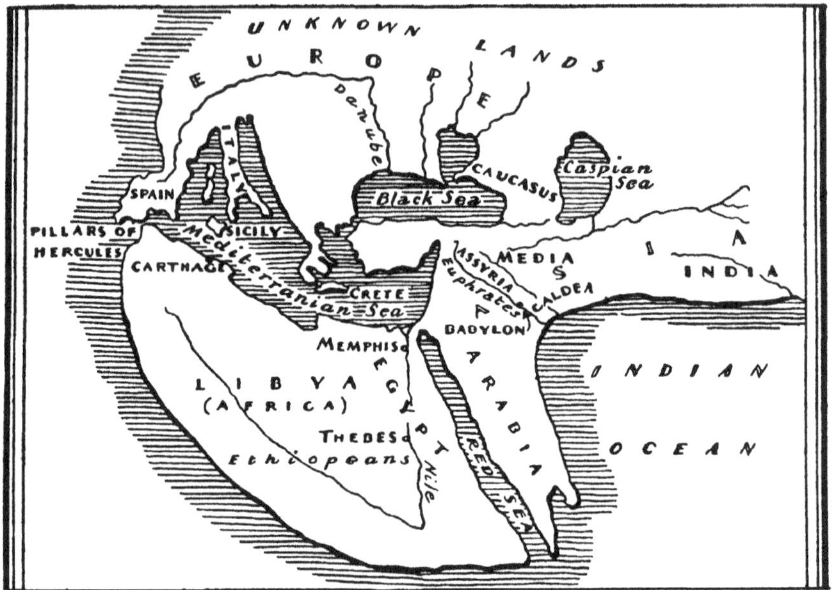

Chapter IV

MAP OF THE WORLD ACCORDING TO HERODOTUS

Science in the Golden Age of Greece

There still remain three studies suitable for freemen. Arithmetic is one of them; the measurement of length, surface, and depth is the second; and the third has to do with the revolutions of the stars in relation to one another . . . there is something in them which is necessary and cannot be set aside . . . if I am not mistaken, [something of] divine necessity. — PLATO, *Laws*, VII, 818 (Jowett).

In biology Aristotle speaks for the first time the language of modern science, and indeed he seems to have been first and foremost a biologist, and his natural history studies influenced profoundly his sociology, his psychology, and his philosophy in general. . . . He must be indeed a dull and muddy-mettled rascal whose imagination is not fired by the . . . picture of the founder of modern biology, whose language is our language, whose methods and problems are our own, the man who knew a thousand varied forms of life, — of plant, of bird, of animal, — their outward structure, their metamorphosis, their early development; who studied the problems of heredity, of sex, of nutrition, of growth, of adaptation, and of the struggle for existence. — WILLIAM OSLER, *The Old Humanities and the New Science*, p. 21.

LITERATURE AND ART

The fifth century B.C. witnessed that astonishing flowering of the Greek genius in literature and military glory which has made it forever famous. The battles of Marathon, 490 B.C., and Salamis, 480 B.C., had flung back the Asiatic hosts which threatened to overrun and enslave Europe, and had transformed the Greeks from a group of jealous and parochial city states into a great democratic nation. Trade prospered, wealth increased, and for about a century letters, art, and science flourished as never before and perhaps never since. History began to be written by Herodotus and Thucydides. The drama was developed by Aeschylus, Sophocles, and Euripides to such a pitch that even today, after the lapse of nearly 2,500 years, we listen with eager interest to the *Oedipus* of Sophocles and the *Iphigenia* of Euripides, while the poetry of Pindar and the wit of Aristophanes have never lost their charm. In architecture and the plastic arts the Parthenon and its sculptures still testify to Greek supremacy.

In science, also, great names are associated with memorable deeds. No such perfection, to be sure, was attained in science as in literature and in sculpture, but vast progress was made in mathematics beyond anything hitherto accomplished, and the foundations were securely laid for a rational interpretation of man and of nature. Literature, architecture, sculpture, and the drama require no special apparatus or equipment. Mathematics also is not dependent upon such externals, and we find geometry and arithmetic moving forward far more rapidly than natural or physical science.

PARMENIDES (*c. 500 B.C.*)

The recognition of the spherical shape of the earth and its division into zones are attributed not only to the Pythagoreans, but also to Parmenides of Elea. He introduced a system of concentric spheres analogous to that soon to be so highly de-

---

NOTE: The map on the facing page is redrawn after Breasted, *Ancient Times*, courtesy of Messrs. Ginn and Company.

veloped by Eudoxus. He identified the evening and the morning stars, and attributed the moon's brightness to reflected light. He regarded the sun as consisting of hot and subtle matter detached from the Milky Way, the moon chiefly of the dark and cold.

EMPEDOCLES (c. 490–c. 435 B.C.)

Passing over the guesses of Heraclitus and Parmenides at the riddle of existence and of man and nature, we may pause for a moment to examine the ideas of Empedocles (about 455 B.C.). A native of Agrigentum in southern Sicily, Empedocles was regarded as poet, philosopher, seer, and immortal god. He appears to have been a close observer of nature, understanding the true cause of solar eclipses and believing the moon to be twice as far from the sun as from the earth. The latter is held in place by the rapidly rotating heavens "as the water remains in a goblet which is swung quickly round in a circle."

Empedocles is mentioned by Aristotle as the first to announce the four elements — *earth*, *air*, *water*, and *fire* as the constituents of all matter in the universe. The four elements possess the attribute of immortality and each of the four is unchangeable in quality throughout all changes. All other substances may be resolved into the four elements. Specific attractions or repulsions, which he called love and hate, produced combinations from these elements. This theory of four elements is the first approach to the conception of what we call a chemical element. Empedocles stated that flesh and blood contain equal quantities of the four elements, bones are half fire, one-fourth earth, and one-fourth water. Such ideas associated with the authority of a great name have impeded the development of chemical theory in later centuries.

ANAXAGORAS (c. 499–c. 428 B.C.)

For the student of science Anaxagoras, a native of Clazomene in Asia Minor, is more important than Empedocles. Turning aside from wealth and civic distinction in his enthusiasm for science, he seems to have occupied himself with the problem of squaring the circle, a problem attacked even by the Egyptians with some degree of success, and destined to exercise great influence on the development of Greek geometry. The beginnings of perspective are also attributed to him, in connection with studies of the stage. He was particularly interested in a great meteorite — the appearance of which he was afterwards said to have predicted — supposing it to have fallen from the sun, and inferring that the latter was a "mass of red-hot iron greater than the Peloponnesus," not very distant from the earth. Like the Pythagoreans he assigned as the order of distances: — moon, sun, Venus, Mercury, Mars, Jupiter, Saturn. The earth's axis was inclined, in order that there might be variations of climate and habitability. He explained the moon's phases correctly, also solar and lunar eclipses, but he misinterpreted the Milky Way as due to the shadow cast by the earth.

Anaxagoras held that the universe originally consisted of infinite space filled homogeneously with small particles or seeds. These particles are the elementary particles of all known material — air, gold, water, flesh, bone, etc. An Intelligence or Will called the "*nous*" acted upon these particles and made substances from them. When any substance is destroyed, it is resolved by the *nous* into its constituents. This theory is of historical interest because it abandons any attempt to account for the evolution of the universe as a result of natural, physical properties of matter, but frankly postulates an external, though perhaps impersonal, intelligence as the organizing and directing force. This idea became widely adopted, diverting the Greek spirit from the physical to the mataphysical. However, the theory was repugnant to those holding the polytheistic dogmas of his time and brought him into popular disfavor. Convicted of impiety, he died in exile, 428 B.C.

HERACLITUS

Heraclitus of Ephesus, who wrote at about 490–478 B.C., considered *fire* as the primeval element and the moving and creative force of the universe. He recognized the existence of universal law and the strict sequence of cause and effect in nature. He attempted to harmonize the evidence of the senses with the requirements of human reason, accounting for the visible universe by natural law rather than by supernatural intervention. These tendencies mark an important step in the development of civilization.

THE ATOMISTS

A very little observation of external nature shows that disintegration is forever going on. Ice turns to water, water to vapor, rocks to sand, and sand to dust — in other words, masses to particles. Furthermore, dust vanishes and vapor disappears, while clouds and fogs, rain and snow, make their appearance without obvious cause, and dust accumulates from invisible sources. What is more reasonable than to suppose that visible things — rocks and ice and water — become gradually resolved into invisible particles, and that these in their turn condense into new visible substances at some later time? For these or similar ideas the material "seeds" of Anaxagoras had, as stated above, paved the way, when later emphasized by Leucippus and his more famous pupil Democritus. Of the life of Leucippus almost nothing is known, but he was probably a contemporary of Empedocles and Anaxagoras, and possibly a pupil of Zeno. Leucippus assumed the existence of empty space as well as of matter, and held that all things are constituted of atoms. Space is infinite in magnitude, atoms infinite in number and indivisible. Atoms are always in activity, and worlds are produced by atoms variously shaped and weighted, falling in empty space and giving rise to an eddying motion by mutual impact.

DEMOCRITUS OF ABDERA (*c. 460–c. 370 B.C.*)

Democritus was a pupil and associate of Leucippus, whose theories of empty space and material atoms he developed and made so famous that his own name alone is often associated with them. Of his life and works, little is certainly known, but he may be regarded as marking the culmination and conclusion of the Ionian school; and his reputation, both in antiquity and in medieval times, was immense. Like contemporary and preceding philosophers, his writings were in verse, and Cicero is said to have deemed his style worthy of comparison with that of Plato.

Democritus regarded all matter as composed of indivisible *atoms*, choosing this name to express that quality, in place of the *seeds* of Leucippus. The atoms differ from each other in form, position, and magnitude. Water is a liquid because its atoms, being smooth and round, glide easily over each other, while atoms of iron are hard and rough. Soul and fire consist of small atoms by inhaling which life is maintained. Respiration introduces new atoms into the body in place of old ones which are rejected. The soul perishes with the body — a doctrine which made Democritus odious to later generations. Dante, for example, places him far down in hell as "ascribing the world to chance."

The atomic theory of perception held that from every object "images" of that object are being given off in all directions, some of which enter the organs of sense and cause "sensations." Democritus further held that sensations are the only sources of our knowledge. He was regarded as one of the extreme sceptics of antiquity, as e.g., in this saying, "We know nothing; not even if there is anything to know." Galileo, himself of a highly sceptical turn of mind, refers with approval to Democritus, and it is probably on this side, i.e., by exemplification of the critical spirit, that Democritus rendered his greatest service. While he praised experiment as a guide to knowledge, there is no record of his performing any, and his positive contributions to science, even in atomism, are apparently neither novel nor important. Democritus explained the Milky Way as composed of a vast number of small stars.

While admiring the penetrating vision of these Greek atomists, the modern student cannot fail to be impressed by the haziness of their conceptions and by the futility of their theories in the prediction of new facts. Their materialism was antithetical to the idealistic tendencies of Plato, but later their ideas were to flower in the poetry of Lucretius.

GREEK GEOGRAPHY AND GEOLOGY

We have very little knowledge of what the Greeks thought of the structure of the earth. Fragments remain of the book, *Circuit of the Earth*, written by Hecataeus of Miletus (*c.* 550–475 B.C.), "the father of geography," who travelled widely and described the world with the Mediterranean at its center. He is said to have improved the map of Anaximander (p. 47). From him Herodotus (*c.* 484–425) drew heavily for his *History*, which also was the result of extensive travel in Europe, Asia, and Africa. It is "one of the most fascinating stories ever written, a prose epic in nine books, each named after one of the nine Muses," and it "is of considerable importance because of the large amount of geographic and ethnographic information that it contains." Having observed fossil seashells in the mountains of Egypt, Herodotus postulated that at one time they had been covered by the sea. Empedocles (p. 62) from his observations of vulcanism on Mt. Etna, concluded that the inside of the earth was composed of molten matter. Practical knowledge of silver ores is shown by the exploitation of the silver mines at Laurium in southern Attica, which, especially during the fifth century B.C., provided a substantial part of the revenue of Athens.

THE MEDICAL SCIENCES BEFORE HIPPOCRATES

Many philosophers of the fifth century B.C. were interested in some aspect of the medical sciences. Anaxagoras (p. 62), the last of the Ionian school, introduced the scientific method into Athens. He dissected animals, including the brain, in which he recognized the lateral ventricles. Empedocles, who founded the medical school in Sicily, discovered the labyrinth in the inner ear. Diogenes of Apollonia in Crete, a younger contemporary of Anaxagoras, wrote a book *On Nature* of which fragments remain. To him is due one of the first Greek descriptions of the blood vascular system, "and it has the great merit of accuracy."

Unlike their anatomical observations, the views of these philosophers on physiology and pathology were mere guesses, but they are important because of their influence upon later writers. Anaxagoras is said to have attributed acute disease to the presence of black [1] bile or yellow bile in the blood and organs. He seems to have had a slight notion of evolution. At any rate, he was quoted by Aristotle as saying "that man is the most intelligent of animals because he has hands." Empedocles was very important for his influence upon the Hippocratic school. He thought respiration to be the ebb and flow of air through bloodless tubes opening by closely packed pores at the surface of the body. Comparing this system to a water-clock, he explained that the blood when surging through the limbs to the interior would permit the inflow of air, and when the blood returns the air would be breathed out again in equal quantity. He attributed importance to the blood vessels because he adopted the common belief that "the blood is the life," the carrier of the innate heat, and he was the first to describe the flow of blood as a surging to and from the heart with the respiratory rhythm. This led to the belief in the heart as the center of the vascular system and the chief organ of the *pneuma*, the soul, the breath of life. Health of the body he believed to be conditioned upon an equilibrium of the four elements, or "roots (*rhizomata*) of all things" —

> For it is with earth that we see Earth, and Water with water; by air we see bright Air, by fire destroying Fire. By love we see Love, and Hate by grievous hate. For out of these are all things formed and fitted together, and by these do men think and feel pleasure and pain. — EMPEDOCLES. (Burnet, *Early Greek Philosophy*, p. 232.)

To Philolaus is due the important distinction between sensory, animal, and vegetative functions, which he located in the brain, the heart, and the belly, respectively.

Empedocles is also the first sanitarian of whom we have any

[1] The idea of "black" bile (*melainan cholen*, melancholia) seems to come from dark discolorations of dejecta during certain states of disease.

record. He is credited with having cut down a hill of his native city and thus to have cured a plague by letting in the north wind, and to have done a similar service to the neighboring city of Selinus (Selinunte) by simply draining a local marsh.

THE BEGINNINGS OF RATIONAL MEDICINE. HIPPOCRATES OF COS

An outstanding feature of the intellectual revolution in Greece during the fifth century B.C. was the medical teaching at Cos in the second half of that century. The chief figure was Hippocrates of Cos, "the Father of Medicine." Aristotle calls him "the Great Hippocrates." In two references by Plato he is classed as an Asclepiad. This does not mean a priest in the temple of Asclepios, but probably a member of a guild of physicians with the lofty ideals set forth in the celebrated *Oath*. There was nothing priestly about Hippocrates. From three ancient and largely fabulous biographies we gather that he was born at Cos in 460 B.C., the son of an Asclepiad, that he travelled, was a friend of Democritus of Abdera, and lived to a very old age.

The aim of Hippocrates was by observation to discover the workings of Nature (in Greek, *physis*) in the body of man, especially as affected by environment and food, the idea of divine interference being totally rejected. His treatment depended upon belief in the healing power of nature (in medieval Latin, *vis medicatrix naturae*), which the physician (note the derivation) must aid by removing obstacles so far as he is able. His success or failure will appear on the day of *crisis*, then the patient definitely improves or declines.

It is the distinction of the Greeks among the nations of antiquity that they practiced a system of medicine based not on theory but on observation accumulated systematically as time went on. . . . Only the Greeks among the ancients could look upon their healers as *physicians* (= naturalists, physis = nature), and the word stands as a lasting reminder of their achievement. — SINGER, *Greek Biology and Greek Medicine*, p. 80.

Significant of the permanent influence of Hippocrates is the large number of modern names of diseases and other scientific terms that are Greek words taken directly from the Hippocratic writings, in which are found not only the highest ideals of practice, but also the germs of a large number of ideas current today in biology and medicine.

What is known of the teaching of Hippocrates is gathered from the *Hippocratic Collection*, or *Corpus Hippocratici*. This includes some 70 works, which, after passing through the hands of many copyists, collectors, and annotators, are represented by a number of manuscripts (the oldest of the 9th or 10th century A.D.) to be found in the libraries of Europe. The works collected in these manuscripts are in a fairly definite order, are anonymous, and vary greatly in character, apparent authorship, and date. All are free from superstition, some are strictly scientific, others exhibit attempts to apply the philosophy of Empedocles.

In the *Collection* are certain great treatises generally believed to be by the hand of the Master. Regarded as of greatest importance was the foretelling of the course of disease. The method, differing fundamentally from the Babylonian (p. 32) was by the study of symptoms, as described in the *Prognostic*. Supplementing this work is the *Regimen in Acute Disease*, in which treatment, chiefly for pneumonia and the three malarial fevers, is prescribed. Our common "zymotic" diseases — measles, etc. — apparently were unknown. The treatment is mild, with few drugs, much attention to a simple diet. The *Epidemics I and III*, a single work, which has been called "the most remarkable product of Greek science," is the record of a physician's observations of the progress, day by day, of 42 cases, 25 of them fatal. Treatment is rarely mentioned. It is pure science, the natural history of disease. *On Airs, Waters, Places* deals with the effect of situation and climate upon health and character, and contains the first effort to classify Man according to physical features. *On Wounds of the Head* is a thoroughly scientific surgical treatise, including description of variations in the skull.

Mention should be made of other treatises, especially the *Aphorisms*, "the most famous book" in the *Collection*. It is a compilation of very brief generalizations, one section being of special interest for its treatment of the incidence of a large number of diseases with reference to age and season. The doctrine of the Four Humors that dominated medical thought for more than two thousand years, is implied in other treatises, but is first clearly stated in the *Nature of Man*, attributed to Hippocrates and his son-in-law, Polybus. The humors are blood (*haema*), phlegm (*phlegma*), yellow bile (*cholen xanthen*), and black (*melainan*); "these make up the nature (*physis*) of his body, and through these he feels pain or enjoys health." The four humors are analogous to, but not identical with, the four elements of Empedocles (p. 67). Health results from a perfect blend. Excess or deficiency of any humor produces disease in the part affected, and may be caused by improper food or impure air. *On the Sacred Disease*, possibly a pupil's thesis, is important for its strictly rational point of view. The "magicians, purifiers, charlatans, and quacks of our day" are criticised, and the disease, epilepsy or similar attacks, is described as an affection of the brain produced by excess of phlegm. This is proved by opening the head of a goat having the disease (the staggers — parasitic) and finding the brain "very full of dropsy and of an evil odour, whereby you may learn that it is not a god but the disease which injures the body." The disease is hereditary, and this is explained by the origin of the "seed" "from every part of the body." "The brain of man, like that of all animals, is double, being parted down the centre by a thin membrane," (earliest mention of the hemispheres). It is the seat of consciousness, of intelligence, and of the emotions, not the heart and the diaphragm as "some people say." *On Diet* (or *Regimen*) is the earliest known book on preventive medicine, and contains the first classification of edible animals into Four-footed (*tetrapoda*), Birds, and Fishes. The *Hippocratic Collection* is weak in anatomy, *On the Heart*, being in that field the only treatise "worthy of those who could so well describe the aspects of disease." It mentions experiments to test the valves. In the *Nature of the Infant* is introduced the basic method of embryology — examination of the hen's egg at successive stages of incubation.

THE SOPHISTS

In the fifth century B.C., after the repulse of the Persians at Salamis (480 B.C.) and at Plataea (479 B.C.), Athens, governed by her citizens under the leadership of Pericles, became the ruler of half of Greece and the finest and richest city in the world. One result was to arouse in the young men of Greece a great interest in politics and a strong desire for intellectual instruction to occupy the four years between the end of their primary education at fourteen, and the beginning of military training at eighteen. They wished to know how to deal with men, to learn the arts of logic, persuasion, and public speaking. Rhetoric and argumentation, human nature and government were the fields of study thought suitable for young men who were really alive and ambitious. In answer to this need appeared teachers, beginning with Protagoras (*c.* 485–411 B.C.), who wandered from city to city, and who taught individually all known subjects wherever they happened to be. They were all included under the general name of Sophists — wise men. About individual Sophists little is known, and that chiefly from hostile critics. The Sophists generally sought honestly to teach wisdom, virtue, and good citizenship. The destructive effect that the critics found in their teachings was not so much in their practical character as in their denial of any truth beyond the practical, their denial of ultimate truth and virtue. This not only irritated the conservatives, but was part of a far-reaching and momentous movement that shook the very foundations of Athenian civilization — an *intellectual revolution.* The more important subjects taught by the Sophists fell into two groups — Rhetoric and Mathematics. The latter could not fail to be esteemed as a means of discipline, and several of the Sophists made notable contributions to its development.

THE CLASSICAL PROBLEMS. HIPPIAS OF ELIS

Hippias of Elis is the first sophist to be mentioned for important mathematical work. About 420 B.C. Hippias invented a curve called the quadratrix, serving for the solution of the first two of the three celebrated "Classical Problems" of Greek geometry; viz., the quadrature of the circle, the trisection of an angle, and the duplication of the cube. By means of straight line and circle constructions, the solution of the quadratic equation had been accomplished, though without algebraic symbolism, or any recognition of negative or imaginary results. The trisection problem, like that of duplicating the cube, was equivalent to the solution of the cubic equation, and could therefore not be accomplished by orthodox line and circle methods. Departing from these, the quadratrix

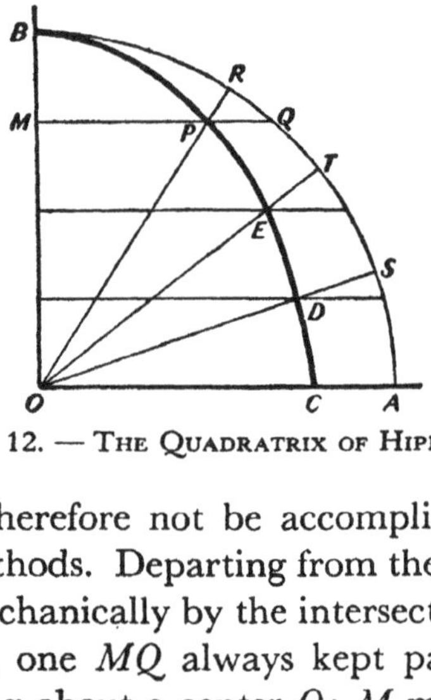

FIG. 12. — THE QUADRATRIX OF HIPPIAS.

was generated mechanically by the intersection *P* of two moving straight lines, one *MQ* always kept parallel to *OA*, the other *OR* revolving about a center *O*; *M* moves at a uniform rate from *O* to *B* while *R* is moving uniformly along the circular quadrant *AB*. By means of this curve the trisection problem is reduced to that of trisecting a straight line, which is elementary.[1] The curve meets the perpendicular lines *OA* and *OB* at *C* and *B*, respectively, so that $OC : OB = 2 : \pi$,

[1] To trisect any angle as *AOR*, draw *MQ* parallel to *OA* and divide *OM* into three equal parts by lines parallel to *OA*, meeting the curve in *D* and *E* respectively. The radii *OS* and *OT* will then trisect the angle *AOP*, by the definition of the curve.

where π is the ratio of the circumference of a circle to its diameter. Dinostratus (2nd half 4th cent.) showed, namely, that the assumptions $OC : OB > 2 : \pi$, and $OC : OB < 2 : \pi$, both lead to contradictions, therefore $OC : OB = 2 : \pi$ — a good example of the Greek *reductio ad absurdum.* To this quadrature solution the name of the curve is due. The study of a problem not capable of solution by elementary means thus led to the invention of this new curve, the first of which we have any definite record.

THE CRITICISM OF ZENO

Zeno of Elea in southern Italy, teaching about 462 B.C., though not himself a mathematician, represents an important phase of philosophical criticism of mathematics. Every manifold, he says, is a number of units, but a true unit is indivisible. Each of the many must thus be itself an indivisible unit, or consist of such units. That which is indivisible, however, can have no magnitude, for everything which has magnitude is divisible to infinity. The separate parts have therefore no magnitude, etc. Again, as to the possibility of motion, he maintains that before the body can reach its destination it must reach the middle point, before it can arrive there it must traverse the quarter, and so on without end. Motion is thus shown to be impossible; so the tortoise, if he have any start, cannot be overtaken by the swift runner Achilles, for while Achilles is covering that distance the tortoise will have attained a second distance, and so on. Such specious criticism was naturally, and in a measure justly, evoked by misguided efforts of certain mathematicians to show that a line consists of a multitude of points, etc. These or similar controversies as to the interpretation of the infinite and the infinitesimal have persisted till our own day, resembling in that respect the classical problems of circle squaring and angle trisection to which reference has been made above. The more or less mystical statements about the new discoveries of the Pythagoreans also invited sceptical epigrams.

Zeno was concerned with three problems. . . . These are the problem of the infinitesimal, the infinite, and continuity. . . . From him to our own day, the finest intellects of each generation in turn attacked these problems, but achieved, broadly speaking, nothing. . . . — B. Russell.

The mathematicians . . . realizing that Zeno's arguments were fatal to infinitesimals, saw that they could only avoid the difficulties connected with them by once for all banishing the idea of the infinite, even the potentially infinite, altogether from their science; thenceforth, therefore, they made no use of magnitudes increasing or diminishing *ad infinitum*, but contended themselves with finite magnitudes that can be made as great or as small *as we please*. — T. L. Heath, *Hist. Greek Math.* I, p. 272.

An excellent discussion of the "History of Zeno's Arguments on Motion" has been given by Cajori (*Amer. Math. Monthly*, **22**, v.p., 1915).

CIRCLE MEASUREMENT: ANTIPHON AND BRYSON; HIPPOCRATES OF CHIOS

Two of the sophists, Antiphon and Bryson (contemporaries of Socrates, 2nd half 5th cent.), made a substantial contribution to the problem of squaring the circle, by means of the inscribed and circumscribed regular polygons. Antiphon started with a regular polygon inscribed in a circle, and constructed by known elementary methods an equivalent square. By doubling the number of sides repeatedly he obtained polygons which become more and more nearly equivalent to the circle — the first correct attack on this formidable problem. Bryson took the important further step of employing both inscribed and circumscribed polygons, making the assumption that the area of the circle may be considered intermediate between them.

Another great step in the development of the theory of the circle was accomplished by Hippocrates of Chios, who had relations with the now dispersed Pythagoreans during the latter half of the fifth century and came to Athens in later life after financial reverses. He is said in the register of mathematicians to have written the first *Elements*, or textbook of mathematics, in which he made effective use of the *reductio ad absurdum* as a method of deriving one proposition from another.

To Hippocrates is probably due the theorem that the areas of circles are proportional to the squares on their diameters. He appears to have employed geometrical figures with letters at the vertices, in the modern fashion. From the theorem in regard to areas of circles follows naturally a general theorem

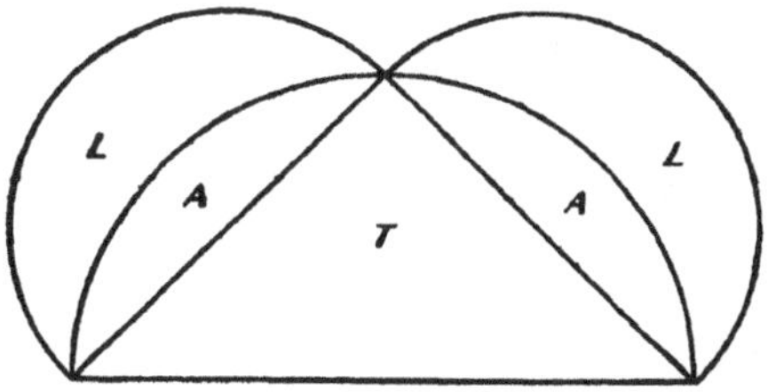

FIG. 13. — THE LUNES OF HIPPOCRATES.

for similar segments and sectors of circles. His work on lunes is remarkable. Starting with an isosceles right triangle, he described a semicircle on each of the three sides. By the theorem just quoted the semicircle on the hypotenuse is equal in area to the sum of the other two. If the larger semicircle is taken away from the entire figure, two equal lunes remain; if the two smaller semicircles are taken away, the triangle remains. Therefore the two lunes are together equivalent to the triangle, and the area of each may be determined. In algebraic notation $T + 2A = 2(A + L)$ whence $T = 2L$. The gulf between rectilinear and curvilinear figures has at last been successfully crossed.

DUPLICATION OF THE CUBE

It is reported that Euripides in one of his tragedies attributed to King Minos of Crete the words referring to a tomb erected at his order:

> Too small thou hast designed me the royal tomb,
> Double it, yet fail not of the cube.

At a somewhat later period it is related that the Delians, suffering from a disease, were bidden by the oracle to double the size of one of their altars, and invoked the aid of the Athenian geometers. Hippocrates transformed the problem of solid geometry into one in two dimensions by observing that it is equivalent to that of inserting two geometric means between given extremes. In our modern algebraic notation, the continued proportion $x : y = y : z = z : a$ leads to the equations $y^2 = xz$, $z^2 = ya$, whence, eliminating z, $y^3 = ax^2$, $y = a^{\frac{1}{3}}x^{\frac{2}{3}}$; y and $z = a^{\frac{2}{3}}x^{\frac{1}{3}}$ are the desired means between x and a, and by putting $a = 2x$ the problem is solved. No such algebraic notation existed at this time, however, and the geometrical methods invented by later Greek mathematicians were necessarily very complicated.

PLATO AND THE ACADEMY

One of the greatest names in the history of philosophy is that of Plato (*c.* 428–348/7 B.C.), and yet with Plato philosophy enters upon a new phase in which it almost parts company with science. Before Plato philosophy was almost wholly devoted to inquiries or speculations touching the earth, the heavens, and the universe, and hence was substantially "nature" or "natural" philosophy. But with Plato and ever since his time the larger part of philosophy has been devoted to observation and speculation upon the human mind and its processes and has accordingly often been called "mental" or "moral" as contrasted with "natural" philosophy. It is Thales and Pythagoras, Hippocrates and Democritus and Aristotle, rather than Plato and his disciples, who are the protagonists of science.

During his early manhood Plato was a devoted friend of Socrates, and it has been said that he found it expedient to leave Athens after the death of his master (399 B.C.). There are traditions of his having travelled widely, engaged in battle, and conversed with Egyptian priests. At any rate, it is known that in middle life he taught in a grove at Athens, where he founded The Academy; and that until his death he remained at the head of this first great philosophical school, which lasted under various forms until A.D. 529. While primarily a philosopher rather than a mathematician, Plato, unlike his master Socrates — who desired only enough mathematics for daily needs — rated highly the importance of mathematics and rendered services of the greatest value in its development. This was doubtless due in part to the influence of Archytas, a friend of the Pythagoreans, with whom he had associated, perhaps during his prolonged exile.

The register proceeds: "Plato . . . caused mathematics in general, and geometry in particular, to make great advances, by reason of his well known zeal for the study, for he filled his writings with mathematical discourses, and on every occasion exhibited the remarkable connection between mathematics and philosophy."

"Let no one ignorant of geometry enter under my roof" was the injunction which confronted Plato's would-be disciples. His respect for mathematics finds interesting expression in the remarks he puts into the mouth of Socrates in the *Dialogues*, and to him it is largely indebted for its place in higher education.

In the *Laws* he advises the study of music or the lyre to last from the age of 13 years to 16, followed by mathematics, weights and measures, and the astronomical calendar until 17. For a few picked boys on the other hand he recommends in *The Republic*, the study before they are 18, of abstract and theoretical mathematics, theory of numbers, plane and solid geometry, kinematics, and harmonics. Of arithmetic he says, "Those who are born with a talent for it are quick at learning, while even those who are slow at it have their general intelligence much increased by studying it." "No branch of education is so valuable a preparation for household management and politics and all arts and crafts, sciences and professions, as arithmetic; best of all by some divine art, it arouses the dull and sleepy brain, and makes it studious, mindful, and sharp."

Under Plato's influence mathematics first acquired its unified significance, as distinguished from geometry, computation, etc. Accurate definitions were formulated, questions of possibility considered, methods of proof criticized and systematized, logical rigor insisted upon. The philosophy of mathematics was begun. The point is the boundary of the line; the line is the boundary of the surface; the surface is the boundary of the solid. Such axioms as "Equals subtracted from equals leave equals" date from this period. The analytical method is developed, connecting that which is to be proved with that which is already known. Another principle carefully observed is to isolate the problem by removing all non-essential elements, and a third consists in proving that assumptions inconsistent with that which is to be proved are in a given problem impossible.

THE ANALYTIC METHOD

The analytic method, proceeding from the unknown to the known, depends for its validity on the reversibility of the steps; the synthetic method on the contrary proceeds from the known to the unknown, with unimpeachable validity. It was characteristic of the Greek geometers to aim at this as the final form for their demonstrations, even if the results had been first obtained analytically. The two methods may be illustrated by the following:

A circle *CDE* is given and two external points *A* and *B*. It is required to draw straight lines *AC* and *BC* meeting the circle

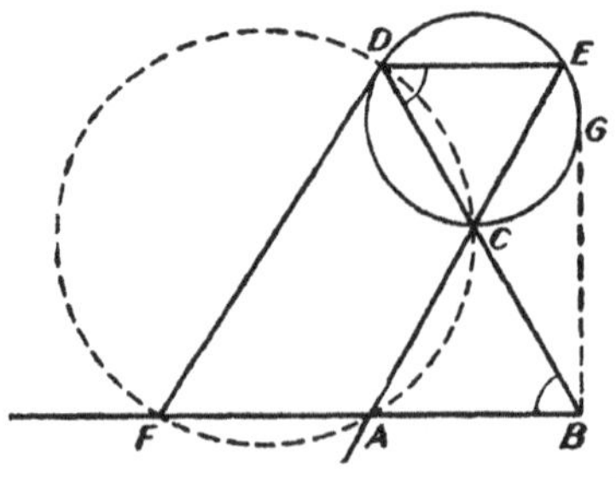

Fig. 14. — Illustration of the Analytic Method.

in *C*, *D*, and *E* so that *DE* shall be parallel to *AB*. It is shown that if the construction can be made, the tangent to the circle at *D* will meet *AB* (produced if necessary) in a point *F* which will lie on a new circle passing through *A*, *C*, and *D*. This *analysis* of consequences is the desired clue on which the following *synthesis* of the construction is then based. Starting again with *A*, *B*, and the circle, we locate *D* by drawing the tangent *BG* and producing *BA* to a point *F* such that $BA \times BF = BG^2$. A tangent from *F* will meet the circle at the required point *D*. *BD* is then drawn determining *C*: *AC* is drawn meeting the circle in *E*. It can then be shown that *DE* is parallel to *AB*.

A solution of the "duplication of the cube" problem is also attributed to Plato, though the mechanical process employed is so much at variance with his usual teachings that the correctness of the attribution is questionable.

To Plato is attributed a systematic method for finding numbers which may be sides of right triangles, his method being essentially an extension of the Pythagorean already described. Plato's Timaeus dialogue is indeed an important source of our information in regard to Pythagorean mathematics. Plato speaks with emphatic scorn of the shameful ignorance of mensuration on the part of his countrymen.

> He is unworthy of the name of man who is ignorant of the fact that the diagonal of a square is incommensurable with its side. — PLATO. (Moritz, *Mem. Math.*, p. 211.)

While predominantly interested in geometry, Plato's arithmetical attainments were considerable for his time. He made, for example, a correct statement about the 59 divisors of 5040, which include all the numbers from 1 to 10.

> This is a kind of knowledge which legislation may fitly prescribe; and we must endeavor to persuade those who are to be the principal men of our State to go and learn arithmetic, not as amateurs, but they must carry on the study until they see the nature of numbers with the mind only; nor again, like merchants or retail-traders, with a view to buying and selling, but for the sake of their military use, and of the soul herself; and because this will be the easiest way for her to pass from becoming to truth and being. . . . Arithmetic has a very great and elevating effect, compelling the soul to reason about abstract number, and rebelling against the introduction of visible or tangible objects into the argument. — PLATO, *Republic*. (Jowett, p. 227.)

PLATONIC COSMOLOGY

The spherical figure of the earth was now generally accepted in Greece, and the older cosmogonies gradually disappeared. To Plato, whose interest in physical science was indeed but secondary, the earth was a sphere at the center of the universe, requiring no support. He supposes the distances of the heavenly bodies from this center to be proportional to the numbers: Moon 1, Sun 2, Venus 3, Mercury 4, Mars 8, Jupiter 9, Saturn 27 — these numbers being obtained by combining the arithmetic and geometric progressions, 1, 2, 4, 8 and 1, 3, 9, 27.

He accepts as a principle that the heavenly bodies move with a uniform circular motion and proposes to the mathematicians this problem: "What are the uniform and regular circular motions which may properly be taken as hypotheses in order that we may save the appearances presented by the planets?"

All the heavenly bodies are looked on as divine beings, the first of all living creatures, the perfection of whose minds is reflected in their orderly motions.

He probably had no real knowledge of those deviations of the planets from uniform circular motion, which were to engross the attention of succeeding philosophers and astronomers. His system is consistently geocentric, and assumes a stationary earth. According to Plutarch, "Theophrastus states that Plato, when he was old, repented of having given the earth the central place in the universe which did not belong to it," this presumably indicating an inclination towards the theories of the later Pythagoreans.

Plato regarded the universe as made up of a material body plus a soul or intelligence. His notion of matter is not easy to understand but resembles that of Pythagoras. It is an indefinite something which does not differ demonstrably from space. When portions of it are enclosed in triangles or squares, elements are formed differing according to the nature of the boundaries. If the bounding surfaces are square, a cube results and the element earth is formed. If the bounding surfaces are triangles bound together to form a tetrahedron, fire is formed and the sharpness of the points where the faces meet characterizes the penetrating power of fire. Air is formed of octahedra, and water of icosahedra. Although Plato accepts the four elements of Empedocles, there is the possibility of one element changing into another by changing the surfaces at the will of a directing intelligence. Water, according to Plato, can be converted into air by heating, and by cooling ice is formed which under the earth may be converted into rocks and stones.

While the theories of Plato contributed little of permanent value to physical science, they had much influence upon ancient and medieval notions and the growth of alchemy.

ARCHYTAS

Archytas of Tarentum in southern Italy, a statesman and several times commander of the military forces of his city, was a Pythagorean philosopher and a close friend of Plato. He applied mathematics to problems of mechanics and is sometimes called the founder of theoretical mechanics. He is said to have invented the pulley and the screw. In his mathematics he came to a remarkable solution of the duplication problem. This very interesting and somewhat elaborate solution involves a combination of three surfaces: a cone of revolution, a cylinder having the vertex of the cone in the circumference of its base, and a surface generated by revolving a semicircle about an axis in its own plane passing through one end of its diameter. It shows remarkable mastery of elementary geometry, both plane and solid, and an interesting tendency to employ a wider range of methods, including motion, which might, but for adverse tendencies, have had important results in connecting mathematics with its possible applications to mechanics, etc. The influence of Plato in avoiding such connections and associating geometry with abstract logic and philosophy, undoubtedly had compensating advantages in promoting elegance and scientific rigor — crystallizing out a more refined product. Vitruvius praises Archytas as an accomplished builder of machines.

MENAECHMUS (c. 350 B.C.): CONIC SECTIONS

Even more interesting in its foreshadowing of future mathematical developments are the solutions of the duplication problem by Menaechmus. The problem which we should express in modern algebraic notation by the continued proportion $\frac{a}{x} = \frac{x}{y} = \frac{y}{b}$, Menaechmus, without any such notation or any system of coordinate geometry, shows to be equivalent to that of determining the intersection either of a parabola and a hyperbola, corresponding, respectively, to the two proportions

$$\frac{a}{x} = \frac{x}{y} \text{ and } \frac{a}{x} = \frac{y}{b}$$

or to the intersection of two parabolas, in case the second proportion is replaced by $\frac{x}{y} = \frac{y}{b}$. The construction of either parabola or the hyperbola naturally required some mechanical device.

The Greeks of this period distinguished three types of the cone formed by the rotation of the right triangle about one of its sides, according as the angle formed by that side with the hypotenuse was less than, equal to, or greater than half a right angle. A plane perpendicular to an element of the cone would cut a cone of the first kind in an ellipse, a cone of the second kind in a parabola, a cone of the third kind in an hyperbola. These curves were named accordingly sections of the acute-angled, the right-angled, the obtuse-angled cone.

The discovery of the conic sections . . . first threw open the higher species of form to the contemplation of geometers. But for this discovery, which was probably regarded . . . as the unprofitable amusement of a speculative brain, the whole course of practical philosophy of the present day, of the science of astronomy, of the theory of projectiles, of the art of navigation, might have run in a different channel; and the greatest discovery that has ever been made in the history of the world, the law of universal gravitation, with its innumerable direct and indirect consequences and applications to every department of human research and industry, might never to this hour have been elicited. —J. J. SYLVESTER, *Coll. Math. Papers*, II, 7.

Many of Plato's followers and disciples in the Academy continued the development of mathematics. This whole period is one of great productivity and importance in the history of mathematics. New theorems and new methods are discovered, former methods are critically scrutinized, loci problems are investigated, these and the study of the three classical problems (p. 72) leading to the introduction of new curves and a general extension of geometrical knowledge. Geometry, with emphasis, indeed, on its philosophical side, predominates over the theory of numbers, and even the latter is given so geometrical a form, for example in the use of such terms as square and cube for powers of a number, that mathematics is unified.

EUDOXUS: A NEW COSMOLOGY

Eudoxus of Cnidos (408?–355 B.C.) was a student both of Archytas and, for a time, of Plato. He was not only mathematician and astronomer, but also physician. In mathematics he is almost a new creator of the science, developing the theory of proportion, making a special study of the "golden section," already mentioned in connection with the regular polygons, and obtaining important results in solid geometry.

To him was formerly attributed the proof that the volume of a pyramid is one third that of the prism having the same base and altitude, as well as the corresponding theorem for cones and cylinders. A manuscript of Archimedes discovered in 1906, called *Method*, shows, however, that for this Democritus deserves the credit. The method of exhaustion, so-called, employed in proving these theorems was expressed in the auxiliary theorem: "When two volumes are unequal, it is possible to add their difference to itself so many times that the result shall exceed any assigned finite volume." This exceedingly useful and important principle, avoiding the difficulties of infinitesimals, was expressed in several approximately equivalent forms, and was already implied in the work of Antiphon and Bryson.

There appear to have been no astronomical instruments at this time except the simple gnomon and sun-dial, but the more obvious irregularities of the planetary motions were beginning to attract attention, and under Eudoxus led to the development of a new and important theory. Nearest to the central earth is the moon, carried on the equator of a sphere revolving from west to east in 27 days. The poles of this sphere are themselves carried on a second sphere, which turns once in every 18½ years about the axis of the zodiac. The angle between the axes of these two spheres corresponds with the moon's variation in latitude. A third outer sphere gives the daily east to west motion. Similarly there are three spheres for the sun. For each of the five planets a fourth sphere is necessary to account for the stations and retrogressions of its apparent orbital motion — thus making with the single sphere of the stars 27 spheres, all having their common center at the center of the earth.

How far these spheres were regarded as having concrete existence, how far they merely expressed in convenient geometrical form the observed relations and motions, we cannot determine from extant evidence. The amount of observational data available was entirely inadequate to serve as a basis for any quantitatively correct theory. For Mercury, Jupiter, and Saturn the theory was reasonably adequate, for Venus less so, and for Mars quite defective.

Callippus, who flourished in Athens in *c.* 330 B.C., a follower of Eudoxus, endeavored with some degree of success to remedy these defects by adding a fifth sphere for each of the refractory planets, and at the same time a fourth and fifth for the sun, in order to account for the recently discovered inequality in the length of the four seasons, using in all 34 spheres.

Reviewing the development of this interesting theory, Dreyer says:

> But with all its imperfections as to detail the system of homocentric spheres proposed by Eudoxus demands our admiration as the

first serious attempt to deal with the apparently lawless motions of the planets. . . . Scientific astronomy may really be said to date from Eudoxus and Callippus, as we here for the first time meet that mutual influence of theory and observation on each other which characterizes the development of astronomy from century to century. Eudoxus is the first to go beyond mere philosophical reasoning about the construction of the universe; he is the first to attempt systematically to account for the planetary motions. When he has done this the next question is how far this theory satisfies the observed phenomena, and Callippus at once supplies the observational facts required to test the theory and modifies the latter until the theoretical and observed motions agree within the limits of accuracy attainable at the time. Philosophical speculation unsupported by steadily pursued observations is from henceforth abandoned; the science of astronomy has started on its career. — DREYER, *Planetary Systems*, pp. 102, 107.

Eudoxus made the first known proposal for a leap-year, and for a star catalogue.

ARISTOTLE AND THE LYCEUM

When we turn from the writings of Plato to those of his pupil Aristotle, we are conscious of a complete change of atmosphere; we have turned from the pulpit to the laboratory. . . . Aristotle is the practical man of the world, interested in the systematic investigation of facts. — W. C. GREENE, *Achievement of Greece*, p. 181.

Aristotle has been called the Stagirite, from Stagira, a town near the Macedonian border, where he is said to have been born in 384 B.C. His father became physician to the king of Macedon. At the age of eighteen Aristotle went to Athens and became prominent in the Academy under Plato.

After the death of Plato, 347 B.C., Aristotle left Athens and carried on extensive researches for two or three years in marine zoology, especially on fishes. Then for about five years he was at the court of Macedon tutoring the prince who was destined to be Alexander the Great. At the age of fifty he returned to Athens, with great prestige, and opened a school in the grounds of the Temple of Apollo Lyceus, hence known as the Lyceum. For teaching he used the covered walks (*peripatoi*) and from them his scholars were named the Peripatetics. When a political upheaval followed the death of Alexander, he retired to Chalcis in Euboea, where he died the next year, 322 B.C.

The outstanding features of Aristotle were originality of thought and keenness of observation. One of the most important of his many services to science is the encyclopedic character of his writings, since from time to time he reviews in them the opinions of his predecessors whose works are sometimes known to us chiefly through his references to them. His observations are often poor, his conclusions often erroneous, but his interest, his curiosity, his zeal are indefatigable.

Aristotle classifies the sciences in three groups: the theoretical, aiming at knowledge for its own sake, the practical, aiming at it as a guide to conduct, the productive, aiming at creating the useful or the beautiful. The first group includes metaphysics, physics, and mathematics. With the others we are not here concerned. Physics deals with "natural bodies"; mathematics with number and spatial figures. Under physics, he deals in a long series of works with the first causes of nature and with natural movement in general; with the order and movement of the stars, the bodily elements and their transformations; with coming to be and passing away, genesis and decay; with animals and plants. That changes happen is established by experience, in spite of the paradoxes of Zeno. There must be at least two first principles and cannot be an infinite number. Animals and plants have in themselves a source of movement or rest, while other objects move on account of their component material. The mathematician abstracts what is quantitive and continuous from such qualities as weight, hardness, etc., dealing in arithmetic with the discrete or unextended, in geometry with the continuous or extended. Such applied sciences as astronomy, optics, mechanics are subordinate to pure mathematics. Physics must take account of both form and matter. It must seek the causes of physical change. "Matter" is the material of a thing as contrasted with its structure. Each type of material has a natural movement, tending towards a definite region of the universe — fire towards the circumference, etc. He maintains the reality and the continuity of movement. Spatial magnitude is not actually infinite but is infinitely divisible, though the actual operation of division into an infinite number of parts will never be made. Time and number, unlike space, are limited. Matter is continuous and there is no void or empty space but we always imagine a body less dense than one given. Time is the measure of movement — and of rest, but necessary truths are not in time. The primary kind of locomotion is that in a circle, as of the celestial sphere. A line cannot be composed of points, which cannot touch unless they coincide. In refutation of Zeno's paradox, while it is impossible to traverse an infinite space in a finite time, it is quite possible to traverse an infinitely divisible space in a finite time. The movement of the heaven as a divine body must be rotation about a center at rest, i.e., the Earth. Since there are Earth and its opposite, fire, there must also be air and water as intermediates. The universe consists of a series of concentric spheres, the outermost containing the fixed stars, carried around the Earth once in 24 hours.

With Plato he accepts the four elements of Empedocles but adds a fifth the *essence* or *quintessence*. He holds that individuals of a certain kind all contain the same essence but different matter, and in this way is able to account for the persistence of a species in spite of the death of the individual. Aristotle does not accept the elements of Empedocles in a purely literal sense but regards them as certain combinations of the fundamental attributes, heat, cold, wetness, and dryness. Fire is hot and dry, water is cold and wet, air is hot and wet, and earth is cold and dry. These properties — heat, cold, wetness, and dryness appear to be antagonistic forces and if one of them overcomes its opposite, the elements themselves are changed. This Aristotelian theory, according to which any element can be changed into another by changing one of its inherent qualities helped keep alive the hopes of the later alchemists in the attempt to convert base metals into gold, and alchemistic writings, of a much later period, are full of allusions to these ideas. Aristotle appears to have believed in the transmutation of metals and in describing the making of bronze from copper and tin he states that the tin vanishes and escapes leaving behind with the copper only a color.

The great works of Aristotle were written while he was at the Lyceum, his aim being to produce an encyclopedia of all known sciences, each reconstructed from his own point of view. He began with an investigation of the principles of reasoning, thus founding the science of Logic — a word invented later. His two chief treatises are entitled *Prior Analytics* and *Posterior Analytics*. "In these works he has produced nothing temporary, or of merely antiquarian interest, but an addition to human knowledge as complete in itself, as permanent, and as irrefragable as the Geometry of Euclid." He invented the name "Syllogism" for the process of deducing a third proposition from two previous assertions put together. The first volume is devoted to the theory of this process; the second deals with the logic of science. Deductive reasoning is the subject of both. The inductive processes are almost entirely unexplored, but Aristotle takes care to say — "in philosophy and in every science or branch of knowledge. *You must study facts.* Experience alone can give you general principles on any subject." (*Prior Anal.* I, xxx.) This is directly opposed to Plato's theory of Ideas. Aristotle argues that you cannot prove everything in science, for each science has its own "primary universal and immediate principle, arrived at by intuition" (like the axioms of Euclid). It is important to remember Aristotle's theory of the Four Causes required for the adequate definition of anything. These are — (1) the material, or antecedent, cause; (2) the formal cause, the whole nature of the thing; (3) the efficient cause, or motive power; and (4) the final cause, the purpose. Of a temple, for example, (1) stone may be the material cause, (2) the formal cause is the idea of a temple, (3) the efficient cause is the work of the architect and builders, and (4) the final cause may be the need of a place in which to worship Apollo. In science the efficient and the final causes are the ones most frequently considered. "Science itself," says Aristotle, "is knowledge of a cause."

Not always ending one treatise before he began another, Aristotle seems to have turned from the subject of method to the "practical sciences" — and wrote the *Ethics*, the *Politics*, and the *Art of Rhetoric*. He then took up constructive science and produced a small work *On Poetics*. Now he was ready to enter his main field, "theoretical science," comprising Natural Philosophy, Biology, and Theology. "The Physical and Natural Sciences occupy 1,447 pages, or fully one half of the writings which are undoubtedly Aristotle's." He began with the *Physical Discourse*, or *Physics*, in which he unfolded the general principles of natural philosophy. Then followed *On the Heavens*, *On Generation and Corruption*, and *Meteorology*. The biological works will be mentioned later. After the death of Aristotle, his ten books or fragments on theology were gathered together under the title *Metaphysics* ("after physics"). The historian of science is greatly indebted to Aristotle for his introduction of the custom, now generally followed, of prefacing a treatise or monograph with a summary of previous works on the subject. The first book of the *Metaphysics* is a conspicuous example.

"Aristotle speaks often of Mathematics as a great and interesting science, capable of affording high mental delight; but he seems to have regarded it as something tolerably finished and settled in his own time, and therefore less requiring his attention than other departments." He left no separate treatise on mathematics, but the theorem that the sum of the exterior angles of a plane polygon is four right angles is attributed to him. He distinguishes sharply between geodesy as an art and geometry as a science; he considers the plane sections of the circular cylinder; he recognizes the physical reason for the adoption of ten as the base number in arithmetic; he designates unknown quantities by letters. Continuity — an idea so important in modern mathematical and physical science — he defines by saying:

A thing is continuous when of any two successive parts, the limits at which they touch, are one and the same, and are, as the word implies, held together. — Gow, p. 188.

ARISTOTLE'S MECHANICS

In mechanics he discusses the composition of motions at an angle with each other. He enunciates the correct relation between the length of the arms of a lever and the loads which will balance each other upon it. He even deals with the central and tangential components of circular motion. He seems almost to recognize the principle of "virtual velocities." He asks such questions as: "Why are carriages with large wheels easier to move than those with small?" "Why do objects in a whirlpool move toward the centre?" etc. He is convinced that the speed of falling bodies is proportional to their weight — a belief credulously accepted until Galileo's time. He illustrates his discussion by geometrical figures.

Of the bearing of Aristotle's physical theories Pierre Duhem [1] says, in substance, "Incapable of any alteration, inaccessible to any violence, the celestial essence could manifest no other than its own natural motion, and that was uniform rotation about the centre of the universe." In his *Physics* he explains the rainbow, attributes sound to atmospheric motion, and discusses refraction mathematically. While he undertakes to deal with motion, space, and time — i.e., with the subject-matter of mechanics — his treatment is too metaphysical to have much real value. He declares, for example, that: — The bodies of which the world is composed are solids, and therefore have three dimensions. Now, three is the most perfect number — it is the first of numbers, for of *one* we do not speak as a number, of *two* we say both, *three* is the first number of which we say *all*. Moreover, it has a beginning, a middle, and an end.

ARISTOTELIAN ASTRONOMY

Only the second of the four books *On the Heavens* is devoted to astronomy. He considers the universe to be spherical, the

[1] *Le Système du Monde*, 1913, I, p. 225.

sphere being the most perfect among solid bodies, and the only body which can revolve in its own space. Rotation from east to west is more honorable than the reverse. He holds that the stars are spherical in form, that they have no individual motion, being merely carried all together by their own sphere.

Again, since the stars are spherical, as our opponents assert and we may consistently admit, inasmuch as we construct them out of the spherical body, and since the spherical body has two movements proper to itself, namely, rolling and spinning, it follows that if the stars have a movement of their own, it will be one of these. But neither is observed. (1) Suppose them to *spin*. They would then stay where they were, and not change their place, as, by observation and general consent, they do. Further, one would expect them all to exhibit the same movement; but the only star which appears to possess this movement is the sun, at sunrise or sunset, and this appearance is due not to the sun itself but to the distance from which we observe it. The visual ray being excessively prolonged becomes weak and wavering. The same reason probably accounts for the apparent twinkling of the fixed stars and the absence of twinkling in the planets. The planets are near, so that the visual ray reaches them in its full vigor, but when it comes to the fixed stars it is quivering because of the distance and its excessive extension; and its tremor produces an appearance of movement in the star; for it makes no difference whether movement is set up in the ray or in the object of vision.
(2) On the other hand, it is also clear that the stars do not *roll*. For rolling involves rotation; but the "face," as it is called, of the moon is always seen. Therefore, since any movement of their own which the stars possessed would presumably be one proper to themselves, and no such movement is observed in them, clearly they have no movement of their own. — ARISTOTLE, *DeCaelo* 290a, J. L. Stocks. (*Works*, W. D. Ross, ed., Vol. II.)

Aristotle adopts the system of homocentric spheres of Eudoxus and Callippus, but seems to suppose these spheres to be concrete, and not merely a geometrical device for interpreting the phenomena or determining the positions. In order, however, to secure what he conceives to be the necessary relation between the motions of the spheres, he is obliged to increase their total number from 34 to not less than 55. The earth is fixed at the center of the universe. That the earth is a sphere is shown logically, and is also evident to the senses. During eclipses of the moon, namely, the boundary line, which shows the shadow of the earth, is always curved. . . . If we travel even a short distance south or north, the stars over our heads show a great change, some being visible in Egypt, but not in more northern lands, and stars are seen to set in the south which never do so in the north. It seems, therefore, not incredible that the vicinity of the Pillars of Hercules is connected with that of India, and that there is thus but one ocean.

The bulk of the earth he considers to be "not large in comparison with the size of the other stars." The estimated circumference of 400,000 stadia — about 39,000 miles — is the earliest known estimate of the size of the earth, and is of unknown origin. While the heavens proper are characterized by fixed order and circular motion, the space below the moon's sphere is subject to continual change, and motions within it are in general rectilinear — a theory destined long to block progress in mechanics. Of the four elements, earth is nearest the center, water comes next, fire and air form the atmosphere, fire predominating in the upper part, air in the lower. In this region of fire are generated shooting stars, auroras, and comets, the latter consisting of ignited vapors, such as constitute the Milky Way.

Against any orbital motion of the earth Aristotle urges the absence of any apparent displacement of the stars. Reviewing his astronomical theories, Dreyer says:

> His careful and critical examination of the opinions of previous philosophers makes us regret all the more that his search for the causes of phenomena was often a mere search among words, a series of vague and loose attempts to find what was "according to nature" and what was not; and even though he professed to found his speculations on facts, he failed to free his discussion of these from purely metaphysical and preconceived notions. It is, however, easy to understand the great veneration in which his voluminous writings on natural science were held for so many centuries, for they were the first, and for many centuries the only, attempt to systematize the whole amount of knowledge of nature accessible to mankind; while the tendency to seek for the principles of natural philosophy by considering the meaning of the words ordinarily used to describe the

phenomena of nature, which to us is his great defect, appealed strongly to the medieval mind, and, unfortunately, finally helped to retard the development of science in the days of Copernicus and Galileo. — DREYER, *Planetary Systems*, p. 122.

At times Aristotle shows consciousness that his theories are based on inadequate knowledge of facts.

"The phenomena are not yet sufficiently investigated. When they once shall be, then one must trust more to observation than to speculation, and to the latter no farther than it agrees with the phenomena."

"An astronomer" he says "must be the wisest of men; his mind must be duly disciplined in youth; especially is mathematical study necessary; both an acquaintance with the doctrine of number, and also with that other branch of mathematics, which, closely connected as it is with the science of the heavens, we very absurdly call geometry, the measurement of the earth."

THE BIOLOGY OF ARISTOTLE

The word "biology" was invented long after Aristotle, but he had the idea. He went beyond the applications to medicine, and was the first individual to undertake the study of the entire range of living things, so far as was possible before the invention of the microscope.

"By insisting on the absolute necessity of anatomical observation, he carried biology at one step from the world of dreams into the world of realities; he set the science on a substantial basis, and indeed may be said to have been its founder, . . ." OGLE, *Aristotle on Youth and Old Age*, p. 29.

His writings on this subject form a consistent whole, the extant works occupying nearly 400 pages of the Berlin edition. Treatises *On Plants* [1] and *On Anatomy* have been lost. *On The Soul* (*peri psyche*, *De Anima*) begins with general biology and contains the first application of the scientific method to the study of psychology. Bodies are divided into lifeless and living and life is defined — "by life we mean the capacity for self-sustenance, growth, and decay." "The soul (*psyche*) may . . . be defined as the first actuality (*entelecheia*) of a natural body

[1] A treatise *On Plants* attributed to Aristotle is probably spurious.

potentially possessing life, and the body must be of a kind that possess sex organs" (*The Soul*, II, i, tr. Hett). In other words — "The soul is the cause and first principle of the living body" (*ibid.*, II, iv). It is the vital principle that gives form to the body. But it is more than that, for it actuates the fundamental functions of nutrition (including growth and reproduction), sensation, thinking, and movement. All living things have the nutritive soul characteristic of plants, from which all animals are distinguished by having also a sensitive soul. The lowest animals possess at least the sense of touch. Higher forms have more complex souls, culminating in "the mind and thinking faculty" of men "which seems to be a distinct species of soul and it alone admits of being separated as the immortal from the perishable" (*ibid.*, II, ii). After expounding his views on the vital principle and its functions, Aristotle devotes the main part of this work to the five senses, vision, hearing, smell, taste, and touch, and their relations to thinking. Appended to this work are a number of short papers (collected under the title *Parva Naturalia*) dealing with special topics in psychology and physiology. Except the important one *On Respiration*, all of these resemble the *Physics* in method.

Now we turn to a very different group of writings in which we see the naturalist at work — the *History of Animals*, *Parts of Animals*, and *Generation of Animals*. Aside from the works on logic, these are the principal works of Aristotle. Here he is at his best — observing, dissecting, or collecting information from farmers, fishermen, hunters, and the like. On almost any page, especially in the *History of Animals*, one finds fascinating glimpses of insight into structure or function of most diverse forms. For example — the regeneration of sponges (*Hist. Anim.* V, xvi); the umbilical blood vessels of the fœtal calf (*ibid.*, VII, ix); the distinction of whales from fishes; the uterine development of the smooth dogfish, that "would in itself be sufficient to establish the claim of Aristotle to a place in the front rank of observing naturalists"; the embryology of the octopus and the cuttlefish, and the structure of the "heterocotylus" arm of the male and the fishermen's story of its copulatory function, which Aristotle doubts but which proved to be true. Such examples could be multiplied indefinitely. We can add only the evidence of illustrations now lost, e.g., the diagram of the mammalian urogenital system that can be restored, so exact is the description. The *History of Animals* treats of comparative anatomy, classification, development, habits, and behavior. Animals are divided into those having blood and those without (red) blood, corresponding to Lamarck's vertebrates and invertebrates, and within these major groups are recognized twelve distinct classes. The words *genus* and *species* are used, but in somewhat different sense from the present usage. The *Parts of Animals* is a treatise on comparative physiology, beginning with principles of classification and the distinction between inorganic matter, tissues, and organs. The aim is to find the purpose, or final cause, of each organ. The *Generation of Animals* is an amazing collection of facts, true or supposed, on the processes of reproduction.

Many of the general ideas of Aristotle are important because of their influence throughout the ages. He was not an evolutionist. He quoted Anaxagoras, but to refute him. Nevertheless, he constructed an ascending *scale of being* from lifeless things through various grades of plants and of animals, from sponges to man. This applied to psychical as well as bodily features, as was proved by the fact that in its mental development the child at first hardly differs from an animal. His definition of the *psyche*, the vital principle, as an *entelecheia*, meaning an inherent quality by virtue of which a specific living thing is what it is, formed the basis for the vitalism that prevailed in physiology to the middle of the nineteenth century and is being revived in so-called neo-vitalism; the word *entelechy* is prominent in relatively recent literature on the factors of embryonic development (e.g., Hans Driesch). Mechanists classify this idea with the errors of Aristotle, some of which are important because of their retarding effect upon scientific progress. Spontaneous generation of certain aquatic animals, parasites, and the like, from mud, flesh, etc., was a popular belief accepted when no other method of reproduction could be proved.

The destructive effect of excessive heat on animal life is evident. Land animals and whales breathe by taking air into their lungs, fishes by passing water over their gills. No other property common to air and water being known to Aristotle, he concluded that the purpose of respiration is to cool the blood. Aristotle's theory of the heart as the place of blood-formation and site of innate heat and intelligence is doubtless founded on his observations of the development of the chick, combined with his principle that the most fundamental features are those first developed. After three days of incubation the first trace of the embryo seen by him in the hen's egg was "the heart . . . like a speck of blood . . . This point beats and moves as though endowed with life." A little *later* he saw the body, especially the head. The brain could not be the seat of intelligence because it is bloodless and cold to the touch and "when it is touched, no sensation is produced" (true, in the living animal). But the cool brain, by tempering the heat of the blood, he supposed to affect consciousness (*On Sleep*, III). The errors of Aristotle are the errors of a pioneer, due mainly to incomplete observation, to the acceptance of folklore, and to construction of theory upon assumed first principles.

THEOPHRASTUS

One of Aristotle's principal pupils, and his successor in his School, was Theophrastus (372–287 B.C.) notable in the history of science chiefly as an early student of plants, and the writer of the most important treatises of antiquity on botany. These are two large works — *On the History of Plants* and *On the Causes of Plants*. "These treatises . . . are in many respects the most complete and orderly of all ancient biological works that have reached our time." In the first, more than 500 species of plants are described, chiefly with reference to their medical uses. But more important are the records of original observation on the anatomy and reproduction of plants, including a clear and accurate account of the germination of the seed, with distinction of monocotyledonous and dicotyledonous forms. In this treatise seventeen general ideas on form and structure of plants appeared for the first time. While not aware of the floral organs of sex, Theophrastus understood the sexual differentiation of the Date-Palm and described the old process of artificial fertilization (cf. p. 31), which he compared to the fertilization of the Fig with the aid of insects, previously described by Aristotle. The work *On the Causes of Plants* deals with physiology and distribution with reference to soil and climate, and with some diseases of plants.

With the death of Theophrastus about 287 B.C. pure biological science substantially disappears from the Greek world, and we get the same type of deterioration that is later encountered in other scientific departments. — SINGER. (*Legacy of Greece*, p. 183.)

EPICURUS AND EPICUREANISM

A few words may be said of another philosopher of the fourth century, a follower to some extent of Democritus and the forerunner and exemplar of the Roman Lucretius. This was Epicurus (341–270 B.C.), who, born in Samos and educated in Athens and Asia Minor, became a famous teacher and the leader of a remarkable community "such as the ancient world had never seen." The mode of life in this community was not that of the so-called "epicures" of today, but very plain — water the general drink, and barley bread the general food. The magnetic personality of Epicurus held the community together, and his chief work was a treatise *On Nature* in thirty-seven books, of which some fragments are known. Epicureanism is of interest in the history of science chiefly because of its effect on its Roman exponent, the poet Lucretius. Much of it was even a negation of science and of the scientific spirit.

TERRESTRIAL MOTION: PHILOLAUS, HICETAS, HERACLIDES

It is of particular interest that later Pythagoreans, in particular Philolaus, about the middle of the fifth century, attributed the apparent daily motion of the heavenly bodies from east to west not to their own actual motion but to a motion of the earth in the opposite direction. This latter motion, however, was thought of, not as a rotation, but as an orbital motion about a so-called "central fire." Just as the moon revolved about the earth, always turning the same face towards the latter, so the earth might revolve about the central fire which would be forever invisible to the inhabitants of the side of the earth away from the central fire. While we say that the moon rotates about its axis in the same time in which it revolves about the earth, to the ancients such a motion was not considered to include rotation at all. A further essentially arbitrary assumption introduced between the earth and the central fire a counter-earth (*antichthon*), which was required to make up the supposed number of the heavenly bodies, and which would hide the central fire from dwellers in the antipodes.

All the other heavenly bodies describe orbits, each in its own hollow sphere about the central fire, the generally adopted order, based on the apparent rate of motion among the stars, being Moon, Sun, Venus, Mercury, Mars, Jupiter, Saturn. Pythagorean speculations as to relative distances of the different planets were naturally mystical notions merely. The sun was said to move around the central fire in an "oblique circle," i.e., the ecliptic. The moon was believed to be inhabited by plants and animals. The moon might be eclipsed either by the earth or by the counter-earth. This remarkable system, admitting the earth to move and not to be the center of the universe, was not generally or long accepted, but had a share in securing the acceptance of the theories of Copernicus nearly 2,000 years later. One at least of the Pythagoreans made the great further step, somewhat loosely described by Cicero in the words:

> Hicetas of Syracuse, according to Theophrastus, believes that the heavens, the sun, moon, stars, and all heavenly bodies are standing still, and that nothing in the universe is moving except the earth, which, while it turns and twists itself with the greatest velocity round its axis, produces all the same phenomena as if the heavens were moved and the earth were standing still.

To Heraclides of Pontus (*c.* 388–*c.* 312 B.C.) belongs the distinction of teaching that the earth turns on its own axis from west to east in 24 hours. He had been connected with the Pythagoreans, and with the schools of Plato and Aristotle. His work is known to us only indirectly, none of his own writings having survived. He is said also to have advanced the hypothesis that Venus and Mercury revolve about the sun, being therefore at a distance from the earth sometimes greater than the sun, sometimes less, an anticipation of the theory of Tycho Brahe (p. 236). Geminus writing in the first half of the first century B.C. of the different fields and points of view of astronomers and physicists, remarks:

> For why do sun, moon, and planets appear to move unequally? Because, when we assume their circles to be excentric, or the stars to move on an epicycle, the appearing anomaly can be accounted for, and it is necessary to investigate in how many ways the phenomena can be represented, so that the theory of the wandering stars may be made to agree with the etiology in a possible manner. Therefore also a certain Heraclides of Pontus stood up and said that also when the earth moved in some way and the sun stood still in some way, could the irregularity observed relatively to the sun be accounted for. In general it is not the astronomer's business to see what by its nature is immovable and of what kind the moved things are, but framing hypotheses as to some things being in motion and others being fixed, he considers which hypotheses are in conformity with the phenomena in the heavens. He must accept as his principles from the physicist, that the motions of the stars are simple, uniform, and regular, of which he shows that the revolutions are circular, some along parallels, some along oblique circles. — DREYER, *Planetary Systems*, p. 131.

This contrast between the physical phenomena and the mathematical theory which corresponds with them, without being true or perhaps even possible in all respects, is of continued and increasing importance in the history of science, as a larger stock of facts was accumulated and as theories still imperfect were more frequently subjected to critical comparison with observed data, instead of being accepted on purely philosophical or metaphysical grounds. Heraclides is not credited with any conception of an orbital or progressive motion of the earth.

REFERENCES FOR READING

ARISTOTLE, *Works in English*, ed. by J. A. Smith and W. D. Ross, Oxford, 1908–31.

BERRY, ARTHUR, *History of Astronomy*, 1898.

BURNET, JOHN, *Early Greek Philosophy*, 1892.

DREYER, J. L. E., *Planetary Systems*, 1906.

FAIRBANKS, ARTHUR, *First Philosophers of Greece*, 1898.

FREEMAN, K. E., *Schools of Hellas*, 1907.

GRANT, (SIR) ALEXANDER, *Aristotle*, 1910.

HEATH, T. L., *Aristarchus of Samos*, 1913.

——, *Manual of Greek Mathematics*, 1931.

HIPPOCRATES, Engl. tr. W. H. S. Jones and E. T. Withington (*Loeb Class. Lib.*), 1923–31.

Legacy of Greece (ed. R. W. Livingstone), 1922.

LUND, F. B., *Greek Medicine* (*Clio Medica*, Vol. 18), 1936.

SMITH, D. E., *History of Mathematics*, 1923–25.

THEOPHRASTUS, *Enquiry into Plants*, tr. (Sir) Arthur Hort (*Loeb Class. Lib.*), 1916.

Books from previous chapters: Allman, *Greek Geom.* I; Cajori, pp. 16–23; Farrington; Gow, Ch. III, IV, VI; Osler, *Evol. Mod. Med.*, pp. 58–71.

NOTE: A sphere and a cylinder marked the tomb of Archimedes as described by Cicero (*Tusculanae disputationes*, v, 23) — hence the design by Elizabeth Tyler Wolcott on the opposite page.

Chapter V

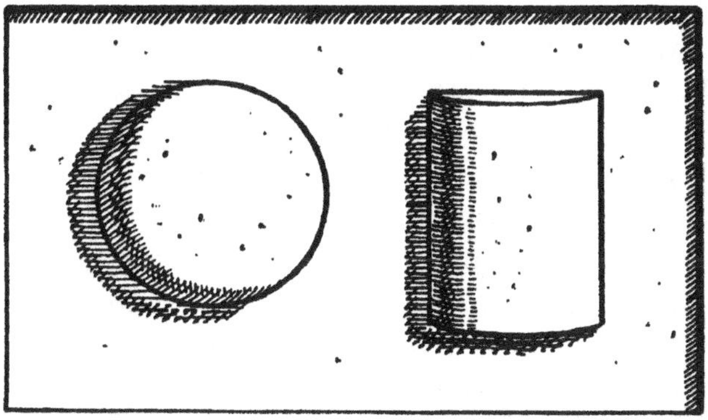

Greek Science in Alexandria

There is an astonishing imagination, even in the science of mathematics . . . We repeat, there was far more imagination in the head of Archimedes than in that of Homer. — VOLTAIRE. (Moritz, *Memo. Math.*, p. 31.)

About the middle of the third century, the famous Museum . . . was founded at Alexandria. The four departments of literature, mathematics, astronomy, and medicine were in the nature of research institutes as well as schools, and the needs of them all were served by the largest library of the ancient world, . . . For some centuries the Library of Alexandria was one of the wonders of the world. — SIR WILLIAM DAMPIER, *History of Science*, p. 51.

THE HELLENISTIC PERIOD

This important epoch in the history of civilization began with the death of Alexander the Great in 323 B.C. and ended when a Roman army destroyed Corinth in 146 B.C. Educated by Aristotle and trained for military command by his father, Alexander at the age of twenty became in 336 B.C. king of Macedon and overlord of Greece. With an army contributed largely by the Greek cities, he carried out his father's plan to expel the Persians from Asia Minor. Success led him along the Phœnician coast to Egypt, where he occupied the island of Pharos, and, finding there a harbor free from Nile silt, in 332 B.C. he ordered his engineer, Deincrates, to lay out on the adjoining mainland the city of Alexandria for a naval base and a commercial center. Leaving a viceroy in command, Alexander soon departed for further conquests, which extended to the confines of India.

The death of Alexander without a competent heir was followed by a prolonged struggle among his generals for supremacy, with the result that in 305 B.C. four of the viceroys proclaimed themselves kings, and after the decisive battle of Ipsus (301 B.C.) the empire of Alexander was divided into a small number of monarchies that were to endure, more or less intact, throughout the Hellenistic period. Cassander took Macedon and Greece; Lysimachus, Thrace and western Asia Minor; Ptolemy founded the Macedonian dynasty in Egypt; and the greater part of Alexander's empire, from Syria to the Indus, went to Seleucus. This unnamed "Seleucid monarchy" is distinguished by being the only one to adopt a system of chronology with consecutively numbered years. This era began with 312 B.C., when Seleucus became governor of Babylon. Later in western Asia Minor another kingdom was established with its capital at Pergamum.

Alexander's conquests and the foundation of these governments made available the wealth of Persia. New lands and markets were opened, trade barriers were lowered, and the Greek aristocratic contempt for commercial enterprise was dispelled. In Asia the Greeks built new cities or rebuilt old ones on a magnificent scale. From southern Italy to the Indus Attic Greek became a universal language, largely replacing local dialects. The court of each sovereign became a center of Greek culture. Literature was endowed as never before. The names of 1,100 Hellenistic writers are on record and the production of 100 volumes by one author was not uncommon. Libraries were established at royal courts. Between 200 and 150 B.C., Pergamum was an important center of literary studies, and there some 200,000 volumes were collected. From Pergamum is derived the name parchment, for skins prepared there as a substitute for papyrus, obtainable only in Egypt.

THE MUSEUM AT ALEXANDRIA

In 294 Ptolemy I invited Demetrius of Phalerum a distinguished Athenian citizen and scholar, a pupil of Aristotle, to Egypt, instructing him to establish what became later the greatest of ancient academies of learning. With him went members of the Peripatetic school who were mostly interested in science, while Athens remained headquarters for humanistic studies, notably historical research. Patterned after the schools of Athens on a much larger scale and well-supplied from the royal treasury, the Museum (Seat of the Muses) was a veritable university of Greek learning. To this were attached a great library, a dining-hall, and lecture rooms for professors. Here for the next 700 years Greek science had its chief abiding place. The library grew so rapidly that by 250 B.C. it contained over 400,000 rolls, and its value was enhanced by scholars who established authoritative texts of standard works with linguistic and historical annotations. The scholar-poet Callimachus (*c.* 250 B.C.) made a catalogue of authors, in 120 volumes, with a full list of works and brief biography of each author. With these facilities and cheap papyrus, Alexandria became the chief center for publication, and exported books to all parts of the Hellenistic world.

The fame of Alexandria soon outshone and eventually eclipsed that of Athens, while Romans journeyed from Rome — never important in ancient times as a scientific center — to study at Alexandria the healing art, anatomy, mathematics, geography, and astronomy. Neither Athens, Rome, nor any other city of the ancient world can boast similar distinction as a home of science.

EUCLID

Euclid's period of activity was about 300 B.C.; his place of birth and even his race are unknown; he is said to have been of a mild and benevolent disposition, and to have appreciated fully the scientific merits of his predecessors. While we know next to nothing of his life and personality, his writings have had an influence and a prolonged vitality almost, if not quite, unparalleled. Continuing the ancient register, Proclus writes:

Not much later than these [of the Academy] is Eulcid, who wrote the "Elements," arranged much of Eudoxus's work, completed much of Theaetetus's, and brought to irrefragable proof propositions which had been less strictly proved by his predecessors. — Gow, *Greek Math.*, p. 137.

. . . It is related that King Ptolemy asked him once if there were not in geometrical matters a shorter way than through the *Elements:* to which he replied that in geometry there was no straight path for kings. . . . — (Cf. Moritz, p. 152.)

EUCLID'S "ELEMENTS"

Scientifically, Euclid is attached to the Platonic philosophy. Thus he makes the goal of his *Elements* the construction of the so-called "Platonic bodies" i.e., the five regular polyhedrons. This treatise, which served as the basis of practically all elementary instruction for the following 2,000 years, is naturally his best-known work, and appears to have been accepted in the Greek world, after many previous attempts, as a finality. It consisted of thirteen books, of which only six are ordinarily included in the school editions of the last centuries. The whole is essentially a systematic introduction to Greek mathematics, consisting mainly of a comparative study of the properties and relations of those geometrical figures, both plane and solid, which can be constructed with ruler and compass. The comparison of unequal figures leads to arithmetical discussion, including the consideration of irrational numbers corresponding to incommensurable lines. Book I deals with triangles and the theory of parallels: Book II with applications of the Pythagorean theorem, many of the propositions being equivalent to algebraic identities, or solutions of quadratic equations, which seem to us more simple and obvious than to the Greeks. It should be noted however that the geometrical treatment is relatively advantageous for oral presentation. Book III deals with the circle, Book IV with inscribed and circumscribed polygons. These first four books thus contain a general treatment of the simpler geometrical figures, together with an elementary arithmetic and algebra of geometrical magnitudes. In Book V, for lack of an independent Greek arithmetical analysis, a theory of proportion (which has thus far been avoided) is worked out, with the various possible forms of the equation $\frac{a}{b} = \frac{c}{d}$. The results are applied in Book VI to the comparison of similar figures. This contains the first known problem in maxima and minima — the square is the greatest rectangle of given perimeter — also geometrical equivalents of the solution of quadratic equations. The next three books are devoted to the theory of numbers, including for example the study of prime and composite numbers, of numbers in proportion, and the determination of the greatest common divisor. They show how to find the sum of a geometrical progression, and prove that the number of prime numbers is infinite.

If there were a largest prime number p then the product $1 \times 2 \times 3 \ldots \times p$ increased by 1 would always leave a remainder 1 when divided by p or by any smaller number. It would thus either be prime itself, or a product of prime factors greater than p, either of which suppositions is contrary to the hypothesis that p itself is the greatest prime number. This hypothesis must therefore be false.

Book X gives a geometrical theory of those irrational numbers which are square roots of integers and square roots of these. A fundamental rôle is played by the theorem: If two unequal magnitudes are given, and if one takes from the greater more than its half, and from the remainder more than its half, and so on, one arrives sooner or later at a remainder which is less than the smaller given magnitude. Books XI, XII, and XIII are devoted to solid geometry, leading up to our familiar theorems on the volume of prism, pyramid, cylinder, cone, and sphere; but in every case without computation, emphasizing the habitual distinction between geometry and geodesy or mensuration — a distinction expressed by Aristotle in the form: "One cannot prove anything by starting from another species, for example, anything geometrical by means of arithmetic. Where the objects are so different as arithmetic and geometry one cannot apply the arithmetical method to that which belongs to magnitudes in general, unless the magnitudes are numbers, which can happen only in certain cases." Book XIII passes from the regular polygons, partly treated in Book IV, to the five regular polyhedra, and proves in conclusion that only the known five are possible.

The extent to which Euclid's *Elements* represents original work rather than compilation of that of earlier writers cannot be determined. It would appear, for example, that much of Books I and II is due to the Pythagoreans, of III to Hippocrates, of V to Eudoxus, and of IV, VI, XI, and XII, to later Greek writers; but the work as a whole constitutes an immense advance over previous similar attempts.

Proclus (A.D. 410–485) is the earliest extant source of information about Euclid, with the exception of Pappus (*c.* A.D. 300), who mentions his honesty, modesty, and kindness. Theon of Alexandria edited the *Elements* nearly 700 years after Euclid, and until the nineteenth century all editions have been based upon his. Only after the discovery in the Vatican Library, in 1810, of a manuscript of older date than Theon's, did it become known that Theon changed the text considerably, probably for the sake of clarity. This Vatican manuscript has been used by Heiberg in his edition, which now is generally used as the basis of all editions, e.g., that of Heath.

Like other Greek learning, Euclid has come down to later times through Arab channels. There is a doubtful tradition that an English monk, Adelhard of Bath, surreptitiously made a Latin translation of the *Elements* at a Moorish university in Spain in 1120. Another dates from 1185, printed copies from 1482 onward, and an English version from 1570. After Newton's time it found its way from the universities into the lower schools.

Different versions vary widely as to the axioms and postulates on which the work as a whole is based. It is believed that Euclid originally wrote five postulates, of which the fourth and fifth are now known as Axioms 11 and 12 — "All right angles are equal"; and the famous parallel axiom: — "If a straight line meets two straight lines, so as to make the two interior angles on the same side of it together less than two right angles, these straight lines will meet if produced on that side." The necessarily unsuccessful attempts which have since been made to prove this as a proposition rather than a postulate constitute an important chapter in the history of mathematics, leading in the last century to the invention of the generalized geometry known as non-Euclidean, in which this axiom is no longer valid.

INFLUENCE OF EUCLID

The *Elements* of Euclid has exerted an immense influence on the development of mathematics. Aside from its substance of geometrical facts, it is characterized by a strict conformity to a definite logical form, the formulation of what is to be proved, the hypothesis, the construction, the progressive reasoning leading from the known to the unknown, ending with the familiar Q.E.D. There is a careful avoidance of whatever is not geometrical. No attempt is made to develop initiative or invention on the part of the student; the manner in which the results have been discovered is rarely evident and is even sometimes concealed; each proposition has a degree of completeness in itself. This treatise translated into the languages of modern Europe has been a remarkable means of disciplinary training in its special form of logic.

An estimate of the circulation of the *Elements* is obtained from the list of editions given by P. Riccardi in 1887, which contains more than a thousand; it seems there is truth in the popular contention that the *Elements* has been, apart from the *Bible*, the most distributed book in Western civilization.

On the other hand, its narrowness of aim, its deliberate exclusion of the concrete, its laborious methods of dealing with such matters as infinity, the incommensurable or irrational, its imperfect substitutes for algebra, as in the theory of proportion, have diminished its usefulness, and have led in comparatively recent times to the substitution of modernized texts. Still, no other mathematical treatise has had even approximately the deservedly far-reaching influence of Euclid. Its author's name is still a current synonym for elementary geometry.

The Greeks occupied themselves with the greatest versatility in all directions, and made in all directions wonderful progress. Nevertheless, from our modern standpoint, they fell short of the possibly attainable in all, and in some directions made only a beginning.

It has become a tradition that Greek geometry reached unique development, while in reality many other branches of mathematics were successfully cultivated. The development of Greek mathematics was particularly hampered by the lack of a convenient number-system and notation as a basis for an independent arithmetic, and by ignorance of negative and imaginary numbers. Euclid's intention in the Elements was by no means to write an encyclopedia of current geometry, which must have included conic sections and other curves, but rather to write for mature readers an introduction to mathematics in general, the latter being regarded in its turn, in the Platonic sense, as necessary preparation for general philosophic studies. Hence the emphasis on formal order and logical method, as well as the omission of all practical applications. He aims at the flawless logical derivation of all geometrical theorems from premises completely stated in advance. — FELIX KLEIN, *Elementar-Mathematik*, II.

OTHER WORKS OF EUCLID

Besides the *Elements* Euclid wrote several other mathematical treatises, some of which exist only in Arabic or Latin translations, others only as fragments. They include one *On Porisms*, a special type of geometrical proposition; and one, the *Data*, containing such theorems as the following:

Given magnitudes have a given ratio to each other.

When two lines given in position cut each other their point of intersection is given.

When in a circle of given magnitude a line of given magnitude is given, it bounds a segment which contains a given angle.

A work *On Fallacies* is designed to safeguard the student against erroneous reasoning. Still other treatises are devoted to *Division of Figures*, *Surface-Loci*, and *Conic Sections;* finally there are works *On Phenomena*, *On Optics*, and *On Catoptrics* dealing with applications of geometry.

The *Phenomena* gives a geometrical theory of the universe, the *Optics* is an unsuccessful attempt to deal with problems of vision on the hypothesis that light proceeds from the eye to the object seen.

The *Catoptrics* deals in 31 propositions with reflections in plane, concave, and convex mirrors. It is remarked that a ring placed in a vase so as to be invisible from a certain position, may be made visible by filling the vase with water. The authenticity of this work is however questionable.

The last two works constitute the earliest known attempt to apply geometry systematically to the phenomena of light-rays. The law of reflection is correctly applied. Just as geometry is based on a definite list of axioms, so Euclid makes his optics depend on eight fundamental facts of experience. For example, the light rays are straight lines. The figure inclosed by the rays is a cone with its vertex at the eye, while the boundary of the object corresponds to the base, etc. This work, though in very imperfect form, continued in use until Kepler's time.

ARCHIMEDES (287?–212 B.C.)

The second great name in the Alexandrian school and one of the greatest in the whole history of science is that of Archimedes. He was both geometer and analyst, mathematician and engineer. He enriched the highly developed Euclidean geometry, made important progress in algebra, laid the foundations of mechanics, and even anticipated the infinitesimal calculus, reaching thus a level which was not surpassed for 2,000 years. Born in Syracuse, probably 287 B.C., the greater part of his life was spent in his native city, to which he rendered on occasion invaluable services as a military engineer. According to Livy it was due to the efforts of Archimedes that the Romans under Marcellus were held in check during the protracted siege of Syracuse. On the fall of the city in 212 B.C. the venerable mathematician, absorbed in a geometrical problem, was killed by a Roman soldier, much to the regret of Marcellus, who appreciated and would have spared him. The conqueror carried out the wish of Archimedes by erecting a monument with a mathematical figure, and this was with some difficulty rediscovered and put in order by Cicero, during his official residence in Sicily, 75 B.C.

Nothing afflicted Marcellus so much as the death of Archimedes, who was then, as fate would have it, intent upon working out some problem by a diagram, and having fixed his mind alike and his eyes upon the subject of his speculation, he never noticed the incursion of the Romans, nor that the city was taken. In this transport of study and contemplation, a soldier, unexpectedly coming up to him, commanded him to follow to Marcellus, which he declined to do before he had worked out his problem to a demonstration; the soldier, enraged, drew his sword and ran him through. Others write, that a Roman soldier, running upon him with a drawn sword, offered to kill him; and that Archimedes, looking back, earnestly besought him to hold his hand a little while, that he might not leave what he was at work upon inconclusive and imperfect; but the soldier, nothing moved by his entreaty, instantly killed him. Others again relate, that as Archimedes was carrying to Marcellus mathematical instruments, dials, spheres, and angles, by which the magnitude of the sun might be measured to the sight, some soldiers seeing him, and thinking that he carried gold in a vessel, slew him. Certain it is, that his death was very afflicting to Marcellus; and that Marcellus ever after regarded him that killed him as a murderer; and that he sought for his kindred and honored them with signal favours. — PLUTARCH, *Life of Marcellus.* (Moritz, *Memo. Math.*, p. 138.)

The known works of Archimedes include the following: two books on the *Equilibrium of Planes*, with an interpolated treatise on the *Quadrature of the Parabola*, two books *On the Sphere and the Cylinder*, the *Circle Measurement*, *On the Spirals*, the book of *Conoids and Spheroids*, *The Sand-Reckoner*, two books on *Floating Bodies*, *Choices*, *The Method.* Unlike Euclid's *Elements*, these are for the most part original papers on new mathematical discoveries, which were also often communicated to his contemporaries in the form of letters.

CIRCLE MEASUREMENT

In this Archimedes proves three theorems.

(1) Every circle is equivalent to a right triangle having the sides adjacent to the right angle equal, respectively, to the radius and circumference of the circle.

(2) The circle has to the square on its diameter approximately the ratio 11 : 14.

(3) The circumference of any circle is three times as great as the diameter and somewhat more, namely less than $3\frac{10}{70}$ but more than $3\frac{10}{71}$ the diameter.

He proves the first theorem by showing that the assumption that the circle is either larger or smaller than the triangle leads to a contradiction. The second he bases on the third, at which he arrives by computing successively the perimeters of both inscribed and circumscribed polygons of 3, 6, 12, 24, 48, and 96 sides. All this is contrary to the spirit of Euclid and essentially modern in its method of successive approximation. The difficulty of the achievement in view of the imperfect arithmetical notation available can hardly be overrated.

QUADRATURE OF THE PARABOLA

Of special interest is his quadrature of the parabola (measurement of the area bounded by an arc of the curve and its chord). A segment is formed by drawing any chord *PQ* of the parabola; it is known that if a line is drawn from the middle point *R* of the chord parallel to the axis of symmetry of the parabola, the tangent at the point *S* where this line meets the curve will be parallel to the chord, and the perpendicular from *S* to the chord is greater than any other which can be drawn from a point of the arc. The triangle formed by joining the same point *S* to the ends of the original chord being wholly contained within the segment, the area of the latter will be greater than that of the triangle and less than that of a parallelogram having the same base and altitude. Now the segment exceeds the triangle by two smaller segments, in each of which triangles *STQ* and *SPU* are again inscribed. It is a known property of the parabola that each of these triangles has one-eighth the area of the triangle *PSQ*. The area of each of the two smaller segments is therefore greater than one-eighth and less than one-fourth that of the triangle *PSQ*. The area of the original segment (consisting of the triangle *PSQ* and the two small segments) is therefore less than three-halves and greater than five-fourths that of triangle *PSQ*. The construction may evidently be repeated any number of times, and the ratio of the segment to the triangle will lie between numbers which may be shown to approach the limit four-thirds — an interesting example of the limit of an infinite series. Archimedes also succeeded in determining the area of the ellipse.

It should be kept in mind, however, that Archimedes did not use the modern approach to a limit. This reasoning — as

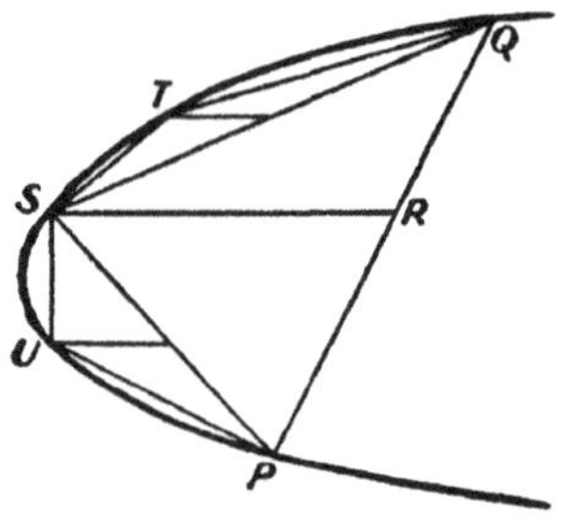

FIG. 15. — ARCHIMEDES'S QUADRATURE OF THE PARABOLA.

was all similar reasoning of Greek mathematicians — was by a *reductio ad absurdum*. He proved that the assumption that the area of the parabolic segment is greater than four-thirds the area of the triangle is wrong, and also the opposite assumption, that the first area is less than the second. Hence, Archimedes reasons, the areas must be equal.

SPIRALS

The discussion of spirals is based on the definition, "If a straight line moves with uniform velocity in a plane about one of its extremities which remains fixed, until it returns to its original position, and if at the same time a point moves with uniform velocity starting at the fixed point, the moving point describes a spiral." With the simple resources at his command, he also succeeds in obtaining the quadrature of this spiral, and in drawing a tangent at any point. In these quadratures he approximates the summation principle of the modern integral calculus. This spiral is called the *spiral of Archimedes* ($r = a\theta$, a = a constant, in modern notation of polar coordinates).

Supplementing Euclid's treatment of the regular polyhedrons, Archimedes investigates the semi-regular solids formed by combining regular polygons of more than one kind. Of these he finds 13, ten of which have two kinds of bounding polygons, the others, three kinds.

SPHERE AND CYLINDER

In his important treatise *On the Sphere and the Cylinder* he derives three new theorems:

(1) That the surface of a sphere is four times the area of its great circle.

(2) That the convex surface of a segment of a sphere is equal to the area of a circle whose radius is equal to the straight line from the vertex of the segment to any point in the perimeter of its base.

(3) That the cylinder having a great circle of the sphere for its base and the diameter of the sphere for its altitude exceeds the sphere by one-half, both in volume and in surface. It was the figure for this last proposition which was at his wish carved upon his tombstone.

In attempting to solve the problem of passing a plane through a sphere so that the segments thus formed shall have either their surfaces or their volumes in an assigned ratio, he is led to a cubic equation; he appears to have given both a solution and a criterion for the existence of a positive root, but the work is lost.

In his *Conoids and Spheroids* he deals with the bodies formed by the revolution of the ellipse, parabola, and hyperbola, by means of plane cross-sections, ascertains the volume of these solids by comparing the portion between two neighboring planes with an inscribed and a circumscribed cylinder — much in the modern manner.

It is not possible to find in all geometry more difficult and more intricate questions or more simple and lucid explanations (than those given by Archimedes). Some ascribe this to his natural genius; while others think that incredible effort and toil produced these, to all appearance, easy and unlabored results. No amount of investigation of yours would succeed in attaining the proof, and yet, once seen, you immediately believe you would have discovered it; by so smooth and so rapid a path he leads you to the conclusion required. — PLUTARCH, *Life of Marcellus*. (Moritz, *Memo. Math.*, p. 139.)

In other branches of mathematical science than geometry the work of Archimedes was relatively not less important.

The so-called Cattle Problem, for example, is a notable performance in the algebra of linear and quadratic equations, involving the use of enormous numbers for their solution.

Again he succeeds in summing the series of squares: 1, 4, 9, 16, 25, 36, etc., to n terms, expressing the result in geometrical form. Both proof and formulation are of course much more complicated by reason of the entire lack of an algebraic symbolism, the same remark naturally applying also to the preceding cattle problem and to the cubic equation referred to above. This last was indeed to Archimedes not primarily an equation at all, but a proportion

$$a - x : b = \tfrac{4}{9}\, a^2 : x^2$$

In his *Circle Measurement*, already outlined, he showed mastery of square root, and the comparison of irrational numbers with fractions, showing for example that

$$\frac{1351}{780} > \sqrt{3} > \frac{265}{153}$$

How these fractions were obtained cannot be certainly determined.

In the *Sand-Reckoner,* Archimedes undertakes to give a number which shall exceed the number of grains of sand in a sphere with a radius equal to the distance from the earth to the starry firmament. The treatise begins: "Many people believe, King Gelon, that the number of sand grains is infinite. I mean not the sand about Syracuse, nor even that in Sicily, but also that on the whole mainland, inhabited and uninhabited. There are others again who do not indeed assume this number to be infinite, but so great that no number is ever named which exceeds this. . . . I will attempt to show however by geometrical proofs which you will accept that among the numbers which I have named . . . some not only exceed the number of a sand-heap of the size of the earth, but also of that of a pile of the size of the universe." He assumes that 10,000 grains of sand would make the size of a poppy-seed, that the diameter of a poppy-seed is not less than one-fortieth of a finger-breadth, that the diameter of the earth is less than a million stadia, that the diameter of the universe is less than $10{,}000^2$ diameters of the earth. To express the vast number which results from these assumptions — 10^{63} in our notation — he employs an ingenious system of units of higher order comparable with the modern use of exponents, an immense advance on current arithmetical symbolism.

MECHANICS OF ARCHIMEDES

In mechanics Archimedes is a pioneer, giving the first mathematical proofs known. In two books *On Equilibrium of Planes,* he deals with the problem of determining the centers of gravity of a variety of plane figures, including the parabolic segment. A treatise on levers and perhaps on machines in general has been lost, as also a work on the construction of a celestial sphere. A sphere of the stars and an orrery constructed by him were long preserved at Rome. He describes an original apparatus for determining the angular diameter of the sun, discussing its degree of accuracy.

The lever and the wedge had been known practically from remote antiquity, and Aristotle had discussed the practice of dishonest tradesmen shifting the fulcrum of scales, but no previous attempt at exact mathematical treatment is known.

Archimedes assumes as evident at the outset:

(1) Magnitudes of equal weight acting at equal distances from their point of support are in equilibrium;

(2) Magnitudes of equal weight acting at unequal distances from their point of support are not in equilibrium, but the one acting at the greater distance sinks.

From these he deduces:

(3) Commensurable or incommensurable magnitudes are in equilibrium when they are inversely proportional to their distances from the point of support.

In a work on *Floating Bodies*, extant in a Latin version by Willem van Moerbeke (1269) from a Greek text which disappeared between 1544 and 1564, and only partly in the original Greek, Archimedes lays down the foundations of hydrostatics as follows: "Let it be assumed that the nature of a fluid is such that, all its parts lying evenly and continuous with one another, the part subject to less pressure is expelled by the part subject to greater pressure. But each part is pressed perpendicularly by the fluid above it, if the fluid is falling or under any pressure." "Every solid body lighter than a liquid in which it floats sinks so deep that the mass of liquid which has the same volume with the submerged part weighs just as much as the floating body." The specific gravity of heavier bodies was of course employed in his solution of the crown problem,[1] which with his achievements as a military engineer gave him a great reputation among his contemporaries.

In his book on *The Method*, which was discovered as recently as 1906 in a Constantinople palimpsest, Archimedes throws a very interesting light on his methods of attacking problems in mechanics, as well as on his use of mechanical methods for geometrical problems. Naturally his mathematical methods are highly developed in comparison with the

[1] A crown of gold was thought to have been made of an alloy of baser metal, which would for the requisite weight, have increased its bulk. Archimedes in his bath realized that this increase of bulk could be tested by immersing the crown in water of which an equal bulk would be displaced.

relatively simple problems of mechanics with which he deals. He actually got hold of the fundamental idea of the calculus and could evaluate some simple integrals.

Certain things first became clear to me by a mechanical method, although they had to be demonstrated by geometry afterwards because their investigation by the said method did not furnish an actual demonstration. But it is of course easier, when we have previously acquired, by the method, some knowledge of the questions, to supply the proof than it is to find it without any previous knowledge. I apprehend that some, either of my contemporaries or of my successors, will, by means of the method when once established, be able to discover other theorems . . . which have not yet occurred to me.

ARCHIMEDES AS AN ENGINEER

His engineering skill, which has gained from an eminent German historian the appellation of "the technical Yankee of antiquity," may be inferred from Plutarch's account of the siege of Syracuse: —

Now the Syracusans, seeing themselves assaulted by the Romans, both by sea and by land, were marvellously perplexed, and could not tell what to say, they were so afraid; imagining it was impossible for them to withstand so great an army. But when Archimedes fell to handling his engines, and set them at liberty, there flew in the air infinite kinds of shot, and marvellous great stones, with an incredible noise and force on the sudden, upon the footmen that came to assault the city by land, bearing down, and tearing in pieces all those which came against them, or in what place soever they lighted, no earthly body being able to resist the violence of so heavy a weight; so that all their ranks were marvellously disordered. And as for the galleys that gave assault by sea, some were sunk with long pieces of timber like unto the yards of ships, whereto they fasten their sails, which were suddenly blown over the walls with force of their engines into their galleys, and so sunk them by their over great weight.

These machines (used in the defense of the Syracusans against the Romans under Marcellus) he (Archimedes) had designed and contrived, not as matters of any importance, but as mere amusements in geometry; in compliance with king Hiero's desire and request, some time before, that he should reduce to practice some part of his admirable speculation in science, and by accommodating the theoretic truth to sensation and ordinary use, bring it more within the appreciation of people in general.

One of his most famous inventions was the water-screw, the idea of which, it is said, occurred to him in Egypt. On occasion of difficulty in the launching of a certain ship he successfully applied a cogwheel apparatus with an endless screw.

Archimedes . . . had stated that given the force, any given weight might be moved, and even boasted, we are told, relying on the strength of demonstration, that if there were another earth, by going into it he could remove this. Hiero being struck with amazement at this, and entreating him to make good this problem by actual experiment, and show some great weight moved by a small engine, he fixed accordingly upon a ship of burden out of the king's arsenal, which could not be drawn out of the dock without great labor and many men; and, loading her with many passengers and a full freight, sitting himself the while far off with no great endeavor, but only holding the head of the pulley in his hand and drawing the cords by degrees, he drew the ship in a straight line, as smoothly and evenly, as if she had been in the sea. —PLUTARCH. (*Memo. Math.*, p. 135.)

ARCHIMEDES AND EUCLID

In contrasting the limitations of Euclid's *Elements* with the broad range of Greek mathematics, Felix Klein characterizes the work of Archimedes somewhat as follows:

(1) Quite in contrast to the spirit controlling Euclid's *Elements*, Archimedes has a strongly developed sense for numerical computation. One of his greatest achievements indeed is the calculation of the ratio π of the circumference of a circle to its diameter, by approximations with regular polygons. There is no trace of interest for such numerical results with Euclid, who merely mentions that the areas of two circles are proportional to the squares of the radii, two circumferences as the radii, regardless of the actual proportionality factor.

(2) A far-reaching interest in applications of all sorts is characteristic of Archimedes, including the most varied physical and technical problems. Thus he discovered the principles of hydrostatics and constructed engines of war. Euclid on the

contrary does not even mention ruler or compass, merely postulating that a straight line can be drawn through two points, or a circle described about a point. Euclid shares the view of certain ancient schools of philosophy — a view unfortunately extant in certain quarters — that the practical application of a science is something mechanical and unworthy. The greatest mathematicians, Archimedes, Newton, Gauss, have combined theory and applications consistently.[1]

(3) Finally, Archimedes was a great investigator and pioneer, who in each of his works carries knowledge a step forward. This affects materially the form of presentation. In *The Method* the procedure is essentially modern as contrasted with the rigid formalism of the *Elements*, though in other works Archimedes remains in the orthodox tradition.

EARTH MEASUREMENT: ERATOSTHENES (*c*. 273–*c*. 192 B.C.)

Eratosthenes, librarian of the great library at Alexandria (235 B.C.), making a systematic quantitative study of data collected from various sources, laid the foundations of mathematical geography — a transformation quite analogous to that taking place in astronomy. After an historical review he gives numerical data about the inhabited earth, which he estimates to have a length of 78,000 stadia [2] and a breadth of 38,000. Pytheas of Massilia (Marseilles) had (*c*. 330) determined latitude by measuring the sun's shadow and Dicaearchus of Massina, about the same time, had suggested the use of two axes perpendicular to each other for reference in plot-

[1] Plutarch, however, says: "Archimedes possessed so high a spirit, so profound a soul, and such treasures of highly scientific knowledge, that though these inventions (used to defend Syracuse against the Romans) had now obtained him the renown of more than human sagacity, he yet would not deign to leave behind him any commentary or writing on such subjects; but, repudiating as sordid and ignoble the whole trade of engineering, and every sort of art that lends itself to mere use and profit, he placed his whole affection and ambition in those purer speculations where there can be no reference to the vulgar needs of life; studies, the superiority of which to all others is unquestioned, and in which the only doubt can be whether the beauty and grandeur of the subjects examined, or the precision and cogency of the methods and means of proof, most deserve our admiration."

[2] The stadium may have been about 488 feet; *Isis*, **28**, 493, 1938.

ting the world's map, but Eratosthenes was the first geographer who was able to draw his map with cross lines indicating latitude and longitude. In connection with this he gives also a remarkably successful determination of the circumference of the earth. This was based on his observation that at Syene (modern Assuan, near the First Cataract of the Nile) at noon

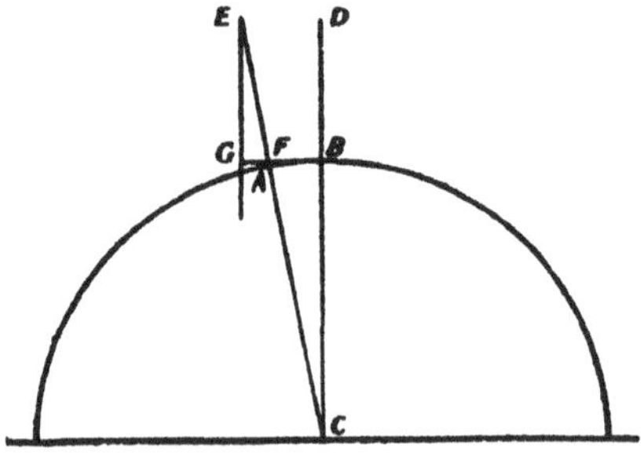

Fig. 16. — Calculation of the Earth's Circumference by Eratosthenes. *C*, center of the earth; *AB*, distance from Alexandria to Syene, 5,000 stadia; *DB*, axis of well; *EF*, vertical rod at Alexandria; *FG*, its shadow. Angle *FEG* = angle *ACB* = 7° 12′. Modified from Cary, *Legacy of Alexandria*, p. 407. Permission of The Dial Press, Inc.

of the summer solstice the sun shone straight down to the bottom of a deep well, while at Alexandria the zenith distance of the sun at noon was $\frac{1}{50}$ of the circumference of the heavens. Assuming the two places to lie in the same meridian, and taking their distance apart as 5,000 stadia, he infers that the whole circumference must be 250,000 stadia. He or some successor afterwards substituted 252,000, perhaps in order to obtain a round number, 700 stadia, for the length of one degree.

This result, subject to some uncertainty as to the length of the stadium, was a close approximation to the real circumference, but we may suppose that this degree of accuracy was to some extent a matter of accident. Eratosthenes is also credited with measuring the obliquity of the ecliptic with an error of about seven minutes.

A student of the Athenian Platonists and a man of extraordinary versatility, philosopher, philologian, mathematician, athlete, Eratosthenes wrote on many subjects. Besides having "contributed more than any other individual to a correct delineation of the earth's map," he also founded the scientific chronology of ancient Greece, and he may well have been responsible for the attempt of Ptolemy III to introduce leap-year into the Egyptian calendar by the "Decree of Canopus" in 238 B.C.,

> in order that the seasons may continually render service according to the present order and that it may not happen that some of the public festivals which are celebrated in the winter come to be observed in the summer . . . ,

but everywhere the people continued to use the inconvenient lunar month of the Greeks.

His so-called "sieve" is a method for systematically separating out the prime numbers by arranging all the natural numbers in order, and striking out first all multiples of 2, then of 3, and so forth, thus sifting out all but the primes, 1, 2, 3, 5, 7, 11, 13, 17, etc.

APOLLONIUS OF PERGA (*c. 260–c. 200 B.C.*)

"The great geometer" was the last of this famous Alexandrian group of mathematicians, and owes his reputation to

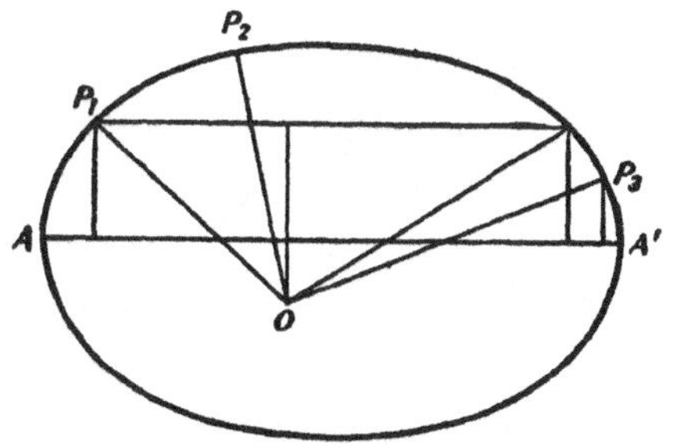

FIG. 17. — THE THREE NORMALS TO THE ELLIPSE. APOLLONIUS.

his important work on the conic sections. His predecessors had in general recognized only those sections formed from right circular cones by planes normal to an element. Archimedes, indeed, and Euclid obtained ellipses by passing other planes through right cones, but Apollonius first showed that any cone and any section could be taken, and introduced the names ellipse, parabola, and hyperbola. In the prefatory letter to Book I, Apollonius says to the friend to whom it is addressed:

Apollonius to Eudemus, greeting. When I was in Pergamum with you, I noticed that you were eager to become acquainted with my *Conics;* so I send you now the first book with corrections and will forward the rest when I have leisure. I suppose you have not forgotten that I told you that I undertook these investigations at the request of Naucrates, the geometer, when he came to Alexandria and stayed with me; and that, having arranged them in eight books, I let him have them at once, not correcting them very carefully (for he was on the point of sailing) but setting down everything that occurred to me, with the intention of returning to them later. Wherefore I now take the opportunity of publishing the needful emendations. But since it has happened that other people have obtained the first and second books of my collections before correction, do not wonder if you meet with copies which are different from this. — Gow, *Hist. of Greek Math.*, p. 47.

In Book I he defines the cone as generated by a straight line passing through a point on the circumference of a circle

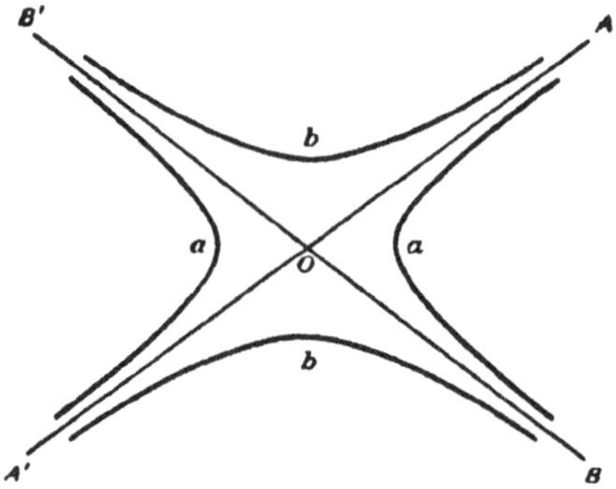

FIG. 18. — CONJUGATE HYPERBOLAS. Hyperbolas *a* and *b* are conjugate. The lines *A'OA* and *B'OB* are asymptotes to both curves.

and a fixed point not in the same plane; he fixes the manner in which sections are to be taken and defines diameters and vertices of the curves, also the latus rectum [1] and center, conjugate diameters and axes. The other branch of the hyperbola

[1] Latus rectum is the chord through a focus normal to the diameter through the two foci. A diameter is a chord through the center of the curve. A diameter bisecting each of a system of parallel chords is conjugate to the diameter included in that system.

is taken due account of for the first time. In Book II asymptote properties and propositions on conjugate hyperbolas are discussed. Book III contains numerous theorems on tangents and secants and introduces foci with the definition: "A focus is a point which divides the major axis into two parts whose rectangle is one-fourth that of the latus rectum and the major axis," or the square on the minor axis. The focus of the parabola, however, is not recognized, nor has he any knowledge of the directrix [1] of a conic section, these omissions being first filled by Pappus in the third century A.D. It is shown that the normal makes equal angles with the focal radii to the point of contact, and that the latter have a constant sum for the ellipse, a constant difference for the hyperbola. This book, he says in the letter quoted above, "contains many curious theorems, most of them are pretty and new, useful for the synthesis of solid loci. . . . In the invention of these, I observed that Euclid has not treated synthetically the locus . . . but only a certain small portion of it, and that not happily, nor indeed was a complete treatise possible at all without my discoveries." On the whole, in some 400 propositions he achieved nearly all the results which are included in our modern elementary analytic geometry, approximating the introduction of a system of coordinates by his use of lines parallel to the principal axes.

It is noteworthy that Fermat (page 318), one of the inventors of modern analytic geometry, was led to it by attempting to restore certain lost proofs of Apollonius on loci.

Of his other mathematical writings little more than the titles are known. Among these are one on burning mirrors, one on stations and retrogressions of the planets, and one on the use and theory of the screw. In astronomy he is believed to have suggested expressing the motions of the planets by combining uniform circular motions, an idea afterwards elaborated by Hipparchus and Ptolemy. How far his mathematical

[1] A conic section may be defined with reference to a fixed line, the *directrix*, and a fixed point (not on that line), the focus, by the statement that the distance from a moving point to the directrix is in a constant ratio to the distance from the moving point to the focus. The term focus is due to Kepler.

results were new, how far he merely compiled and coordinated the work of others, notably Euclid and Archimedes, cannot precisely be determined, but the proportion of original work is certainly very large.

In a treatise on *Contacts*, described by Pappus, a famous problem of plane geometry is solved, namely, the construction of a circle tangent to three given circles and its modifications by substituting points or tangent lines for one or more of the circles. This problem has received the attention of many mathematicians of later date, as Vieta and Newton.

On the arithmetical side Apollonius obtained, it is said, a closer approximation than Archimedes for the value of π, invented an abridged method of multiplication, and employed numbers of higher order in the manner of Archimedes.

APOLLONIUS AND ARCHIMEDES

With Apollonius and Archimedes the ancient mathematics had accomplished whatever was possible without the resources of analytic geometry and infinitesimal calculus, which, though already foreshadowed, were not fully realized until the seventeenth century, after an interval, that is, of nearly 2,000 years.

It is not only a decided preference for synthesis and a complete denial of general methods which characterizes the ancient mathematics as against our newer science (modern mathematics): besides this external formal difference there is another real, more deeply seated, contrast, which arises from the different attitudes which the two assumed relative to the use of the concept of *variability*. For while the ancients, on account of considerations which had been transmitted to them from the philosophic school of the Eleatics, never employed the concept of motion, the spatial expression for variability, in their rigorous system, . . . modern geometry dates from the instant that Descartes left the purely algebraic treatment of equations and proceeded to investigate the variations which an algebraic expression undergoes when one of its variables assumes a continuous succession of values. — H. HANKEL. (Moritz, *Memo. Math.*, p. 115.)

The works of Archimedes and Apollonius marked the most brilliant epoch of ancient geometry. They may be regarded, moreover, as the origin and foundation of two questions which have occupied geometers at all periods. The greater part of their works are connected with these and are divided by them into two classes, so that they seem to share between them the domain of geometry.

The first of these two great questions is the quadrature of curvilinear figures, which gave birth to the calculus of the infinite, conceived and brought to perfection successively by Kepler, Cavalieri, Fermat, Leibnitz and Newton.

The second is the theory of conic sections, for which were invented first the geometrical analysis of the ancients, afterwards the methods of perspective and of transversals. CHASLES, *Aperçu*. (Gow, p. 260.)

ORBITAL MOTION OF THE EARTH. ARISTARCHUS (*c.* 280 B.C.)

Before dealing with certain other mathematical developments, we have to consider the highly interesting and signifi-

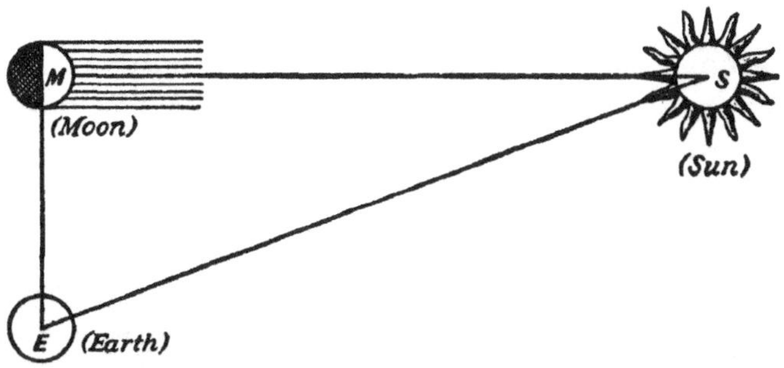

FIG. 19. — METHOD OF ARISTARCHUS FOR DETERMINING THE RELATIVE DISTANCE OF SUN AND MOON. After Jeans, *Through Space and Time*, p. 81. Permission of The Macmillan Company.

cant astronomical theories of Aristarchus of Samos, the author of a treatise *On the Dimensions and Distances of the Sun and Moon.* He endeavored to determine these distances relatively by ascertaining or estimating the angular distance between the two bodies when the moon is just half illuminated, that is, when the lines joining sun, earth, and moon form a right angle at the moon. The difficulties of this determination are so serious, however, that no high degree of accuracy could be attained, the actual result of Aristarchus $\frac{29}{30}$ of a right angle — against the true $\frac{539}{540}$ — corresponding to a ratio of about 1 to 19 of the two distances. Aristarchus had no trigonometry, and no other method of attacking this problem seems to have been known to the Greeks.

In his *Sand-Reckoner* already mentioned, Archimedes says of Aristarchus,

> He supposes that the fixed stars and the sun are immovable, but that the earth is carried round the sun in a circle which is in the middle of the course; but the sphere of the fixed stars, lying with the sun round the same centre, is of such a size that the circle, in which he supposes the earth to move, has the same ratio to the distance of the fixed stars as the centre of the sphere has to the surface. But this is evidently impossible, for as the centre of the sphere has no magnitude, it follows that it has no ratio to the surface. It is therefore to be supposed that Aristarchus meant that as we consider the earth as the centre of the world, then the earth has the same ratio to that which we call the world, as the sphere in which is the circle, described by the earth according to him, has to the sphere of the fixed stars.

Aristarchus meets the objection that motion of the earth would cause changes in the apparent positions of the stars by assuming that their distances are so great as to render the motion of the earth a negligible factor. Another reference to Aristarchus, in Plutarch, mentions an opinion that he

> ought to be accused of impiety for moving the hearth of the world, as the man in order to save the phenomena supposed that the heavens stand still and the earth moves in an oblique circle at the same time as it turns round its axis.

How far this remarkable anticipation of the Copernican theory was a conviction rather than a mere fortunate speculation cannot be known, but at any rate it failed of that acceptance necessary to its permanence. In the next century the rotation of the earth on its axis and the revolution of the earth around the sun was indeed taught by Seleucus, an astronomer in Babylonia, but it was 1,700 years before these daring theories were again advanced. Seleucus also observed the tides, saying "that the revolution of the moon is opposed to the earth's rotation, but the air between the two bodies being drawn forward falls upon the Atlantic Ocean, and the sea is disturbed in proportion."

PLANETARY IRREGULARITIES

The earlier theory of homocentric spheres, while accounting more or less successfully for the apparent motions of the heavenly bodies, had maintained each of them at a constant distance from the earth, and thus quite failed to explain the differences of brightness which were soon discovered, as well as the variations in the apparent size of the moon. The conception of motion in any other path than a straight line or a circle was repugnant to the Greek philosophers, and the difficulty was therefore met, first by supposing the earth not to be exactly at the center of the circular orbits about it, second by introducing subsidiary circles or epicycles.

EXCENTRIC CIRCULAR ORBITS

The planetary system was conceived somewhat as follows. Around the Earth at the center of the universe revolved the Moon in a lunar month, the Sun in a year, each in a circular path. Mercury and Venus each moved in a circular path with its center in the revolving straight line joining Earth and Sun. The Earth is always outside these circles and the two planets are never more than a limited angular distance from the Sun. The three other known planets, Mars, Jupiter, and Saturn, also move in circular paths, the centers of which are on the same revolving straight line, but both Earth and Sun lie within these larger circles.

It seems probable that Aristarchus was led through this theory to conceive of heliocentric orbits, and then to reflect that the earth, too, might revolve about the sun as easily as the sun and planets round the earth.

EPICYCLES

Progress in observational astronomy increased the number and magnitude of planetary irregularities beyond the stationary points, retrograde motions, and variations, known to Aristarchus, and apparently far beyond possible explanation by the simple theory of excentric circles. The system was therefore superseded by, or combined with, that of epicycles, not necessarily as physically realized, but as at least a geometrical working hypothesis, which should conform to and explain the observed phenomena.

The system of epicycles consists in superimposing one circular motion upon another, and repeating the process to any needful extent. The motion of the moon about the earth, for example, is explained by assuming first a circle (later called the deferent) on which moves the center of a second smaller circle called the epicycle, on which the moon itself travels. By varying the dimensions of both circles and the velocities of the two motions, the observed changes, both of position and brightness of the moon, may be more or less satisfactorily accounted for and even computed in advance. In particular, the apparent retrograde motions of the planets in certain parts of their orbits may be explained.

In the figure *E* denotes the earth, the large circle is the deferent of the Moon, *C* the center of the epicycle, P_1, P_2, P_3, P_4 dif-

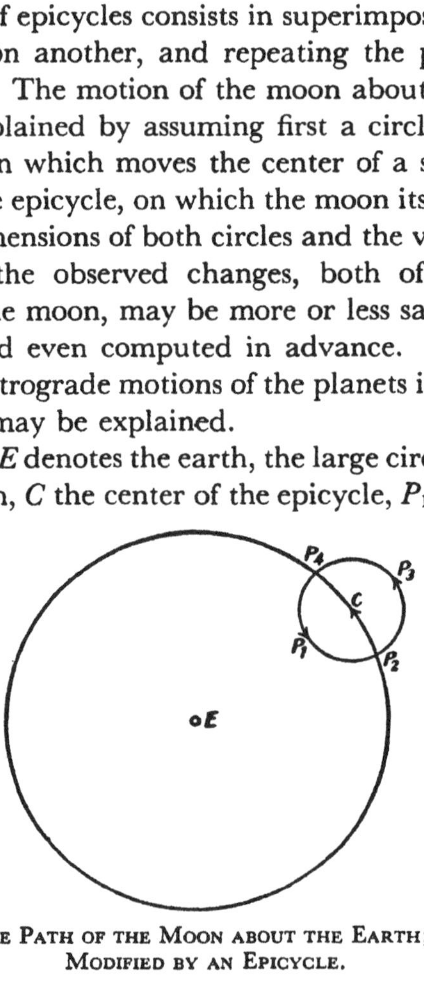

FIG. 20. — THE PATH OF THE MOON ABOUT THE EARTH; A DEFERENT MODIFIED BY AN EPICYCLE.

ferent possible positions of the Moon in its epicycle. The distance of *P* from *E* obviously varies; the apparent motion of *P* as seen from *E*, being compounded of a forward motion of *C* and a circular motion of *P*, will appear at P_1 to be slower and at P_3 faster than the average. By suitable adjustment of the dimensions and velocities there may be retrogression for a certain length of arc near P_1, bounded by stationary points where the two motions seem to an observer at E to neutralize each other.

How far this complicated scheme really departed from the original postulate of uniform circular motion is sufficiently indicated by Proclus's remark, "The astronomers who have presupposed uniformity of motions of the celestial bodies were ignorant that the essence of these movements is, on the contrary, irregularity." While in point of fact the theory of epicycles and that of excentric circles have much in common, the former gradually displaced the latter on account of its greater simplicity. Had Aristarchus worked out the earlier system in full detail, the history of astronomy might have been considerably modified.

MEDICAL SCIENCE AT ALEXANDRIA. BEGINNINGS OF HUMAN ANATOMY

The school of medicine, one of the four departments, of the Museum, began its first century with brilliant success and important contributions to science through the labors of two men — Herophilus of Chalcedon and his younger contemporary and rival, Erasistratus of Chios. Their writings are lost, but a good idea of their results has been extracted from the works of Galen (p. 167). Structure and function are inseparable in their studies, but Herophilus was regarded as the greatest anatomist of antiquity, while Erasistratus is called the father of experimental physiology. Both men made important additions to knowledge of the anatomy of the heart and blood vessels, and of the nervous system and brain. Both distinguished motor nerves from sensory, and, disagreeing with Aristotle, recognized the brain as the central organ of the nervous system and the seat of intelligence.

Herophilus, an eminent physician and teacher, a Platonist, is said by Galen to have been the first to dissect (publicly?) both animal and human bodies (favored doubtless by the Egyptian practice of embalming). He named the prostate gland and gave the name duodenum (twelve fingers) to the first part of the intestine. He was the first to distinguish clearly between arteries and veins, and the first to count the pulse with the aid of a clock (clepsydra, p. 40), and he made an elaborate analysis of its rhythm.

Erasistratus was a rationalist, adopting the materialistic philosophy of Epicurus. He made the important generalization that every organ is supplied with a three-fold system of "vessels" — vein, artery, and nerve — and he followed their ramifications to the limit of visibility. But his physiology was dominated by the theory of the *pneuma* (cf. Empedocles, p. 62). which he elaborated as follows: Air from the lungs becomes in the heart a peculiar *pneuma*, the *vital spirit*, and passes from there through the arteries to the organs. In the brain it is changed to *animal spirit*, which is propelled hence through motor nerves, supposed to be tubular, to the muscles, causing them to swell and shorten and thus to produce motion in the part — an idea that persisted till the time of Descartes (p. 291). He definitely recognized the pumping of the heart and the action of its valves, but, led astray by the doctrine of the *pneuma*, he supposed the blood to flow outward in the veins, from which it could pass to the arteries, but only after the escape of *pneuma* (air) through a wound — thus narrowly he missed the circulation. He thought the chief cause of disease to be *plethora*, excess of blood, but he opposed frequent blood-letting, then a common practice. The accusation by Celsus (p. 166) that he performed human vivisection is improbable, chiefly because not mentioned by Galen, who violently opposed his views. Eudemus of Alexandria (*c.* 250 B.C.), a disciple of Herophilus and Erasistratus, made a deeper study of the skeleton, nervous system, pancreas, and the female sexual organs, and of embryology.

After one generation, medicine at Alexandria became dominated by a sect that was founded by Seraphion. They were called Empirics, and claimed to be practical, scorned theory, abhorred dissection, and treated symptoms with a multitude of drugs. Teaching flourished for centuries, but the contributions to science were unimportant.

REFERENCES FOR READING

BALL, W. W. R., *Short History of Mathematics*, 1901, Ch. IV to p. 84.
CARY, M., *Legacy of Alexander*, 1932.
HEATH, T. L., *Apollonius of Perga*, 1896.
——, *Euclid's Elements*, 1908, Vol. 1.
——, *Works of Archimedes*, 1897.
MACH, E., *Science of Mechanics*, Ed. 4, 1919 (on Archimedes).
SINGER, C., *Evolution of Anatomy*, 1925, pp. 28–36.

Books from previous chapters: Berry, Art. 31–36; Breasted, *Ancient Times*, pp. 489–554; Farrington; Gow, Ch. VII; Heath, *Aristarchus;* Sarton, *Introd.*, I, pp. 149–200. Singer, *Hist. Med.*, pp. 36–41.

Chapter VI

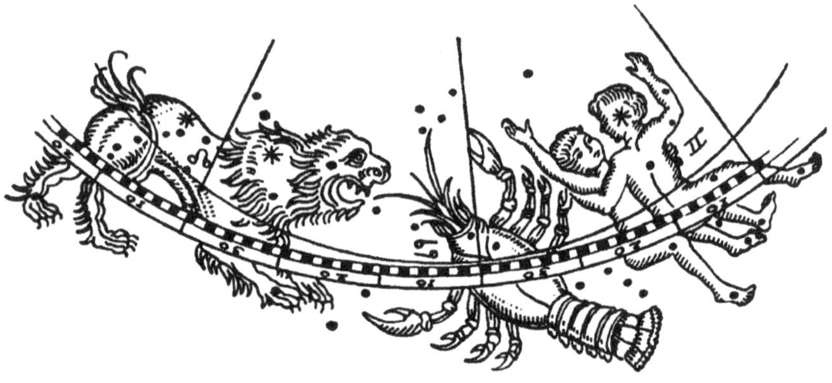

The Decline of Alexandrian Science

The force of nature could go no further in the same direction than the ingenious applications of exhaustion by Archimedes and the portentious sentences in which Apollonius enunciates a proposition in conics. A briefer symbolism, an analytical geometry, an infinitesimal calculus were wanted, but against these there stood the tremendous authority of the Platonic and Euclidean tradition, and no discoveries were made in physics or astronomy which rendered them imperatively necessary. It remained only for mathematicians, as Cantor says, to descend from the height which they had reached and "in the descent to pause here and there and look around at details which had been passed by in the hasty ascent."— Gow, *History of Greek Mathematics*, p. 265.

The Greek astronomers of the scientific period, such as Aristarchus, Eratosthenes, and above all, Hipparchus, appear, moreover, to have followed in their researches the method which has always been fruitful in physical science — namely, to frame provisional hypotheses, to deduce their mathematical consequences, and to compare these with the results of observation. There are few better illustrations of genuine scientific caution than the way in which Hipparchus, having tested the planetary theories handed down to him and having discovered their insufficiency, deliberately abstained from building up a new theory on data which he knew to be insufficient, and patiently collected fresh material, never to be used by himself, that some future astronomer might thereby be able to arrive at an improved theory. — BERRY, *History of Astronomy*.

THE GRAECO-ROMAN PERIOD (146 B.C. to A.D. 476)

"By the first century before Christ, the Romans had conquered the world, and Greek learning had conquered the Romans." In the interval between 146 B.C., when Corinth fell, and 30 B.C., when Cleopatra died and Egypt became a Roman province, the remaining Hellenistic dynasties became extinct and their lands were annexed to Rome, except Babylonia, taken by the Parthians in 141 B.C. Latin remained the language of Rome and Greek influence at first was resisted. But in 156 B.C. Diogenes the Babylonian, while on a mission from Athens, lectured to large audiences in Rome on Stoic philosophy, and a century later Cicero did much to popularize Greek philosophy and literature. Greek became a second language for cultivated Romans, as French or German is now for many English-speaking persons.

In the mean time Greek learning had been influenced by contact with remnants of the ancient cultures of the East. In Mesopotamia, Babylonian learning was maintained by a small group of conservative priests, astronomers, and lawyers. A cuneiform astronomical tablet was dated at Babylon in the Seleucid year 305, that is 7 B.C. Under the shadow of the Museum, the practical arts and the priestly mysteries of the Egyptians were in close contact with Greek learning, and the Jews also had made Alexandria their intellectual and commercial center. The influence of these contacts is seen in the spread westward of astrology mingled with valuable astronomical data, in the rise of alchemy in Egypt, and in the development of a mystical Neo-Platonism in place of the pure philosophy of the classic period.

While medical science had declined much earlier in Alexandria, notable contributions to the mathematical sciences were made there till late in the Graeco-Roman period, which, also, saw the birth of experimental chemistry.

NOTE: The figure on the opposite page contains three of the Signs of the Zodiac from a map of the heavens in a 16th century Latin version of Ptolemy's works, *Opera Omnia*, Basel, 1551. Redrawn by Elizabeth Tyler Wolcott.

MATHEMATICS AND ASTRONOMY

In the second century B.C. Hypsicles developed the theory of arithmetical progression and added two books of *Elements* to Euclid's thirteen, but the chief mathematical work of this century was due to Hipparchus, a great astronomer, and to Hero, an engineer.

At the Museum of Alexandria a school of observers, of whom Aristyllus and Timocharis were notable members, instituted systematic astronomical observations with graduated instruments and made a small star catalogue. Thus was laid a foundation for the brilliant discoveries of Hipparchus and Ptolemy, while astronomy, which had in the work of Eudoxus assumed the character of true science, though with a too slender observational basis, now became an exact science, gradually shedding its encumbrances of speculation and vague generalization.

HIPPARCHUS. STAR CATALOGUE

The next great astronomer and much the greatest of antiquity is Hipparchus, probably a native of Bithynia, but long resident at Rhodes, a city which rivalled Alexandria itself in its intellectual activity. All his works but one are lost, but his great successor and disciple, Ptolemy, has based his famous treatise on the work of Hipparchus and it is possible to determine in a general way how much is to be credited to each. Having at his disposal the primitive star catalogue of Aristyllus and Timocharis, Hipparchus was profoundly impressed — as was Tycho Brahe centuries later — by the sudden appearance in 134 B.C. in the supposedly changeless starry firmament of a new star of the first magnitude. He accordingly set himself the heavy task of making a new catalogue, which ultimately included more than 1,000 stars, for the part of the sky visible to him, and "remained, with slight alterations, the standard for nearly sixteen centuries." His list of constellations and his classification of stars in six "magnitudes" according to their brightness are the bases of our own.

PRECESSION OF THE EQUINOXES

While this great piece of routine work was deliberately planned by Hipparchus, not so much as an end in itself as a necessary basis for future investigators, it nevertheless led to his most remarkable discovery, that of the precession of the equinoxes.[1] In comparing, namely, the positions of certain stars with those observed about 150 years earlier, he detected a change of distance from the equinoctial point — where the celestial equator and the ecliptic meet — amounting in one case to about 2°. By an inspiration of genius, he interpreted this correctly as due to a slight progressive shifting of the equinoctial points, corresponding to a slow rotation of the earth's axis, by means of which the celestial pole in many thousand years describes a complete circle. His estimate of 36″ per year was considerably below the actual value, which is about 50″.

OTHER ASTRONOMICAL DISCOVERIES. PLANETARY THEORY

Striving always for greater accuracy and completeness of data, he determined the length of the year within about six minutes. In attempting to explain the annual motion of the sun, he was aware that the change of direction is not uniform, and its distance from the earth, as shown by its apparent size, not constant. He determined the length of spring as 94 days, that of summer as $92\frac{1}{2}$, and by a somewhat complicated calculation arrived at the value $\frac{1}{24}$ as the eccentricity of the earth's position in the sun's orbit. These determinations were naturally very difficult and imperfect on account of the entire lack of accurate time measurement. Following Apollonius, Hipparchus devised a combination of uniform circular motions which should account for the observed facts within the limits of probable error of observation, and in this undertaking he was successful, the degree of accuracy of his theory corresponding to that of which his instruments were capable.

With the more complicated lunar theory he was naturally

[1] Recently discovered evidence indicates the anticipation of this discovery in Babylonia. (See page 23.)

less successful. He is believed, however, to have discovered the more important irregularities of the moon's motion, supposing it to have a circular orbit in a plane making an angle of 5° with that of the sun's orbit — the ecliptic. The earth is not at the center, but the latter revolves about the earth in a period of nine years.

Extending his study of eclipses to the ancient records of the Chaldeans, he made substantial improvements in the theory of both solar and lunar eclipses, and obtained a close approximation for the distance of the moon. He estimated the sun's radius at about twelve times that of the earth, its distance from the earth at about 2,550 earth-radii, the moon's radius $\frac{29}{100}$ that of the earth, its distance about 60 earth-radii. The comparison of these figures with Ptolemy's and with the actual are (in earth-radii) —

	Sun's Radius	Sun's Distance	Moon's Radius	Moon's Distance
Hipparchus . . .	12	2550	.29	60
Ptolemy	5.5	1210	.29	59
Actual	109	23,000	.273	$60\frac{1}{3}$

Hipparchus realized that he had no adequate method for determining these numbers for the sun.

The generally accepted order of the planets had now become Moon, Mercury, Venus, Sun, Mars, Jupiter, Saturn — an order adopted very early in Babylonia, and received as a more or less probable hypothesis from this time until that of Copernicus. In attempting to deal with the motions of the other planets as he had done with that of the sun and moon, Hipparchus was soon baffled by lack of adequate data, and set himself steadfastly to supply the need, resigning to more fortunate future astronomers the task of interpretation.

Eudoxus, more than two centuries earlier, had developed a logical mathematical theory of the planetary motions. The more exact methods and data of Hipparchus brought out the entire inadequacy of existing theory to furnish anything better than a crude approximation to the motions of the planets, and showed the necessity both of a better theory and of more complete observational data. It is interesting to speculate on the consequences which might have resulted for astronomical science had the genius of Hipparchus adopted the daring heliocentric theories of Aristarchus instead of adhering to the traditional geocentric ideas.

INVENTION OF TRIGONOMETRY

Not least important among the services of Hipparchus to science was his laying the foundations of trigonometry, by constructing for astronomical use a table of chords, equivalent to our tables of natural sines. He gave also a method for solving spherical triangles. It is said that he first indicated position on the earth by latitude and longitude — the germ of coordinate geometry — Eratosthenes having merely given the latitude by means of the height of the pole-star. For mapping the sky he used stereographic projection, for mapping the earth orthographic.[1]

To sum up the chief work of Hipparchus: — he made very effective use of extant records of earlier astronomers with critical consideration of their value; he made a prolonged and systematic series of observations with the best available instruments; he worked out a consistent mathematical theory of the motions of the heavenly bodies so far as his data warranted; he made a new catalogue of 1,080 stars, with the classification by magnitude still in use; he discovered the precession of the equinoxes; he laid the foundations of trigonometry.

Delambre, the great French historian of astronomy, says:

When we consider all that Hipparchus invented or perfected and reflect upon the number of his works and the mass of calcula-

[1] In stereographic projection every point in the object or area projected is supposed joined by a straight line to a certain central point; the point where the joining line meets a certain interposed plane is the projection of the original point. The projection of an object is the aggregate of the projection of the points. If one holds a sheet of glass in front of the eye, rays of light from any object to the eye form a stereographic projection on the glass. In orthographic projection the rays are parallel and perpendicular to the plane on which the projection is made.

tions which they imply, we must regard him as one of the most astonishing men of antiquity, and as the greatest of all in the sciences which are not purely speculative, and which require a combination of geometrical knowledge with a knowledge of phenomena, to be observed only by diligent attention and refined instruments. — *Hist. de l'Astron. Ancienne*, I, 185 (tr. by Berry, p. 61).

In spite of these brilliant achievements, the position of Apollonius and Hipparchus had become relatively isolated under the prevalent Stoic philosophy, which was attended with a reversion to primitive cosmical notions. Even in Hipparchus a somewhat critical attitude, excellent in its immediate results, has been regarded by some as foreshadowing the period of decadence which actually followed. Astronomy is to remain nearly stationary for sixteen centuries.

INVENTIONS. CTESIBIUS AND HERO

In the period of civil war following the death of Alexander and followed in turn by Roman conquest, much attention naturally was devoted to the invention and improvements of military engines. Compressed air came into use as a motive power and the foundations of pneumatics were laid.

Ctesibius, a mechanic of Alexandria, who probably lived about the beginning of the second century B.C., was distinguished by his mechanical inventions (hydraulic clock, hydraulic organ). We will also discuss here Hero (or Heron), though there is no agreement as to the period in which he lived (2nd century B.C.–2nd century A.D.). Hero made notable inventions and some real contributions to mathematical science. The works attributed to Hero, on the basis of a great quantity of confused and doubtful material, include: — a *Mechanics*, treating of centers of gravity and of the lever, wedge, screw, pulley, and wheel and axle; various works on military engines and mechanical toys, a *Pneumatics* — the oldest work extant on the properties of air and vapor — describing many machines, among others a fire-engine, a water-clock, organs, and in particular a steam-engine which we may regard as a remote precursor of our modern steam turbine.

Many of the machines depend for their action on the flow of water into a vacuum, which Hero, having no conception of atmospheric pressure, attributed to nature's "abhorrence" of a vacuum. He arrived at the important law for the lever and

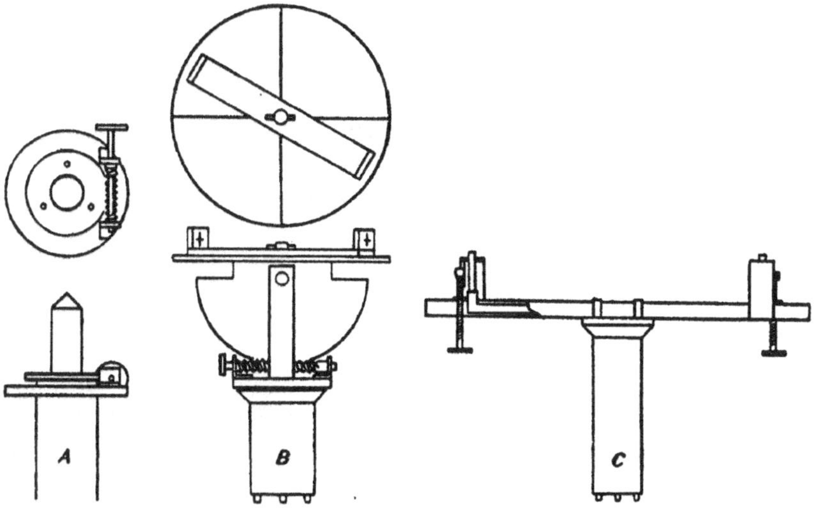

FIG. 21. — HERO'S DIOPTRA AND ATTACHMENTS. *A*, the dioptra, a stand with revolving disk perforated to receive *B* or *C; B*, the theolodite, its lineal swings above a bronze disk marked by right angles or by 360 degrees; *C*, the level, its lineal, at least 1.85 meters long, was grooved for a bronze tube with glass tubes cemented into upturned ends, and had sights adjustable to the water level. After Drachmann, *Pauly-Wissowa-Kroll*, Suppl. VI.

the pulley: "The ratio of the times is equal to the inverse ratio of the forces applied." The *Dioptra*, a treatise on a kind of rudimentary theodolite, discusses such engineering problems as finding differences of level, cutting a tunnel through a hill, sinking a vertical shaft to meet a horizontal tunnel, measuring a field without entering it, etc. The instrument employed is described as a straight plank, 8 or 9 feet long, mounted on a stand but capable of turning through a semicircle. It was adjusted by screws, turning cog-wheels. There was an eye-piece at each end and a water level at the side. With it two poles, bearing disks, were used, exactly as by modern surveyors. A cyclometer for a carriage is also described, with a series of cog-wheels and an index.

In optics he shows that under the law of equal angles of

incidence and reflection, the path described by the ray is a minimum.

His *Geodesy*, — also the *Dioptra* — contains the well-known formula for the area of a triangle

$$K = \sqrt{s(s - a)(s - b)(s - c)}, \quad 2s = a + b + c,$$

which, since it involves the multiplication of *four* lengths together, is heterodox from the Euclidean standpoint. A triangle with sides 13, 14, 15 is selected as an illustration. Its area is

$$\sqrt{21 \times 6 \times 7 \times 8} = 84.$$

This work seems to have become a standard authority for generations of surveyors, and thus in course of time to have lost much of its identity by successive changes. The whole spirit of the work is rather Egyptian than Greek, that of the practical engineer as distinguished from that of the mathematician, thus in a measure a reversion to the aims of the Ahmes manuscript. "Let there be a circle with circumference 22, diameter 7. To find its area. Do as follows. $7 \times 22 = 154$ and $\frac{154}{4} = 38\frac{1}{2}$. That is the area."

Some of Hero's methods indicate knowledge of the new trigonometry of Hipparchus and of the principle of coordinates. Thus he finds areas of irregular boundary by counting inscribed rectangles, a process corresponding to the modern use of coordinate paper for such a purpose.

From Hero date such time-honored problems as that of the pipes. A vessel is filled by one pipe in time t_1, by another in time t_2. How long will it take to fill it when both pipes are used?

He defines spherical triangles and proves simple theorems about them: — for example, that the angle-sum lies between 180° and 540°. He determines the volume of irregular solids by measuring the water they displace. Having by a blunder introduced $\sqrt{-63}$ he confuses it with $\sqrt{63}$.

INDUCTIVE ARITHMETIC. NICOMACHUS

As in the case of astronomy, progress in geometry now lags and finally ceases altogether. About A.D. 100 a final era of Greek mathematical science, predominantly arithmetical in character, begins with the Neo-Pythagorean mathematician Nicomachus of Gerasa, whose work remained the basis of European arithmetic until the introduction of the Arabic arithmetic a thousand years later. He enunciates curious theorems about squares and cubes, for example: — In the series of odd numbers from 1, the first term is the first cube, the sum of the next two is the second, of the next three the third, etc. — doubtless simple observation and induction. He refers to proportion as very necessary to "natural science, music, spherical trigonometry, and planimetry," and discusses various cases in great detail.

Mathematics had passed from the study of the philosopher to the lecture-room of the undergraduate. We have no more the grave and orderly proposition, with its deductive proof. Nicomachus writes a continuous narrative, with some attempt at rhetoric, with many interspersed allusions to philosophy and history. But more important than any other change is this, that the arithmetic of Nicomachus is *inductive*, not deductive. It retains from the old geometrical style only its nomenclature. Its sole business is classification, and all its classes are derived from, and are exhibited in, actual numbers. But since arithmetical inductions are necessarily incomplete, a general proposition, though *prima facie* true, cannot be strictly proved save by means of an universal symbolism. Now though geometry was competent to provide this to a certain extent, yet it was useless for precisely those propositions in which Nicomachus takes most interest. The Euclidean symbolism would not show, for instance, that all the powers of 5 end in 5 or that the square numbers are the sums of the series of odd numbers. What was wanted, was a symbolism similar to the ordinary numerical kind, and thus inductive arithmetic led the way to algebra. — Gow, pp. 94, 95.

PTOLEMY AND THE PTOLEMAIC SYSTEM

With Claudius Ptolemy, in the second century of our era, Greek astronomy reaches its definitive formulation. In the

260 years which had elapsed since Hipparchus no progress of consequence had been made.

Of Hipparchus, from whom he inherited so much, Ptolemy writes:

> It was, I believe, for these reasons and especially because he had not received from his predecessors as many accurate observations as he has left to us, that Hipparchus, who loved truth above everything, only investigated the hypotheses of the sun and moon, proving that it was possible to account perfectly for their revolutions by combinations of circular and uniform motions, while for the five planets, at least in the writings which he has left, he has not even commenced the theory, and has contented himself with collecting systematically the observations, and showing that they did not agree with the hypotheses of the mathematicians of his time. He explained in fact not only that each planet has two kinds of inequalities but also that the retrogradations of each are variable in extent, while the other mathematicians had only demonstrated geometrically a single inequality and a single arc of retrograde motion; and he believed that these phenomena could not be represented by excentric circles nor by epicycles carried on concentric circles, but that, by Jove, it would be necessary to combine the two hypotheses. — DREYER, pp. 165, 166.

The instruments used by Ptolemy for his astronomical observations included: — the "Ptolemaic rule," consisting of a rod with sights pivoted to a vertical rod, the angle at the junction being measured by the subtended chord; the armillary circle, a copper or bronze ring marked in degrees and mounted in the meridian plane on a post. A second movable ring is fitted into this with pegs diametrically opposite each other, by means of which the sun's midday heights could be measured; the armillary sphere, similar in principle but somewhat more complicated; the astrolabe or astronomical ring for measuring either horizontal or vertical angles. Like the Chaldeans, Ptolemy also used meridian quadrants of masonry. Time was still measured by the flow of water, with apparatus considerably improved by Ctesibius and Hero. The numerous observations of Ptolemy were made during the period A.D. 127–151 and he was in Alexandria in 139.

THE ALMAGEST

In his celebrated *Syntaxis*, better known from Arabic translations as the *Almagest*, Ptolemy undertakes to present for the first time the whole astronomical science of his age.

In Book I he reviews the fundamental astronomical data thus:

> The earth is a sphere, situated in the centre of the heavens; if it were not, one side of the heavens would appear nearer to us than the other and the stars would be larger there; if it were on the celestial axis but nearer to one pole, the horizon would not bisect the equator but one of its parallel circles; if the earth were outside the axis, the ecliptic would be divided unequally by the horizon. The earth is but as a point in comparison to the heavens, because the stars appear of the same magnitude and at the same distances *inter se*, no matter where the observer goes on the earth. It has no motion of translation, first, because there must be some fixed point to which the motions of the others may be referred, secondly, because heavy bodies descend to the centre of the heavens which is the centre of the earth. And if there was a motion, it would be proportionate to the great mass of the earth and would leave behind animals and objects thrown into the air. This also disproves the suggestion made by some, that the earth, while immovable in space, turns round its own axis, which Ptolemy acknowledges would simplify matters very much.[1] — DREYER, p. 192.

Chapter IX explains the calculation of a table of chords. Starting with the chords of 60° and 72°, already known as sides of regular polygons, he devises ingenious geometrical methods for finding chords of differences and of half-angles. Thus he computes the chords for 12°, 6°, 3°, $1\frac{1}{2}$°, and $\frac{3}{4}$°. Hipparchus had already computed such a table, but Ptolemy completes it by showing that

$$\tfrac{2}{3}\text{ chord }1\tfrac{1}{2}^\circ < \text{chord }1^\circ < \tfrac{4}{3}\text{ chord }\tfrac{3}{4}^\circ$$

[1] "For Ptolemy more geometer and astronomer than philosopher, the astronomer who seeks hypotheses adapted to save the apparent movements of the stars knows no other guide than the rule of greatest simplicity: It is necessary as far as possible to apply the simplest hypotheses to the celestial movements, but if they do not suffice, it is necessary to take others which fit better." — PIERRE DUHEM.

and thence deriving close approximations for the chords of 1° and $\frac{1}{2}$° and constructing a table for each half-degree up to 180°. His results are expressed in sexagesimal fractions of the radius (of which they are thus numerically independent) and are equivalent in accuracy to five decimals in our notation. He also employs our present method of interpolation skilfully, and in his table he uses a symbol similar to our zero to indicate a vacant place in the sexagesimal notation — thus, $o\alpha'\beta''o''' = 0° 1' 2'' 0'''$. This chapter is the culmination of Greek trigonometry, which owed its further development to Indian and Arabic mathematicians.

In Books III, IV, and V, Ptolemy discusses the apparent motions and distances of the sun and moon by means of excentrics and epicycles, his method for determining the moon's distance being substantially the same as the modern. Book V describes the construction and use of his chief instrument, the astrolabe. Book VI deals with eclipses, using a value of π equivalent to our 3.1416. He determines the distance of the sun, following Hipparchus, by observing the breadth of the earth's shadow when the moon crosses it during an eclipse. Books VII and VIII contain a catalogue of 1,028 stars based on that of Hipparchus, and a discussion of precession of the equinoxes, with a close determination of the unequal intervals between successive vernal and autumnal equinoxes. The remainder of the treatise is devoted to the planets, containing Ptolemy's chief original contributions.

While Ptolemy did not take advantage of the better data at his command to improve the theory of the sun's motion, he did make substantial progress with that of the moon, the discrepancies for which rarely exceed 10′, which represented about the maximum precision of his instruments. Hipparchus had assumed the moon to have a motion representable by one circle with the earth as a center and by an epicycle with its center upon this. Discrepancies between observed and computed positions led Ptolemy, bound as he was by the Aristotelian dictum that celestial bodies can move only in circular paths, to modify this by making the first circle excentric to the earth, the line joining the centers of the circle and the earth being itself assumed to revolve. This theory, while giving results of sufficient accuracy for the observations at certain positions of the moon, exaggerated considerably the variation of its distance from the earth, making this at times almost twice as great as at others.

For the five planets, or "wandering stars," he also assumed excentric deferents, and as a further means of accounting for discrepancies, an additional point, in line with the centers of earth and deferent, called the "equant," with respect to which the center of the epicycle would have uniform angular velocity. The planes of the epicycles were slightly inclined to that of the ecliptic.

Thus in the figure, C is the center of the circular deferent, E the earth, and E' the equant. The center A of the epicycle

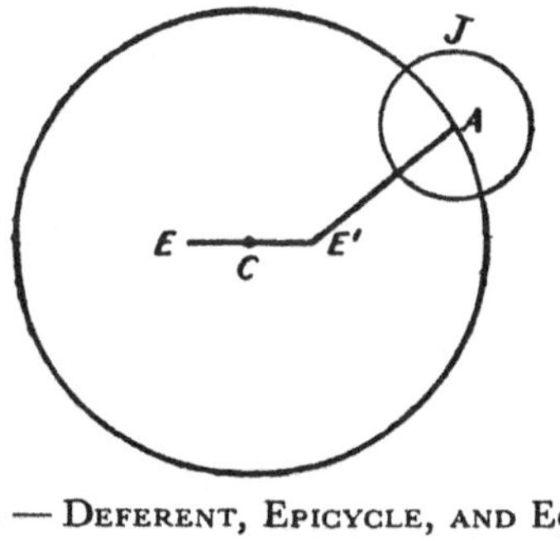

FIG. 22. — DEFERENT, EPICYCLE, AND EQUANT.

travels at such a rate that the line $E'A$ has uniform angular velocity. The planet J travels in an epicycle about A. These assumptions afforded the needful freedom for a fairly close approximation to observed planetary motions, the mathematical computations involved becoming naturally quite elaborate. Ptolemy disclaimed the power of determining the distances or even the order of the planets.

To the modern mind, accustomed to the heliocentric idea, it is difficult to understand why it did not occur to a mathematician like Ptolemy to deprive all the outer planets of their epicycles, which were nothing but reproductions of the earth's annual orbit transferred to each of these planets, and also to deprive Mercury and Venus of their deferents and place the centres of their epicycles in the sun, as Heraclides had done. . . . The system of Ptolemy was a mere geometrical representation of celestial motions, and did not profess to give a correct picture of the actual system of the world. . . . For more than 1400 years it remained the Alpha and Omega of theoretical astronomy, and whatever views were held as to the constitution of the world, Ptolemy's system was almost universally accepted as the foundation of astronomical science. — Dreyer, *op. cit.*, pp. 201, 202.

After Ptolemy we have no record of any important advance in astronomy for nearly 1,000 years.

OTHER WORKS OF PTOLEMY

In spite of his scientific attainments Ptolemy did not disdain to write an elaborate treatise on astrology. In a lost work on geometry, Ptolemy made the first known of the interminable series of attempts to give a formal proof of Euclid's parallel postulate, an attempt naturally foredoomed to failure.

In a great treatise on geography, hardly less important than the *Almagest*, Ptolemy gave a description of the known earth, locating not less than 5,000 places by latitude and longitude. He even gave in addition to position the maximum length of day for 39 points in India, a land probably better known at this period than in the time of Mercator, near the end of the sixteenth century. Ptolemy reckoned longitude from the "Fortunate Isles," — the western boundary of the known world. Various methods of projection were discussed in connection with directions for map drawing.

Ptolemy also wrote on sound and on optics, dealing particularly in the latter with refraction, with what has been called "the oldest extant example of a collection of experimental measures in any other subject than astronomy." He discovered by careful experiment and induction the law that light-rays passing from a rarer to a denser medium are bent towards the perpendicular, and invented a simple apparatus for measuring angles of incidence and reflection. He accepted Plato's idea that vision is due to the meeting of rays from the eye with those from the object.

PAPPUS

The last two of the great Greek mathematicians were Pappus and Diophantus, who lived in Alexandria about A.D. 300.

The most important work of Pappus is his *Collection*, in eight books, of which all but the first and a part of the second are preserved. In this he comments fully on the most important Greek mathematical works known to him, making his treatise of the highest historical value, particularly in its careful summaries of books which have been lost. He reviews the various solutions of the duplication of the cube, and discusses the regular inscribed polyhedrons; the higher curves, spirals, conchoid, quadratrix, etc., the problem of describing a circle tangent to three given circles which touch each other. Book VII contains his well-known theorems (sometimes attributed to Guldin), that the volume of a solid generated by the revolution of a plane area about an exterior axis is equal to the product of the revolving area and the length of the path of its center of gravity, and that the area generated by the bounding curve is equal to the product of its length times that of the circular path described by its center of gravity. The whole is somewhat weak on the arithmetical side.

With the political decline of Greece and the awakening to intellectual activity of great Semitic and Egyptian populations, mathematical science changed radically from the traditional deductive geometry, to an arithmetical and algebraic science in harmony with the aptitudes which have characterized these races.

BEGINNINGS OF ALGEBRA. DIOPHANTUS

Diophantus was active in Alexandria in the second half of the third century A.D., though we know so little about him that even his precise name is doubtful. Aside from fragments of his work, all that is known of him is found in an epigram, probably of the sixth century, to the effect that his boyhood lasted $\frac{1}{6}$ of his life, his beard grew after $\frac{1}{12}$ more, after $\frac{1}{7}$ more he married, and his son was born 5 years later; the son lived to half his father's age, and the father died 4 years after the son. This gives 84 years as his span of life. His chief work is his *Arithmetic*, which is extant, however, only in somewhat mutilated form. It is the first known treatise on algebra, and is devoted to the solution of equations, employing algebraic symbols and analytical methods. Euclid had given the geometrical equivalent of the solution of a quadratic equation, and Hero could solve the same problem algebraically but lacked a satisfactory symbolism. The algebra of Diophantus was therefore not a sudden invention, but the result of gradual evolution during several centuries of increasing interest in arithmetical problems, and declining vogue of the abstract Euclidean geometry.

Writers on the history of algebra distinguish three classes or methods of algebraic expression:

(*a*) the *rhetorical*, where no symbols are used, but every term and operation is described in full. This was the only method known before Diophantus, and was later in vogue in western Europe until the fifteenth century;

(*b*) the *syncopated*, which replaces common words and operations by abbreviations, but conforms to the ordinary rules of syntax. This was the style of Diophantus;

(*c*) the *symbolical* or modern, using symbols only, without words.

The syncopated method may be illustrated by the following passage from Heath's *Diophantus:*

> Let it be proposed then to divide 16 into two squares. And let the first be supposed to be $1S$; therefore the second will be $16U - 1S$. Thus $16U - 1S$ must be equal to a square. I form the square from any number of N's minus as many U's as there are in the side of 16 U's. Suppose this to be $2N - 4U$. Thus the square itself will be $4S\ 16U - 16N$ etc.

In his *Arithmetic*, which is really a treatise on algebra, Diophantus represents the (single) unknown by the Greek *sigma* — all the other letters of the Greek alphabet standing for definite numbers — with successive powers to the sixth inclusive. If he requires two unknowns he admits only one at a time. His originality and power in the solution of problems are amply shown, though the solutions are rarely complete. For quadratic equations, for example, he gives but one of the two roots, even when both are positive. Negative numbers are for him unreal, and he also avoids the irrational. He admits fractional results however, and is indeed the first Greek for whom a fraction is a number rather than a mere ratio of two numbers. For the solution of pure equations his rule is: "If a problem leads to an equation containing the same powers of the unknown on both sides but not with the same coefficients, you must deduct like from like till only two equal terms remain. But when on one side or both some terms are negative, you must add the negative terms to both sides till all the terms are positive and then deduct as before stated."

The modern so-called Diophantine equations involving the solution in integers of one or more indeterminate equations, do not occur in his own extant work.

With the very important process of reducing problems to equations he is relatively successful and often highly ingenious. For example, "to find three numbers, so that the product of any two plus the sum of the same two shall be given numbers, for example, 8, 15, and 24." We should write: $xy + x + y = 8$; $yz + y + z = 15$; $zx + z + x = 24$.

Hence, by subtraction, $x(z - y) + z - y = 16$, $x + 1 = \frac{16}{z - y}$, $z(x - y) + x - y = 9$, $z + 1 = \frac{9}{x - y}$, etc. He, on the other hand, takes $a - 1$ for one of the numbers and readily obtains $\frac{9}{a} - 1$ and $\frac{16}{a} - 1$ for the others, and $a = \frac{12}{5}$. The three numbers are therefore, $\frac{7}{5}, \frac{17}{3}, \frac{11}{4}$.

He employs tentative assumptions with great effect. For example, "To find a cube and its root such that if the same number be added to each, the sums shall also be a cube and its root." If $2x$ is the original number and x the number added (an arbitrary and presumably erroneous assumption), $8x^3 + x = 27x^3$, giving $19x^2 = 1$. The coefficient 19 not being a square, he now seeks to find two cubes whose difference is a square. If $(x + 1)^3 - x^3$ is equated to $(2x - 1)^2$ the special solution $x = 7$ is easily obtained. Returning to the original problem, the new assumption is made: — let x = number to be added, $7x$ = original number.

$$(7x)^3 + x = (8x)^3 \text{ whence } x = \frac{1}{13}.$$

Summarizing his methods of dealing with equations we may say that:

(1) he solves completely equations of the first degree having positive roots, showing remarkable skill in reducing simultaneous equations to a single equation in one unknown;

(2) he has a general method for equations of the second degree but employs it only to find one positive root;

(3) more remarkable than his actual solutions of equations are his ingenious methods of avoiding equations which he cannot solve.

How far his work was original, how far like Euclid in his *Elements* it was the result of compilation, cannot be definitely ascertained. For him the earlier Greek distinction between computation and arithmetic has lost its force.

In reviewing the work of Pappus and Diophantus Gow says:

the *Collection* of Pappus can hardly be deemed really important. . . . But among his contemporaries Pappus is like the peak of Teneriffe in the Atlantic. He looks back, from a distance of 500 years, to find his peer in Apollonius. . . . His work is only the last convulsive effort of Greek geometry which was now nearly dead and was never effectually revived. It is not so with Ptolemy or Diophantus. The trigonometry of the former is the foundation of a new study which was handed on to other nations indeed but which has thenceforth a continuous history of progress. Diophantus also represents the outbreak of a movement which probably was not Greek in its origin, and which the Greek genius long resisted, but which was especially adapted to the tastes of the people who, after the extinction of Greek schools, received their heritage and kept their memory green. But no Indian or Arab ever studied Pappus or cared in the least for his style or his matter. When geometry came once more up to his level, the invention of analytical methods gave it a sudden push which sent it far beyond him and he was out of date at the very moment when he seemed to be taking a new lease of life. — Gow, p. 308.

A melancholy interest attaches to the fate of Charles Kingsley's well-known heroine, Hypatia, daughter of Theon, an Alexandrian mathematician, herself a teacher of Greek philosophy and mathematics, who was torn to pieces by a Christian mob, doubtless as a representative of pagan (Greek) learning, at Alexandria in A.D. 415.

ALEXANDRIAN ALCHEMY

Probably the earliest of the Alexandrian authors on alchemy was Comarius, whose fragmentary treatise of the first

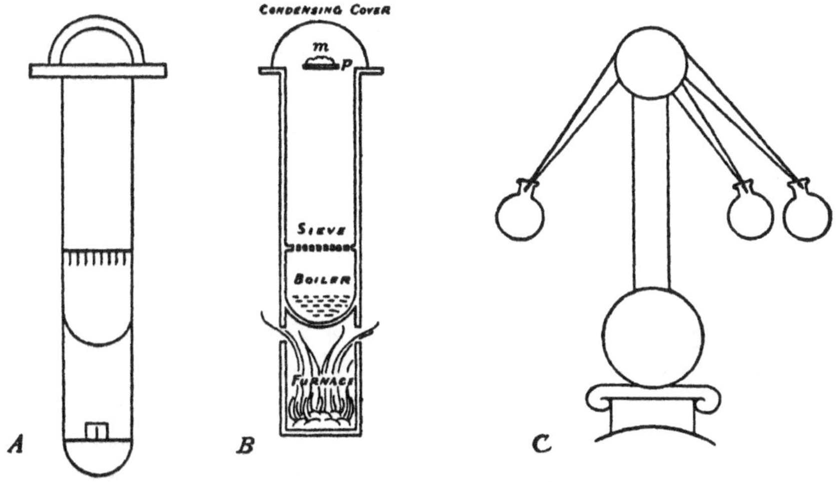

FIG. 23. — EARLY ALEXANDRIAN ALCHEMICAL APPARATUS. *A*, percolator, long form; *B*, conjectural restoration of *A; m*, material to be treated resting on *p*, the *kerotakis*, or palette; *C*, distilling apparatus. *A* and *C* from a Greek Manuscript of the 10th or 11th century at Venice. After F. S. Taylor, *Jour. Hellenic Studies*, **50**, 132 and 136. Permission of the Hellenic Society.

century A.D. is chiefly mythical and symbolic, suggesting Egyptian influence. Very different were two outstanding practical alchemists — Mary the Jewess, of about the same time, and Democritus (Pseudo-Democritus), not later than A.D. 250. Mary, whose works are known from numerous quotations, was quite the most remarkable of early alchemists for her clear descriptions of apparatus, with details such as the making of copper tube from sheet metal, as well as for her inventions, some of them now common laboratory and household utensils. They include a percolator, the hot-ash bath, the dung-bed ("hot-bed"), and the water-bath (double boiler, "bain marie"). Her improvements of the distilling apparatus remained unsurpassed for two thousand years.

Mary was especially interested in alloys of copper and lead. Democritus dealt with the superficial coloring of metals and the preparation of alloys by fusion.

In the most important Greek texts the word *alchemia* is not used. The subject was referred to as "The Work." "The divine and Sacred Art," "The Making of Gold," rarely as *chemia* or *chymia*. In all the alchemical texts there is a religious atmosphere and an entire absence of scientific spirit — no interest in phenomena not directly helpful toward the object in view. Much space is devoted to the preparation of "divine water." All the definite recipes deal with the production of gold, silver, and purple, or in a few instances, with precious stones. Success was not theoretically improbable because of the old theory that all matter is composed of the four elements — air, fire, earth, and water. If the proportions of the elements could be altered, it would be possible to change the qualities of the material. The Alexandrian alchemists [1] adopted an ancient doctrine of two contrary principles — a positive and active principle identified with the sun, male, dry, light, and fiery; and a negative and passive principle identified with the moon, female, moist, heavy, and cold. Zosimos and the Alexandrians of about the third century A.D. called these two principles Sulphur and Mercury, respectively. The chief contribution of Zosimos is an elaborate theory and technique for the preparation of silver and gold from the base metals. The metals are defined by their colors, and could be produced by combinations of the *spirits* of Sulphur and Mercury on an inert sub-stance (note the etymology), the *body* common to all metals — part of Plato's universal matter. With this idea was coupled the Neo-Platonic belief in a tend-

[1] See *Isis* **28**, 73–86 and 424–431, Feb. and May, 1938.

ency toward perfection, and gold was the "perfect metal." At an early period each of the seven metals was associated with one of the seven planets. The alchemist, with a reverence for ancient science that paralyzed research, tried to rediscover lost or hidden knowledge supposed to have been possessed by Greek philosophers or Egyptian priests.

In spite of the order of the Emperor Diocletian suppressing alchemy in Alexandria in A.D. 292, all of the books on the subject were not destroyed, and its influence upon the progress of science in later periods was profound.

CONCLUSION AND RETROSPECT

Intellectual interests in the Greek world (now really Roman) were by this time so completely alienated from mathematics, and indeed from science in general, that the brilliant work of Pappus and Diophantus aroused but slight and temporary interest. Geometry had reached within the possible range of the Euclidean method a relatively complete development. Algebra under Diophantus attained in spite of hampering notation a level not again approached for many centuries.

Little need be said of sciences other than those already dealt with. These, even more than mathematics and astronomy, shrank under Roman autocracy and Christian hostility. With the works of Galen, Strabo, and Pliny we deal in the next chapter.

The torch of science now passes from the Greeks to the Indians of the far East, to be in turn surrendered to the Mohammedan conquerors of Alexandria, A.D. 641. By them it is kept from extinction until in later ages it is once more fanned to ever increasing radiance in western Europe.

In attempting a retrospective estimate of Greek science it is fundamentally important to judge the whole background fairly. In science the Greeks had to build from the foundations. Other peoples had extensive knowledge and highly developed arts. Only among the Greeks existed the true scientific method with its characteristics of free inquiry, rational interpretation, verification or rectification by systematic and repeated observation, and controlled deduction from accepted principles.

In asking ourselves why these extraordinary beginnings seemed after a time to lose their power of continued development, we must not forget the effect of external conditions. It is conceivable indeed that scientific progress should continue from age to age, through the genius of individual teachers and students, regardless of political and social conditions. Such, however, is not the historic fact. For progress in science men of genius are indispensable, but in no country or age have they alone been able to make science flourish under conditions so unfavorable as were those of the early centuries of the Christian era.

Greek science, however, did not "fail," learned and elaborate as are the explanations that have been given of its alleged failure. Under "the chill breath of Roman autocracy" its growth was indeed checked, its animation suspended, for a full thousand years. Then in the Renaissance it renewed its vitality and has ever since been advancing more and more magnificently. This is not to say that criticisms as to the imperfections of the Greek scientific method are invalid, but rather to assert, as most critics must agree, that its merits outweighed its defects, and that the latter would not have proved disastrous but for the development of political, economic, and military conditions under which the free Greek spirit could not continue its wonderful achievements.

We have to remember that the Greeks, preeminently among all the nations of the world, possessed three gifts which are essential to the initiation and development of philosophy and science. They had in the first place a remarkable power of accurate observation; and to this were added clearness of intellect to see things as they are, a passionate love of knowledge for its own sake, and a genius for speculation which stands unrivalled to this day. Nothing that is perceptible to the senses seems to have escaped them; and when the apparent facts had been accurately ascertained, they wanted to know the *why* and the *wherefore*, never resting satisfied until they had given a rational explanation, or what seemed to them to be

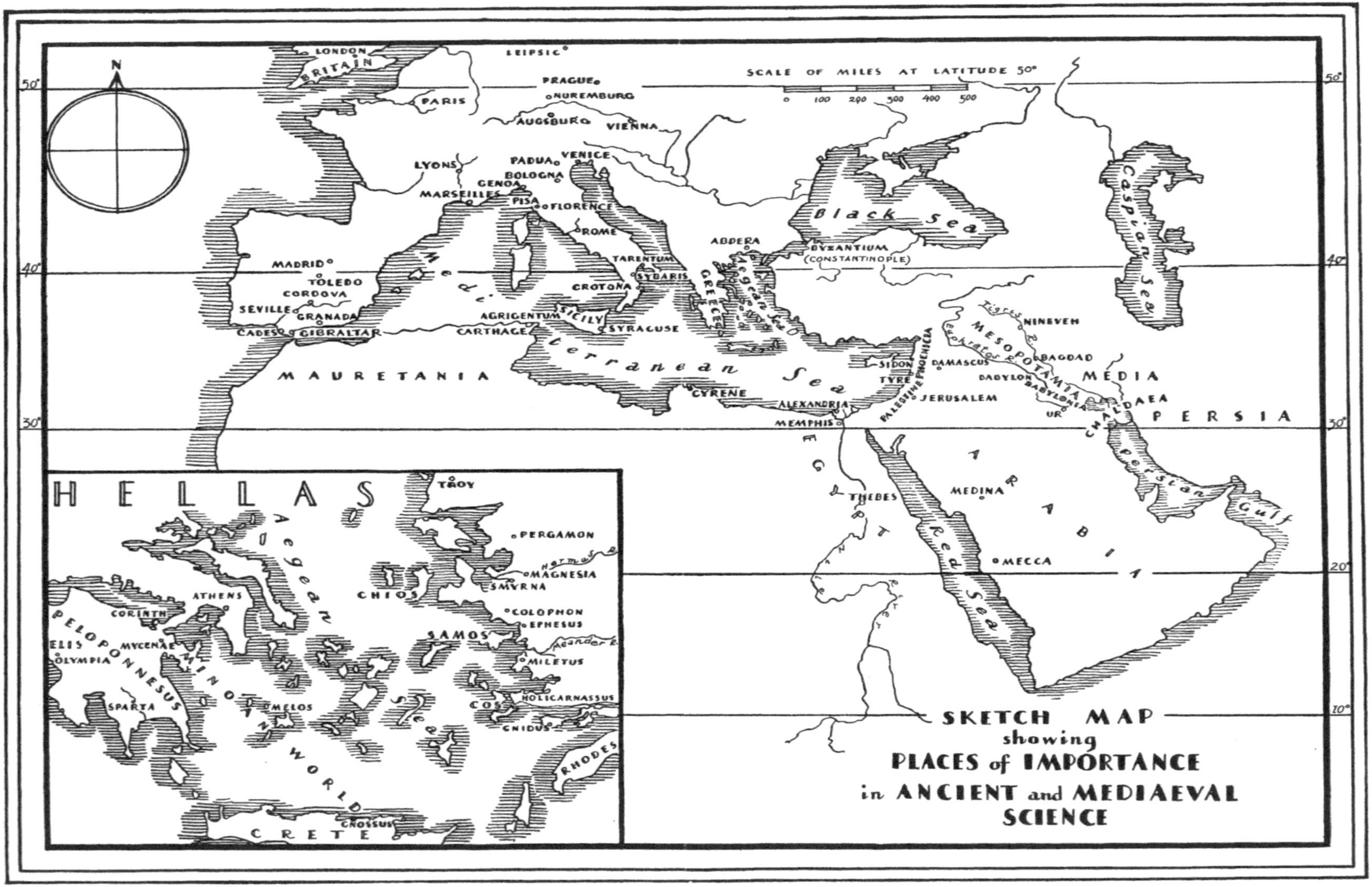

SKETCH MAP
showing
PLACES of IMPORTANCE
in ANCIENT and MEDIAEVAL
SCIENCE

such, of the phenomena observed. Observation or experiment and theory went hand in hand. So it was that they developed such subjects as medicine and astronomy. In astronomy their guiding principle was, in their own expressive words, to "save the phenomena." This means that, as more and more facts became known, their theories were continually revised to fit them. — HEATH, *The Copernicus of Antiquity*, p. 2.

The Greek spirit, that disinterested love of truth which is the very spring of knowledge, was finally smothered by the combination of Roman utilitarianism and Christian sentimentality. — G. SARTON, *History of Science and the New Humanism*, p. 93.

REFERENCES FOR READING

FARRINGTON, B., *Science in Antiquity*, 1936.
HEATH, T. L., *Aristarchus of Samos*, 1913.
——, *Diophantus of Alexandria*, 1910.
HOPKINS, A. J., *Alchemy, Child of Greek Philosophy*, 1934.

Books from previous chapters: Ball, Ch. IV, V; Berry, Art. 37–54; Dreyer, Ch. VI–IX; Gow, Ch. IV, VIII, IX, X.

Chapter VII

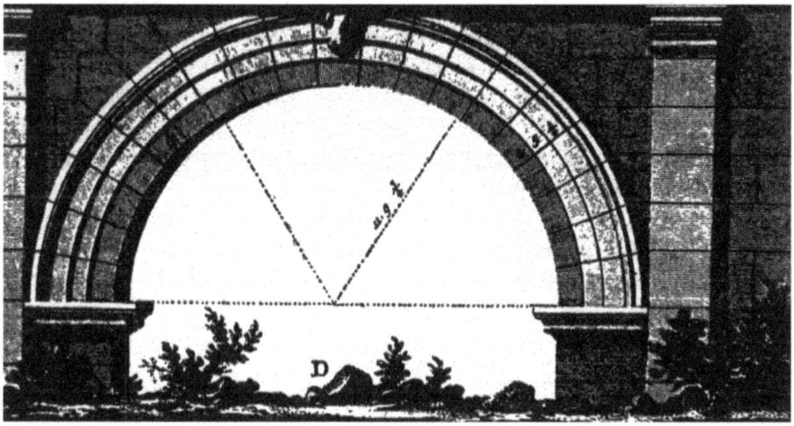

The Roman World. The Dark Ages

The Romans, with their narrow, rustic horizon, their short-sighted, practical sobriety, had always in their heart of hearts that mixture of suspicion and contempt for pure science which is still the mark of the half-educated. — J. L. HEIBERG, *Math. and Phys. Sci.*, p. 80.

Despite the stimulus that followed on the contact with Alexandrian thought, Rome produced no great creative scientist. . . . The general scientific principles of the Greeks were seldom understood even by educated Romans.

Ancient science can be clearly traced as an active process up to the end of the second century of the Christian era. . . . We may fix the end of Antiquity and the beginning of the Middle Ages for science at the end of the fourth or the beginning of the fifth century. — SINGER. (*Science and Civilization*, p. 114.)

THE ROMAN WORLD-EMPIRE

For several centuries, during the decline of Greek learning both in Greece itself and in Alexandria, two new and powerful States were developing; one having its center at Carthage on the northern shore of Africa, almost opposite Sicily, the other — the Roman Empire — on the western shore of Italy in the valley of the Tiber. The latter, at first comparatively insignificant, rapidly rose to a position of world-wide power, conquering in turn Carthage, Greece, and the East, and eventually extending over the greater part of the then known world, from Britain on the north to the Cataracts of the Nile on the south, from the river Tigris in the east to the Pillars of Hercules in the west.

THE ROMAN ATTITUDE TOWARDS SCIENCE

One of the most striking facts in the history of science is the total lack of any evidence of real interest in science or in scientific research among the Roman people or any people under Roman sway. Alexandrian science, even, though previously flourishing, languished and went steadily to its fall after the submission of that city to the Romans in the first century B.C. The truth seems to be that the Roman people, while highly gifted in oratory, literature, and history, were not interested and therefore not successful in scientific work. This is the more impressive when we reflect upon their military genius, and their preeminence in world-wide dominion and influence. In vain do we look for any Roman scientist or philosopher of such originality or range as Aristotle or Plato; for any Roman astronomer, like Aristarchus or Hipparchus or Ptolemy; for any Roman natural philosopher, like Democritus; for any Roman pioneer in medicine, like Hippocrates. The Romans used the knowledge accumulated by the Greeks without replenishing it; they made little or no distinction between astrology and astronomy, between the real and the imaginary.

ROMAN ENGINEERING AND ARCHITECTURE

There is, however, one marked feature of Roman civilization in which extraordinary ability was displayed and peculiar excellence achieved and in which the Romans were unquestionably far superior to all their predecessors and, until very

NOTE: The figure of a Roman arch facing this page is a detail from Piranesi, *Lapides Capitolini* (*Works*, Vol. IX). Courtesy of the Metropolitan Museum of Art.

recent times, to all their successors. This feature is the characteristic Roman genius for both military and civil engineering. It is only necessary to mention the extant remains of Roman walls, fortresses, roads, aqueducts, theaters, baths, and bridges. Never before and never since has any empire built so many, so splendid, and so enduring monuments for the service of its peoples in peace and in war. The surface of southern Europe, western Asia, and northern Africa is still covered after the lapse of twenty centuries with Roman remains which bid fair to resist decay and destruction for another two thousand years. Roman engineering is almost as distinguished as is Roman law. The Emperor Constantine in the fourth century wrote: "We need as many engineers as possible. As there is lack of them, invite to this study persons of about 18 years, who have already studied the necessary sciences. Relieve the parents of taxes and grant the scholars sufficient means." The land surveyors formed a well-organized guild, but they were merely practitioners of a traditional art, perpetuating the errors of their ancient Egyptian predecessors, not dreaming of new discoveries, nor even of imparting such knowledge as they had — outside the ranks of their own guild.

VITRUVIUS ON ARCHITECTURE

The most famous ancient work on building and kindred topics, including building materials, is that entitled *De Architectura*, by Vitruvius, a Roman architect and engineer living (about 14 B.C.) in the age of Augustus. This celebrated work was the only one of importance on architecture known to the Middle Ages, and was the guide and text-book of the builders of that period as well as of those of the Renaissance. The book (accessible in English translation) is in part a compilation from earlier, and especially Greek, authors, and in part original. Vitruvius uses for π the value $3\frac{1}{8}$ — less accurate than that of Archimedes, but displaced later by the crude approximation 3. Of Vitruvius's life and work almost nothing is known, but no other ancient treatise of a similar technical nature has had in its own field so much influence on posterity.

A certain modernity in his points of view may be illustrated by a quotation:

> In the famous and important Greek city of Ephesus there is said to be an ancient ancestral law, the terms of which are severe, but its justice is not inequitable. When an architect accepts the charge of a public work, he has to promise what the cost of it will be. His estimate is handed to the magistrate, and his property is pledged as security until the work is done. . . .
>
> Would to God that this were also a law of the Roman people, not merely for public, but also for private buildings. For the ignorant would no longer run riot with impunity, but men who are well qualified by an exact scientific training would unquestionably adopt the profession of architecture. Gentlemen would not be misled into limitless and prodigal expenditure, even to ejectments from their estates, and the architects themselves could be forced, by fear of the penalty, to be more careful in calculating and stating the limit of expense, so that gentlemen would procure their buildings for that which they had expected, or by adding only a little more. — VITRUVIUS, *The Ten Books on Architecture*, trans. by M. H. Morgan, p. 257.

FRONTINUS ON THE WATERWORKS OF ROME (*c.* A.D. *40–103*)

At about the end of the first century of our era, Sextus Julius Frontinus, a Roman soldier and engineer, wrote a highly interesting and valuable account of the waterworks of Rome. Frontinus served as *prætor* under Vespasian; was afterwards sent to Britain as Roman governor of that island; was superseded by Agricola in A.D. 78 and was appointed in A.D. 97 *Curator Aquarum*, an office only conferred upon persons of very high standing. "Who," he asks, "will venture to compare with the mighty aqueducts of Rome the pyramids or the useless though famous works of the Greeks?"

SLAVE LABOR IN ANTIQUITY

It must never be forgotten that throughout antiquity, and to a great extent even until very recent times, the labor question was wholly different from what it is today. Instead of the labor-saving machinery which is so remarkable a feature of our time, but which was practically non-existent before the end of the eighteenth century, the slave was the machine for all heavy labor. It is not likely that he was ever a particularly cheap machine, but in the mass he was powerful, and it was largely by his labor that the fields were cultivated and irrigated, and that dams and ditches, walls and towers, roads and bridges, and pyramids and temples were built and fortified. It is notorious that the so-called "ships" of war, the galleys, were manned by slaves, even down to modern times. It is difficult to determine the efficiency of labor of this kind because we are generally ignorant as to the time factor, but whether from our modern point of view efficient or not, the results were often remarkable and sometimes, as in the case of the Pyramids, stupendous.

JULIUS CAESAR AND THE JULIAN CALENDAR

Julius Caesar himself undertook two great problems of practical mathematical science: — the rectification of the calendar and a survey of the whole Roman empire. The Romans, like the Babylonians and the Greeks, had been reckoning the year as twelve lunar months, 355 days beginning with March, with an extra month when needed for adjustment to the seasons. The Senate, for political reasons, having refused to order intercalary months, the difference between solar and official dates had reached some 85 days in 47 B.C. when Julius Caesar was at Alexandria in defense of Cleopatra. It is probable that he learned there of the Egyptian solar year of 365¼ days and secured the aid of Sosigenes to calculate a calendar that should be free from politics. The result was the decree of 46 B.C., whereby that year was lengthened by three months and a new era began on the first of January, 45 B.C.[1] The Julian Calendar, thus established, ignored the moon, and fixed a year of 365 days divided into twelve nearly equal months, the first to begin 8 days after the winter solstice. Every fourth year an additional day, called *bis-sextus*, was interpolated before the

[1] This is the year 709 of the old Roman Era of the Founding of the City. Our current distinction between A.D. and B.C. was established in the sixth century of our era.

24th of February (sixth day before the first of March). The heliacal rising of Sirius was dated the 20th of July, and from this other dates were calculated.

The survey, of which the results were to be embodied in a great fresco map showing the marching routes for the Roman armies, was not carried out until the reign of Augustus.

ROMAN NATURAL SCIENCE AND MEDICINE

Among the Roman workers and authors of importance in the history of natural science and medicine only a few require more than passing notice. This is the more remarkable when we reflect upon the vast extension of the Roman empire and the novel and hitherto unequalled opportunities afforded for observation and collection in natural history, and for the study of anthropology, geography, geology, meteorology, climatology, zoology, botany, and the like — not to mention military surgery, and the hygiene and sanitation of camp-life.

LUCRETIUS (c. 98–c. 55 B.C.)

Lucretius is today regarded not only as a great Roman poet but also as the most perfect exponent in his time of the atomistic, that is, materialistic, school of Greek philosophy. He was a contemporary (a few years junior) of Cicero and Julius Caesar. The first two books and the fifth of his *De Rerum Natura* (*On the Nature of Things*) are of interest to the modern scientific student, because of their dealing with problems of permanent importance to mankind. He was a disciple of Epicurus, and apparently also well acquainted with the works of Empedocles, Democritus, Anaxagoras, and many other of the great Greek writers. The title of his famous poem shows his interest in natural philosophy, and there is evidence that he was also a teacher and reformer. He is antagonistic to superstition and a strong advocate of rationalism, but he is neither irreverent nor revolutionary.

His point of view is illustrated by his address to Epicurus and his discussion of eclipses:

O thou who first from so great a darkness wert able to raise aloft a light so clear, illuminating the blessings of life, thee I follow, O glory of the Grecian race, and now on the marks thou has left I plant my own footsteps firm, not so much desiring to be thy rival, as for love, because I yearn to copy thee; for why should a swallow vie with swans, or what could a kid with its shaking limbs do in running to match himself with the strong horse's vigour? (III, 1–8.)

Eclipses of the sun also and hidings of the moon you must suppose to have several possible causes. For why should the moon be able to shut off the earth from the sun's light, and from the side of the world to push her head in his way on high, obstructing his burning rays with her dark orb, and yet at the same time some other body gliding along ever without light not be thought able to do the same thing? And the sun, why should not he also be able to lose his fires and faint at a fixed time and to renew his light, when he has passed through regions of air that are hurtful to his flames, making the fires to be quenched and to perish for a time? And why should earth be able in turn to rob the moon of light and herself passing above the sun to keep him in subjection, while the moon in her monthly course glides through the clear-cut conical shadow; yet at the same time some other body not be able to pass beneath the moon, or glide above the sun's orb to intercept the rays and flood of light? (V, 751–767.) — LUCRETIUS, *De Rerum Natura* (trans. W. H. D. Rause, pp. 171, 395).

Unmatched among the ancients or moderns is the vision by Lucretius of continuity in the workings of Nature. . . . The description of the wild discordant storm (Book V) which led to the birth of the world might be transferred verbatim to the accounts of Poincaré or Arrhenius of the growth of new celestial bodies in the Milky Way. . . . Book II, a manual of atomic physics with its marvelous conception of —

> . . . the flaring atom streams and
> torrents of her myriad universe

can only be read appreciatively by pupils of Roentgen or of J. J. Thomson. — OSLER, *Old Humanities*, p. 22.

ROMAN GEOGRAPHY AND GEOLOGY: STRABO

Discussions of geological phenomena are to be found in the works of Strabo and Seneca. Strabo, Greek traveller, historian, and geographer, was born in the province of Pontus about

63 B.C. and died after A.D. 20. His *Geography* is the most important work on that subject surviving from antiquity and, while apparently building on the foundation laid by Eratosthenes, is plainly an original work devoted largely to his own explorations and observations during years of travel and study in different countries, including Italy, Greece, Asia Minor, Egypt, and Ethiopia. He himself says:

Westward I have journeyed to the parts of Etruria opposite Sardinia; towards the South from the Euxine to the borders of Ethiopia, and perhaps not one of those who have written geographies has visited more places than I have between those limits.

Geography unfolds to us the celestial phenomena, acquaints us with the occupants of the land and ocean, and the vegetation, fruits, and peculiarities of the various quarters of the earth, a knowledge of which marks him who cultivates it as a man earnest in the great problem of life and happiness. On the one hand, it embraces the arts, mathematics, and natural science; on the other, history and fable. — STRABO, *Geography*, II, 5.

His work is invaluable as a picture of the limited geographical knowledge of the time, but he had no such mathematical knowledge of geography as had his great predecessors, Eratosthenes, Hipparchus, and Ptolemy.

In Strabo's *Geography* the evidence of marine fossils found far inland is adduced to prove the ancient philosophical belief in the frequent interchange of land and sea. He believed not only islands but entire continents to have sunk and risen. Islands far from continents were thrown up by subterranean fires; Strabo cited several islands in the Mediterranean where such upheavals had been observed while taking place. Because of these observations, he is often called the father of modern theories of mountain building. Noting that Sicily had less frequent earthquakes in his time than it had experienced in previous ages before volcanic eruptions were known on the island, he concluded that volcanic outbursts act as safety valves for pent-up vapors below, a view held today. He tells also of observations on erosion by water and on the formation of alluvial deposits, and gives information on the mining of salt and its extraction from salt-springs.

SENECA

Lucius Annaeus Seneca (4 B.C.–A.D. 65) was born in Spain, but lived in Rome from an early age. He was a brilliant Stoic and statesman. In his *Natural Questions* physical phenomena are explained on atomistic principles, with a moral at the end of each book. The material is mostly borrowed, but he gives the earliest detailed account of an earthquake, the one of A.D. 63 that did much damage in Campania. Earthquakes he thought due to the expansion of gases accumulating within the earth, or, at times, to the collapse of subterranean cavities. He is very modern in his views on the action of water. He wrote of it as dissolving and carrying away solid rock material and depositing the residue in the form of deltas. The earth in the beginning was a watery chaos.

PLINY THE ELDER (A.D. 23–79)

Pliny the Elder, sometimes called Pliny the Naturalist, was a Roman general, or an admiral, and an omnivorous reader. His great work entitled *Natural History*, published A.D. 77, is a monument to his industry and is useful as a collection of the current knowledge of his time. He quotes 400–500 authors; a hundred are given special attention. This treatise, like the works of Herodotus, is a landmark in the history of civilization. It consists of thirty-seven books (accessible in English translation). Pliny deals with the universe, God, nature, and natural phenomena; with earth, stars, earthquakes; with man, beasts, shells, fishes, insects, trees, fruits, gums, perfumes, timber, the diseases of plants, metals, stones, precious stones, etc. More space is given to plants than to animals. The accounts of animals are uncritical and rich in curious anecdotes. There is throughout the work much of truth and much of fable. Pliny argues for the spherical shape of the earth because masts of ships are seen before their hulls.

"Ebb and flow of the sea" he says "have with all their alternations their cause only in sun and moon. . . . The great open surface of the ocean feels the might of the far-reaching heavenly bodies more than the shut in spaces" of lakes and rivers. Yet he thinks mussels expand with the waxing moon. Of a certain Mediterranean fish, Pliny alleges, "Let the storms rage and the waves dark, this little creature scorns their wrath, tames their power and compels a ship to stand still when no rope and no anchor can do so." And indeed it checks the storm and controls the elements not through its own labor or opposition, but simply and solely by attaching itself to the ship. "Man alone knows ambition and greed, cares for his grave, yes, even for the future after his death. No creature is so robbed of his wits by fear. In none is rage more violent. All others live in peace with their kind. . . . "

The author met his death in that eruption of Vesuvius which overwhelmed Pompeii in A.D. 79 and because of his scientific curiosity which led him to approach too near to the volcano.

ROMAN CHEMISTRY

Knowledge of the Roman chemical industries may be found in the *Architecture* of Vitruvius (p. 158) as well as in the *Natural History* of Pliny. The former treats incidentally of building materials, pigments (including India ink), and even hygiene. He mentions a remarkable natural cement, which when mixed with lime and rubble will set under water. He describes the method of obtaining mercury from cinnabar and of amalgamating gold. He recommends earthern ware pipes for water, instead of lead pipe, because of the danger from the latter of lead-poisoning, the symptoms of which he describes.

Pliny likewise gives details of some of the chemical industries of his time, with a great deal of information concerning unguents, medicines, and poisons; oil, wine, and beer; pigments, dyes, and ink; metals and metallurgy, glass, etc. So far as they treat of chemical theory, both authors are dominated by the doctrine of the four elements of Empedocles (p. 62), which also furnished the philosophical basis for Alexandrian alchemy (p. 151).

HELLENISTIC BOTANY IN ROME

The earliest known botanical illustrations are the work of Cratevas, a Greek, physician to Mithridates, king of Pontus 120–63 B.C. "His drawings are particularly interesting as introducing a new art that was combined with the real study of nature." Copies of them have come down to us; of his text we have only quotations by Dioscorides.

Dioscorides, a military surgeon under Nero (A.D. 54–68), is the author, in Greek, of the *Materia Medica*, concerned chiefly with the medical properties of about 600 plants. These are arranged first by their uses and then by their similarities of structure. Some descriptions are very good; including root, stem, leaf, flower, fruit, and seed, with comments on habitat and distribution. Not so scientific as the works of Theophrastus, this treatise is important because it was "the most practically serviceable book of botany that the world of learning knew of during sixteen centuries."

GREEK MEDICINE IN ROME. GALEN

Until the first ideas of Greek medicine were brought to Rome by slaves attendant upon the athletes, medicine was a family affair based on folk-lore and superstition. The first eminent Greek physician in Rome was Asclepiades (91 B.C.). A disciple of Erasistratus, his physiology and pathology were founded on the atomistic theory of Democritus, rejecting the doctrines of Hippocrates. The earliest systematic treatise and "the greatest medical work of antiquity," excepting those of Hippocrates and of Galen, also "a fundamental source of knowledge of Alexandrian medicine," is the *De re medica* of Celsus, compiled about A.D. 30 in elegant Latin. The four extant works of Rufus of Ephesus, who flourished in Rome and in Egypt a century later, include the first treatise *On the Naming of the Parts of the Body*. His dissections of animals added to the knowledge of the nervous system and the eye. "The greatest Greek physician of the Roman Empire" before Galen, his important treatise on the pulse is the earliest attempt to base pathology upon anatomy and physiology, and of many diseases and symptoms his descriptions are the first.

Galen (A.D. 130–200), "the greatest physician of antiquity after Hippocrates," is of interest to us for two reasons: (1) his contributions to anatomy and physiology and (2) his dominance in the history of medicine during the subsequent centuries. Born in Pergamum and educated in philosophy and in Alexandrian medicine, he arrived at Rome in A.D. 166 and quickly achieved great success, becoming physician to the Emperor, Marcus Aurelius. Of his many works, written in polished and precise Attic Greek, 83 are extant, but only the one *On the Natural Faculties* is completely available in English. In philosophy he emphasized the teleological and vitalistic views of Aristotle, and his physiology and pathology are based on Hippocratic doctrine combined with the theory of the *pneuma* of Erasistratus, whose materialistic philosophy and theory of disease he opposed. He clearly recognized certain fundamental principles of general biology and his physiological experiments, carefully made and accurately recorded, deserve a high place in the history of science. Of special value are his observations on the excretion of urine, on the nervous control of the larynx, and on the effects of the removal of part of the brain and of cutting the spinal cord at different levels. He gives a correct description of muscular movement, which he believes to be actuated from the brain by means of "animal spirits," *pneuma*, flowing through *hollow* nerves.

Discussing the anatomy and physiology of almost every organ of the animal body and unwilling to confess ignorance on any question, Galen was led by false analogies and by authority of the ancients into many errors. Some are particularly important because "Galen's influence upon men's views concerning the vascular system was greater than that of anyone else in antiquity," Figure 24. Although by experiment he knew of the peripheral connections between arteries and veins and was well acquainted with the anatomy of the heart, lungs, and blood vessels, he missed the circulation of the blood by a false analogy with the ebb and flow of the breath and through an ancient theory of the liver as the center of the venous system and site of the "natural spirits."

Disease Galen defined as impairment of normal function, and he recognized four types, each characterized by the preponder-

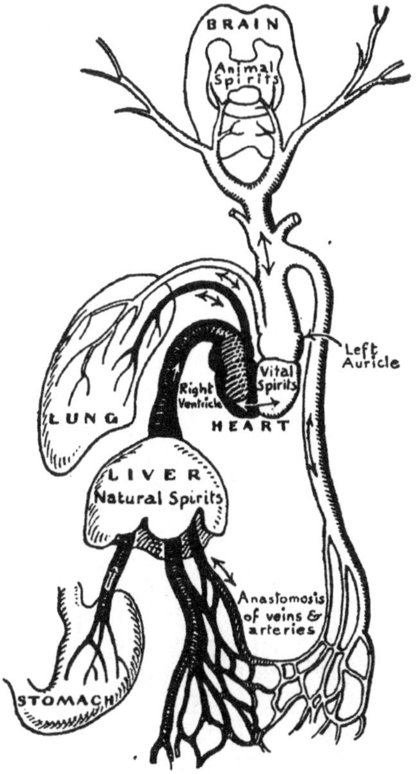

FIG. 24. — THE BLOOD VASCULAR SYSTEM AS DESCRIBED BY GALEN. After Fulton *Clio Medica*, V. Permission of Paul B. Hoeber, Inc.

ance of one of the fundamental qualities: warmth, cold, dryness, and moisture. Treatment was by opposites. He made the important suggestion that phthisis (consumption of the lung) is infective, and he frequently insisted upon the importance of seeking the cause of disease. After Galen we find no great name in anatomy until we come to Vesalius, some 1,400 years later.

LATE ROMAN MATHEMATICAL SCIENCE

Two periods may be distinguished in ancient mathematical science, the first beginning with Pythagoras and ending with

Hero. To these four to five centuries belong all the original works in geometry, astronomy, mechanics, and music. The period closes with the extension of the *pax Romana* over the Orient. The second extends to the sixth century, when Hellenism is proscribed by the new religion, the genius of invention is extinct, and men merely study the older works, commenting and coordinating. Astronomy gradually reverts to astrology, the mathematical geography well begun under Eratosthenes and Ptolemy becomes superficial and descriptive, with Strabo and even with Posidonius.

Whatever the eminence of the Romans in the practical arts of war, politics, and engineering, their interest in abstract sci-

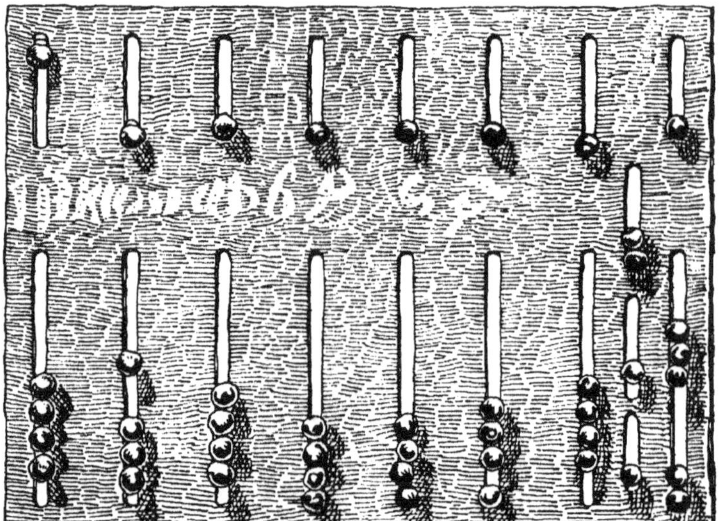

FIG. 25. — A ROMAN ABACUS. After Smith and Ginsburg, *Numbers and Numerals*. Permission of Professor David Eugene Smith and of Professor W. D. Reeve.

ence was almost nil. On the other hand, commercial arithmetic, which had been studiously neglected by Greek mathematicians, now had the place of honor. The Roman numerals, clumsy as they seem to us, were superior to the Greek, and a useful system of finger-reckoning [1] was used, supplementing the skilful use of the abacus. If no abacus was at hand,

[1] L. J. Richardson, *American Mathematical Monthly*, 23, 7–13, 1916.

the corresponding lines were quickly traced on sand or dust, small stones or *calculi* — whence our words calculation and calculus — serving as counters. A complete Roman abacus — of which no example has come down to us — seems to have had eight long and eight short grooves. Of the former, one held five counters or buttons, each of the others four, each of the short grooves one, these last counting as five units each. The grooves with six counters served for computations with fractions. Geometry — but of Hero rather than of Euclid — was valued for its utility in surveying and architecture. Preparation for the engineering art included mathematics, optics, astronomy, history, and law. There were also teachers of mechanics and architecture. (See Vitruvius, above.)

CAPELLA

The Roman writer, Martianus Capella, who lived in Carthage, wrote *c.* A.D. 470 a compendium of grammar, dialectics, rhetoric, geometry, arithmetic, music, and astronomy, of great and lasting educational influence. His classification of these "seven liberal arts" maintained itself throughout the Middle Ages. Gregory of Tours for example says: — "If thou wilt be a priest of God, then let our Martianus instruct thee first in the seven sciences."

SCIENCE AND THE EARLY CHRISTIAN CHURCH

In the earlier centuries of our era the history of science gradually enters upon a new phase. The more highly developed civilization of Greece and Rome, weakened by corruption, has finally yielded to the attacks on the one hand of barbarous or semi-civilized races — Goths, Vandals, Huns, and Arabs — and on the other hand to a moral revolution of humble Jewish origin. These changes were adverse to the development, or even the survival, of Greek science. The destructive relation of the northern barbarians to scientific progress may be easily imagined. The policy of official Christianity was based on antecedent antipathy for the unmoral intellectual attitude and the degenerate character which the early Christians found in close association with Greek learning, and on a too literal interpretation of the Jewish scriptures, with their primitive Chaldean theories of cosmogony and the world.

Justin Martyr, born *c.* A.D. 100, says that what is true in the Greek philosophy can be learned much better from the Prophets. Clement of Alexandria (born *c.* A.D. 150) calls the Greek philosophers robbers and thieves who have given out as their own what they have taken from the Hebrew prophets. Tertullian (*c.* 155–*c.* 220) insists that since Jesus Christ and his gospel, scientific research has become superfluous. St. Isidore (*c.* 560–636), a bishop of Seville, an author of works of encyclopedic character, in which he shows considerable interest in actual science, nevertheless declares it wrong for a Christian to occupy himself with heathen books, since the more one devotes himself to secular learning, the more is pride developed in his soul. Lactantius early in the fourth century includes in his "Divine Institutions" a section,

"On the false wisdom of the philosophers," of which the 24th chapter is devoted to heaping ridicule on the doctrine of the spherical figure of the earth and the existence of antipodes. It is unnecessary to enter into particulars as to his remarks about the absurdity of believing that there are people whose feet are above their heads, and places where rain and hail and snow fall upwards. — DREYER, p. 209.

It was natural that Augustine (354–430), . . . should express himself with . . . moderation, as befitted a man who had been a student of Plato as well as of St. Paul in his younger days. With regard to antipodes, he says that there is no historical evidence of their existence, but people merely conclude that the opposite side of the earth, which is suspended in the convexity of heaven, cannot be devoid of inhabitants. But even if the earth is a sphere, it does not follow that that part is above water, or, even if this be the case, that it is inhabited; and it is too absurd to imagine that people from our parts could have navigated over the immense ocean to the other side, or that people over there could have sprung from Adam. With regard to the heavens, Augustine was, like his predecessors, bound hand and foot by the unfortunate water above the firmament. He says that those who defend the existence of this water point to Saturn being the coolest planet, though we might expect it to be much hotter than the sun, because it travels every day through a much greater orbit; but it is kept cool by the water above it. The water may be in a state of vapor, but in any case we must not doubt that it is there, for the authority of Scripture is greater than the capacity of the human mind. — DREYER, p. 213.

Arguing elsewhere that the soul perceives what the bodily eye cannot, Augustine avails himself of the geometrical analogy of the ideal straight line which shall have length without breadth or thickness, but he lapses into mysticism when he passes to the circle.

On the other hand, St. Augustine sought a naturalistic interpretation of the book of Genesis. Creation he regarded as potential rather than special, like the growth of a tree from a seed. "All development takes its natural course through the powers imparted to matter by the Creator," a theory of evolution under divine control.

Such was the attitude of the Fathers of the Church. But it should be remembered that the Christians during the first two centuries were a small and obscure sect. Whereas, the decline of science began before the Christian era; and zeal for observation died with Galen, A.D. 200, not to be revived for a thousand years. A more effective cause of the decline was the "practical" habit of the Roman people, who despised any science not applicable to their immediate needs, as we have seen in the case of mathematics (p. 169). During the golden age of Rome the Stoic philosophy prevailed.

The Roman had forsaken his early gods. . . . He had abandoned the faith of his fathers and had flung himself into the arms of what he believed to be a lovelier god, and lo! he found himself embracing a machine! His soul recoiled and he fled into Christianity. — SINGER. (*Legacy of Rome*, p. 322.)

There can be no doubt that it was defense by the Church and the monasteries that preserved in Western Europe what was left of our civilization during the periodic anarchy and corruption of the Dark Ages.

The following is a broad survey of the whole period:

The soft autumnal calm . . . which lingered up to the Antonines over that wide expanse of empire from the Persian Gulf to the Pillars of Hercules and from the Nile to the Clyde . . . was only a misleading transition to that bitter winter which filled the half of the second and the whole of the third century, to be soon followed by the abiding dark and cold of the Middle Ages. The Empire was moribund when Christianity arose. Rome had practically slain the ancient world before the Empire replaced the Republic. The barbarous Roman soldier who killed Archimedes absorbed in a problem, is but an instance and a type of what Rome had always done and everywhere by Greek art, civilization, and science. The Empire lived upon and consumed the capital of preceding ages, which it did not replace. Population, production, knowledge, all declined and slowly died. . . .

The sun of ancient science, which had risen in such splendour from Thales to Hipparchus, was now sinking rapidly to the horizon; and when it at last disappeared, say, in the fifth century, the long night of the Middle Ages began. . . . —J. C. MORISON, *The Service of Man.*

THE DARK AGES

After the tottering Roman Empire of the West had come to its end in 476, when the Germans expelled the last emperor from Rome, the peoples of Christian Europe and of

the Graeco-Roman world descended into the great hollow which is roughly called the Middle Ages, extending from the fifth to the fifteenth century, a hollow in which many great and beautiful and heroic things were done and created, but in which knowledge, as we understand it and as Aristotle understood it, had no place. The revival of learning and the Renaissance are memorable as the first sturdy breasting by humanity of the hither slope of that great hollow which lies between us and the ancient world. The modern man, reformed and regenerated by knowledge, looks across it and recognizes on the opposite ridge, in the far-shining cities and stately porticoes, in the art, politics and science of antiquity, many more ties of kinship and sympathy than in the mighty concave between, wherein dwell his Christian ancestry in the dim light of scholasticism and theology. — MORISON.

The "great hollow" here so graphically portrayed may be described as the Middle or Medieval Age (*c.* A.D. 450–1450) and of these ten centuries the first three, or thereabouts, are often called the Dark — as they certainly were the darkest — Ages.

However, in Western Europe the intellectual and even scientific darkness was not complete. The works of Pliny and of Seneca never were lost and were read during the Dark Ages. Dioscorides, translated into Latin and copied unintelligently, was the favorite author on plants. Before the fall of Rome, Capella in Carthage had written his encyclopedia on the *Seven Liberal Arts*, and St. Patrick (*c*. 389–461) had carried the Latin culture to Ireland, which became one of the most civilized countries of the West, sending missionaries to northern and central Europe.

BOETHIUS (*c. 480–524*)

Despite the intellectual darkness that had fallen on Western Europe, a few scholars tended the feeble flame of ancient learning. The most prominent of these was Boethius, born at Rome just after its fall, he is the author not only of the famous *Consolations of Philosophy*, the "last work of Roman literature," but also of works *On Music* and *On Arithmetic* which long-served to represent Greek mathematics to the medieval world. In the course of his public-spirited career, Boethius interested himself in the reform of the coinage and in the introduction of water-clocks and sun-dials. His *Geometry* consists merely of some of the simpler propositions of Euclid, with proofs of the first three only, with applications to mensuration. Yet the intellectual poverty of the age was such that this remained long the standard for mathematical teaching. Boethius's *Arithmetic* begins:

> By all men of old reputation who following Pythagoras's reputation have distinguished themselves by pure intellect it has always been considered settled that no one can reach the highest perfection of philosophical doctrines, who does not seek the heights of learning at a certain crossway — the quadrivium.

For him the things of the world are either discrete (multitudes), or continuous (magnitudes). Multitudes are represented by numbers, or in their ratios by music; magnitudes at rest are treated by geometry, those in motion by astronomy. These four of the seven liberal arts form the *quadrivium;* grammar, dialectics, and rhetoric, the *trivium.* His treatise *On Music* continued in use as a textbook until the 18th century. He translated works of Ptolemy, Nicomachus, Euclid, Archimedes, and Aristotle. Porphyry, who came from Palestine to Rome in 264, wrote a commentary on Aristotle's *Categories.* This, translated into Latin by Boethius, was widely used as a textbook of logic (cf. p. 88). A Christian in faith, a pagan in culture, Boethius has been called the "bridge from antiquity to modern times," "the last of the Romans, the first of the Scholastics."

> It is a world-misfortune that Boethius did not see his way to prepare versions of those works of the Peripatetic school that display powers of observation . . . Boethius repaired the omission, to some extent, by handing on certain mathematical treatises of his own compilation, . . . Thanks to them we can at least say that during the long degradation of the human intellect, mathematics, the science last to sink with the fall of the Greek intellect, was not dragged down so low as the other departments of knowledge. — SINGER. (*Science and Civilization,* p. 120.)

The scholars of the time were almost without exception men whose first interests were theological. Mathematics, having no direct moral significance, seemed to them in itself unworthy of attention. On the other hand, they attached exaggerated importance to all sorts of mystical attributes of numbers and to the interpretation of scriptural numbers. Thus Augustine says the science of numbers is not created by man, but merely discovered, residing in the nature of things.

> Whether numbers are regarded by themselves or their laws applied to figures, lines or other motions, they have always fixed rules, which have not been made by men at all, but only recognized by the keenness of shrewd people.

In 529 St. Benedict founded his first monastery at Monte Cassino (between Rome and Naples) and from there the Benedictine rule, with its emphasis on useful work, spread through Europe. The monasteries became centers of education, using the textbooks of Boethius and of Cassiodorus (*c.* 490–580), who founded a monastery in southern Italy and employed the monks to copy the ancient manuscripts that he collected.

THE EASTERN EMPIRE

While Western Europe suffered the poverty and confusion of the Dark Ages, Constantinople, the New Rome on the Bosporus, stood fast as the capital of the first Christian empire, giving security to government, great wealth, and intellectual and religious life for 900 years. Created by the enlargement of Byzantium, an ancient Greek city, and dedicated in A.D. 330, it quickly became the center of a new civilization founded on Greek culture, Roman law, and Christian faith.

The Byzantine Era, thus inaugurated, was characterized, among other things, by a great respect for ancient learning. Works of the ancients were preserved by constant copying. With an inherited disdain for manual labor, observation and experiment were inhibited, even in alchemy. Productive scholarship took the form of commentaries and compilations. While Theon of Alexandria (p. 106) was editing Euclid, Oribasios (*c.* 325–*c.* 400) in Constantinople, physician of emperor Julian the Apostate, was compiling a medical encyclopedia of historical value for its exact quotations, many from works now lost; and he made Galen available for the ordinary physician. Aëtius of Amida (502–575) made a compilation largely of the writings of Rufus of Ephesus. *Original observations* in physics may be found in the commentaries of Philoponos (*c.* 525), and in pathology in the works of Alexander of Tralles (525–605), especially in his treatise on parasitic worms, the first on that subject. "Paul of Aegina (625–690) wrote an *Epitome* of medicine of great and lasting influence." During the earlier centuries there were many similar works in philosophy, mathematics, etc.; but after the seventh century even this degree of activity almost ceased. Nestorius, a learned Patriarch, was expelled in 431; and the Academy at Athens was closed by order of Justinian in 529, the same year that he made his great contribution to civilization in his *Code of Roman Law*. Although there was a revival of science and philosophy in the ninth century and classical studies continued till the City was pillaged by the army of the Fourth Crusade in 1204, the intellectual center of the world had moved farther east — to Mesopotamia and Persia — whither the Greek learning was carried by the Nestorian Christians.

THE ESTABLISHMENT OF SCHOOLS BY CHARLEMAGNE

We have seen above how the schools of Athens were closed by Justinian in 529. Such schools as existed after that time were chiefly ecclesiastical and their teachings opposed to pagan or heathen (i.e., Greek) learning. At length, however, in 787 Charlemagne, moved, it is said, by the troublesome variety of writing as well as the general illiteracy of his people, ordered the establishment of schools in connection with every abbey of his realm, and summoned to take charge of them Peter of Pisa and Alcuin of York (735–804), whose names stand among the highest in a revival of learning thus begun in Western Europe.

Alcuin himself taught rhetoric, logic, mathematics, and divinity, becoming master of the great school St. Martin at Tours. Of his arithmetic the following problem is an illustration:

> If 100 bushels of corn are distributed among 100 people in such a manner that each man receives 3 bushels, each woman 2, and each child half a bushel; how many men, women and children are there?

Of six possible solutions Alcuin gives but one.

The mathematics taught in Charlemagne's schools would naturally include the use of the abacus, the multiplication table, and the geometry of Boethius. Beyond this, a little Latin with reading and writing sufficed for the needs of the church and her servants, and was supplemented by music and theology for her higher officers. The recognized intellectual needs of the world were indeed but slight. The civilization of

Rome had been gradually submerged by successive waves of barbaric invasion from the north, as a similar fate was soon to be met by the still higher culture of Alexandria. The best intellect of the times was perforce drawn into other forms of activity, while such scholars as remained found no favorable environment for fruitful study. The Benedictine monasteries, indeed, sheltered a few studious monks whose scientific interest scarcely extended beyond the mathematics necessary for their simple accounts, and the computation connected with the determination of the date of Easter.

Near the close of the tenth century Gerbert of Aurillac (Auvergne), devoted his versatile genius in part to mathematical science. Born *c.* 930, he spent a few years in Christian Spain, taught at the monasterial school at Rheims, and from 999 till his death in 1003 was pope, under the name of Sylvester II. He constructed not only abaci, but terrestrial and celestial globes, and collected a valuable library. To him were also attributed a clock, and an organ worked by steam. He wrote works on the use of the abacus, on the division of numbers, and on geometry. The last named contains a solution of the relatively difficult problem to find the sides of a right triangle whose hypotenuse and area are given. He was probably the first Christian to give an account of the so-called ghubar numerals (Spanish-Arabic numerals), which are a transition to the Arabic numerals introduced into Western Europe some centuries later. These numerals are connected with counters and do not have a special sign for zero. Unfortunately the latter part of his life was absorbed in political intrigue, and death cut short his plans for attempting the recovery of the Holy Land.

Out of the schools of Charlemagne gradually grew up that subtle, minute, and over-refined learning of the later Middle Ages which has come to be known as Scholasticism. Based as it was upon authority instead of experiment, and magnifying, as it did, details more than principles, it sharpened rather than broadened the intellect, and was indifferent if not unfavorable to science.

REFERENCES FOR READING

ALLIBUTT, T. C. *Greek Medicine in Rome*. 1921.
BAYNES, N. H. *The Byzantine Empire*. 1925.
CREW, HENRY, *The Rise of Modern Physics*. 1935. Ch. II.
Legacy of Rome. Ed. by Cyril Bailey. 1923, pp. 9–43; 265–324.
LOCY, W. A. *Growth of Biology*. 1925, pp. 47–80.
LONG, E. R. *History of Pathology*. pp. 18–40.
NORDENSKIÖLD, ERIK, *History of Biology*. 1928, pp. 45–67.
SINGER, CHARLES, *From Magic to Science*. 1928.

Books from previous chapters: Farrington; *Science and Civilization*, pp. 112–160; Sarton, *Introd.* I, pp. 201–459; Singer, *Hist. Med.*, pp. 41–60.

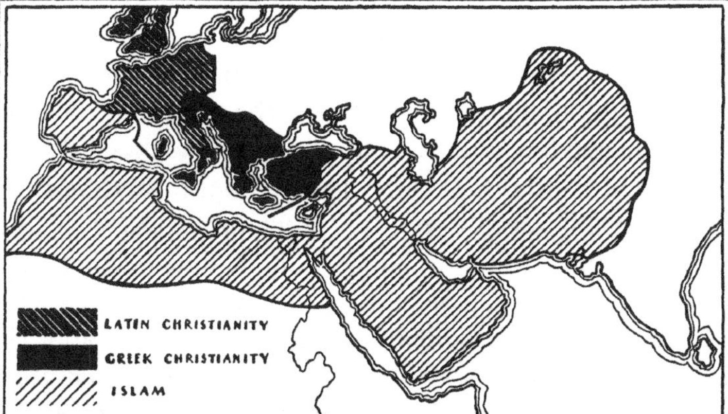

Chapter VIII

SPHERES OF INFLUENCE ABOUT A.D. 750

Hindu and Arabian Science. The Moors in Spain

With the great prophets of the East — Moses, Isaiah, Mahomet — the word was, Thus saith the Lord; with the great seers of the West, from Thales and Aristotle to Archimedes and Lucretius, it was, What says Nature? — DAMPIER.

In the ninth century the School of Bagdad began to flourish, just when the Schools of Christendom were falling into decay in the West, and into decrepitude in the East. The newly-awakened Moslem intellect busied itself at first chiefly with Mathematics and Medical Science; afterwards Aristotle threw his spell upon it, and an immense system of orientalized Aristotelianism was the result. From the East Moslem learning was carried to Spain; and from Spain Aristotle reëntered Northern Europe once more, and revolutionized the intellectual life of Christendom far more completely than he had revolutionized the intellectual life of Islam. — RASHDALL, *Universities of Europe*, I, 351.

TRANSMISSION OF SCIENCE TO WESTERN EUROPE

Alexandria fell to the Arabs in A.D. 641. As a matter of historical perspective it is noteworthy that the interval between its foundation by Alexander the Great and its capture by the Mohammedans — during most of which period with its great library and learned society of scholars it was the intellectual center of the world — is almost equal to that between Charlemagne's time and our own.

The preservation and transmission of portions of Greek science through the Dark Ages to the dawn of science in Western Europe about A.D. 1200 was mainly effected through three distinct, though not quite independent, channels. First, there was to a limited extent a direct inheritance of ancient learning within the Italian peninsula, through all its political and military turmoil. Second, a substantial legacy was received indirectly through the Moors in Spain and Sicily; while, third, additions of importance came later through Italy from Constantinople. Before following the direct Latin-Italian line a brief sketch of Hindu and Arabic science is desirable.

HINDU MATHEMATICS

The far-reaching conquests of Alexander the Great (330 B.C.) immensely stimulated the communication of ideas between the Mediterranean world and Asia, and the East was able to make certain great contributions to mathematical science just where the Greeks were relatively weakest, namely in arithmetic and the rudiments of algebra and trigonometry. Several centuries before our era the Pythagorean theorem and an excellent approximation for the square root of 2, i.e., the length of the hypotenuse of a right triangle of unit sides were known in India in connection with the rules for the construction of altars.

Probably the most important contribution of Hindu mathematics is our modern decimal position system for numbers, which implies the introduction of a sign for zero. Already centuries before the Christian Era, Hindu writers were used to calculating with large numbers in the decimal notation. For different powers of 10 different names were used, but, knowing the complete sequence, the value of any unit was given by

---

NOTE: The map facing this page represents the distribution of Latin, Byzantine, and Islamic power about A.D. 750. Redrawn after Singer (*Legacy of Israel*, p. 178) by permission of The Clarendon Press, Oxford, publishers.

its position in the sequence. A special sign for a vacant spot in the sequence was occasionally used in Greece or Babylon, but the Hindus carried the idea out to its full logical development in a system of calculation. The numeral forms of the Hindus had even then a considerable resemblance to our own forms, they are found in inscriptions from the third century B.C. to the second century A.D. The first definite reference to Hindu numerals outside of India is in a writing by a bishop in Western Syria of 662. Early in the ninth century the numerals became known to Arabic scholars (p. 187).

The mathematicians, however, from whom we trace the later development of mathematics date from the sixth and following centuries. Aryabhata in A.D. 499 wrote a book in four parts dealing with astronomy and the elements of spherical trigonometry, and enunciating numerous rules of arithmetic, algebra, and plane trigonometry. He gives the sums of the series

$$1 + 2 + \dots + n$$
$$1^2 + 2^2 + \dots + n^2$$
$$1^3 + 2^3 + \dots + n^3,$$

solves quadratic equations, gives a table of sines of successive multiples of $3\frac{3}{4}°$ — i.e., twenty-fourths of a right angle — and even uses the value $\pi = 3\frac{177}{1250} = 3.1416$, correct to five places. His geometry is in general inferior.

About A.D. 628, Brahmagupta composed a system of astronomy in verse, with two chapters on mathematics. In this he discusses arithmetical progression, quadratic equations, areas of triangles, quadrilaterals and circles, volume and surface of pyramids and cones. His value of π is $\sqrt{10} = 3.16+$.

Five centuries later Bhaskara (*c.* A.D. 1150) also wrote an astronomy containing mathematical chapters, and the contents of this work soon became known through the Arabs to Western Europe. He acknowledges his indebtedness to the more diffuse work of Brahmagupta. The following problems are typical:

Say quickly, friend, in what portion of a day will [four] fountains, being let loose together, fill a cistern, which, if severally opened, they would fill in one day, half a day, the third, and the sixth part, respectively?

Four jewellers, possessing respectively eight rubies, ten sapphires, a hundred pearls, and five diamonds, presented, each from his own stock, one apiece to the rest in token of regard and gratification at meeting; and they thus became owners of stock of precisely equal value. Tell me, severally, friend, what were the prices of their gems, respectively?

In a pleasant, spacious and elegant edifice, with eight doors, constructed by a skillful architect, as a palace for the lord of the land, tell me the permutations of apertures taken one, two, three, &c. Say, mathematician, how many are the combinations in one composition, with ingredients of six different tastes, sweet, pungent, astringent, sour, salt, and bitter, taking them by ones, twos, or threes?

A snake's hole is at the foot of a pillar, and a peacock is perched on its summit. Seeing a snake, at the distance of thrice the pillar, gliding towards his hole, he pounces obliquely upon him. Say quickly at how many cubits from the snake's hole do they meet, both proceeding an equal distance? — H. T. COLEBROOKE, *Algebra from the Sanscrit*, pp. 42, 45, 50, 65.

While the preceding writers had no algebraic symbolism, but depended laboriously on words and sentences, Bhaskara made considerable progress in abbreviated notation. A partial list of subjects, treated in his first book, includes weights and measures, decimal numeration, fundamental operations, addition, etc.; square and cube root, fractions, equations of the first and second degrees, rule of three, progressions, approximate value of π, volumes. Applications are made to interest, discount, partnership, and the time of filling a cistern by several fountains. While there is reason to believe that the decimal system (without decimal fractions) was known as early as the time of Brahmagupta, this work contains the first systematic discussion of it, including the so-called Arabic numerals and zero.

As an intermediate stage between the earlier use of entire words and our modern use of single letters, he employs abbreviations, but multiplication, equality, and inequality have still to be written out. The divisor is written under the dividend without a line, one member of an equation under the other with verbal context to insure clearness. Polynomials are arranged in powers, though without our exponents, coefficients follow the unknown quantities. In his "rules of cipher" he gives the equivalent of $a \pm 0 = a$, $0^2 = 0$, $\sqrt{0} = 0$, $a \div 0 = \infty$.

In comparison with Greek mathematics, power and freedom are gained at the cost of some sacrifice of logical rigor. Among the Greeks, only the greatest appreciated the possibility and the importance of an unending series of numbers; but the Hindu imagination tended naturally in this direction. A notable achievement of the Hindus was the introduction of the idea of negative numbers and the illustration of positive and negative by assets and debts, etc.

On the whole, the Hindus, having received a part of their mathematics originally from the Greeks, made great contributions on the arithmetical and algebraic side, their influence on European science with which they had little or no direct contact being exerted mainly through the Arabs.

HINDU ASTRONOMY

In astronomy a parallel development took place. It seems probable that Greek planetary theory was introduced into India between the times of Hipparchus and Ptolemy, but Hindu astronomy is characterized as "a curious mixture of old fantastic ideas and sober geometrical methods of calculation." Aryabhata says indeed "The sphere of the stars is stationary, and the earth, making a revolution, produces the daily rising and setting of stars and planets."

MOHAMMED AND THE RISE OF ISLAM

During the sixth and following centuries great events were happening in Arabia, long inhabited by wandering tribes and up to that time a blank in the history of civilization and of science. In A.D. 569 or thereabouts was born, probably in Mecca, that extraordinary man Mohammed, whom millions still regard as the Prophet of the Almighty (Allah). Mecca was the principal commercial town near the eastern shore of the Red Sea, and it was a place of pilgrimage because of its shrine containing many fetishes and idols, above all the "Black Stone" symbolic of *Allah* ("The God"). At the age of forty, Mohammed heard a "call" to preach "belief in Allah and the Last Day." He named his religion *Islam*, "Submission" to God, and a believer, *Muslim*, "one who surrenders" to God. In 622, Mohammed and a few disciples fled to Medina, an agricultural town 250 miles north of Mecca, where he continued to issue his "revelations" of law and doctrine that he called the *Koran* ("reading"), and which after his death was collected into one manuscript. Here he died in A.D. 632.

In Mecca Mohammed was "the despised preacher of a small congregation," but after the flight (*Hegira*) to Medina, he became the leader of a powerful party and ultimately the autocratic ruler of Arabia. At the battle of Bedr in 624 with about 300 Moslems he repulsed 1,000 Meccans. This small engagement is one of the decisive battles of the world and of tremendous consequence. After it thousands confessed Islam, and the religious zeal with which they were fired has never been surpassed. Mohammed taught that idolaters should be converted or killed, but the "People of the Book," chiefly Christians and Jews, were to be undisturbed if they would submit and pay additional taxes. This concession was later of great importance in the history of science. The first two Caliphs, Successors of the Prophet, with great ability started a holy war that was continued with a combination of fanatical fury and greed which overcame all obstacles. Within one century Islam swept like a tidal wave from the barren valleys of western Arabia northward through Syria to the Byzantine border in Asia Minor, eastward through Mesopotamia over all of Persia, and westward along the African shore of the Mediterranean to the Strait of Gibraltar. Egypt, Alexandria, and Carthage fell before the Moslems, and the Islamic empire soon rivaled in extent its great predecessor, the Roman.

In 711, Moslems crossed the Strait of Gibraltar to Spain. Pushing northward into western France as far as Poitiers, their great western and northern movement was checked finally by Charles Martel at the battle of Tours in 732. This extraordinary onrush, occurring within a single century, naturally left the Moslems little time for the development of learning or for the arts and sciences. But after it was over, the Mohammedan invaders settled down in their various conquered countries and in some of them cultivated the arts of peace. The Caliphate became hereditary with the Umayyad Dynasty (A.D. 661–749), and the simple life of the Caliphs of the heroic age was replaced by oriental luxury at the court in Damascus. The intellectual life of Islam began after the overthrow of this dynasty (A.D. 750) by the Abbasids and the removal of the capital to Baghdad, founded in 762.

The Arab invaders had left undisturbed at Jundishapur in southwest Persia a center of learning established there upon a Persian foundation by the Nestorians early in the sixth century. Here were faculties of Christian Theology and of Medicine with a hospital; and from here Caliphs at Baghdad drew Christian physicians to attend their families. Previously there had been translations of Greek and Hindu works into Persian and Syriac. Now ensued a great period of translation into Arabic, the language of the Koran and hence the official language of Islam, fortunately well adapted to scientific writing. Successive Arabian rulers (beginning with al-Mansur, in 754) patronized learning, and to this end collected Greek manuscripts. During the reign of al-Mamun (813–33) the new learning reached its first climax. He founded in Baghdad a kind of academy called the "House of Wisdom" with an observatory and a library. The most celebrated of the translators was Hunain (*c.* 809–877), a Nestorian physician. He and others made available to the Islamic world nearly all that was important in Greek science and medicine. The period of translation was followed by the Golden Age of Arabian Science, about A.D. 900 to 1100. It was "Arabian," however, chiefly in language, not many of the scholars were Arabs and some were not even Moslems. They were mostly Syrians, Persians, and Jews with Arab names. With the exception of some notable advances in mathematics and physics, their positive contributions to science were not great, but they did an enormous service in preserving and arranging the ancient learning of Greece, Persia, and India,[1] and in keeping alive the spirit of science and the arts of civilization while Christian Europe was engaged in a desperate struggle against barbarism.

ARABIAN MATHEMATICAL SCIENCE

We may accept the traditional dictum attributed to the Caliph Omar (634–644) that whatever in the library of Alexandria agreed with the Koran was superfluous, whatever disagreed was worse, and all should therefore be destroyed — as typical of only the earliest phase of Islam. In later periods it was inevitable that individuals among these active-minded peoples should fall under the spell of Greek mathematical science. Their religion was in fact more tolerant towards science than was contemporary Christianity.

It would appear that by A.D. 900 some Arabs were familiar on the one hand with Brahmagupta's arithmetic and algebra, including the decimal position system with the zero, and on the other hand with the chief works of the great Greek mathematicians, some of which have come down to us only through Arabic translations.

The monumental *Algebra* of al-Khwarizmi written about 825 was based on much older sources, perhaps Hindu, perhaps Babylonian. It served in turn as the foundation for many later treatises. From its title is derived our word "algebra," from the author's name our "algorism." The Introduction reads in part:

> That fondness for science, by which God has distinguished the Imam al-Mamun, . . . that affability and condescension which he shown to the learned, that promptitude with which he protects

[1] For a graphic and picturesque account of the transmission of Hindu science into Arabic, see D. E. Smith and J. Ginsberg, *Amer. Math. Monthly*, March, 1918: "Rabbi Ben Ezra and the Hindu-Arabic Problem."

and supports them . . . has encouraged me to compose a short work on Calculating by (the rules of) Completion and Reduction, confining it to what is easiest and most useful in arithmetic, such as men constantly require in cases of inheritance, legacies, partition, law-suits, and trade, and in all their dealings with one another, or where the measuring of lands, the digging of canals, geometrical computation, and other objects of various sorts and kinds are concerned . . . — L. C. KARPINSKI, *Robert of Chester's Latin Trans. of the Algebra of Al-Khowarizmi*, p. 46.

The first book contains a discussion of five types of quadratic equations:
$ax^2 = bx$, $ax^2 = c$, $ax^2 + bx = c$, $ax^2 + c = bx$, $ax^2 = bx + c$; only real positive roots are accepted, but, unlike the Greeks, he

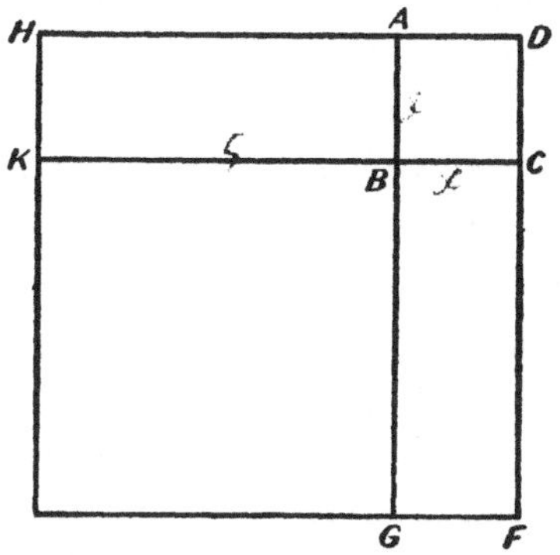

FIG. 26. — GEOMETRICAL SOLUTION OF THE QUADRATIC EQUATION.

recognizes the existence of two roots. He gives a geometrical solution of the quadratic equation analogous to those of Euclid. Suppose $x^2 + 10x = 39$ and let $AB = BC = x$, $AH = CF = 5$; then the areas are $AC = x^2$, $AK = BF = 5x$.

The sum of these is $x^2 + 10x$. Complete the square HF by adding $KG = 25$. $HF = (x + 5)^2 = 64$, whence $x = 3$.

With him analysis begins to be on a level with geometry and mathematical science in Western Europe is more influenced by him than by any other writer from the Greeks to Regiomontanus. Al-Khwarizmi also wrote an arithmetic, of which a Latin translation of the twelfth century was instrumental in bringing to Europe the Hindu system of numeration.

Another great mathematician was the Persian, Omar

Khayyam (*c.* 1043–1131/32) of Naishapur, known to the larger public of Western Europe as poet of the *Rubaiyat.* He is equally entitled to fame for his work on algebra, which contains a discussion of cubic equations, which he solved by geometrical constructions. Omar obtained a root by intersecting a conic section with another conic section. He rejected negative roots and did not find all positive roots. This work, nevertheless, was a notable achievement, because it seems that he set himself for the first time the problem: How can cubic equations with numerical coefficients be solved? He is said to have asserted the impossibility of finding two cubes whose sum should be a cube. This problem was generalized many years later by Fermat, who affirmed that not only were there no two cubes whose sum should be a cube, but neither were there two powers of any order, except the second, whose sum would equal that same power. The proof of this famous theorem has defied all analysis up to the present time.

There is in Arabic science no definite separation of algebra and arithmetic, and the former, in spite of relatively rapid development, remains entirely rhetorical.

In Physics, Ibn al-Haitham, or Alhazen, (*c.* 965–1020) wrote a work on optics enunciating the law of reflection and making a study of spherical and parabolic mirrors. He also devised an apparatus for studying refraction, being probably the first physicist to note the magnifying power of spherical segments of glass — i.e., lenses, and also the first to prove that the ratio between the angle of incidence and the angle of reflection (rather than that between their sines) is *not* constant. He gave a detailed account of the human eye and maintained, in opposition to Euclid, that rays of light have their origin in the object seen, not in the eye. He attempted to explain the change of apparent shape of the sun and the moon when approaching the horizon. In view of the difficulty and the fundamental nature of the problems he attacked, Alhazen may be considered the most outstanding of Mohammedan physicists.

The Arabs employed the pendulum for time measurement,

and tabulated specific gravities of metals, etc. Paraphrasing the words of a modern physicist:

> The Arabs have reproduced what came down to them from the Greeks in thoroughly intelligible form, and applied it to new problems, and thus built up the theorems, at first only obtained for particular cases, into a greater system, adding much of their own. They have thus rendered an extraordinarily great service, such as would correspond in modern times to the investigations which have grown out of the pioneer work of such men as Newton, Faraday, and Röntgen. — E. WIEDEMANN, "Ueber das Experiment in Altertum und Mittelalter," *Unterrichtsblätter*, 1906 (4–6).

ARABIAN ASTRONOMY

In the ninth century Ptolemy's *Syntaxis* (p. 143) was translated into Arabic acquiring its since familiar title *Almagest*. New astronomical tables were calculated, eclipses observed, and improvements made in trigonometry, introducing sines and other functions since current. Masonry quadrants of large size were used, and even a combination of a horizontal circle with two revolving quadrants mounted upon it, foreshadowing the modern theodolite. Better and more complete observational data demanded improved mathematical methods, while the necessary computations were accomplished much more economically, through the use of the decimal number system. Harun al-Rashid sent to Charlemagne an ingenious water-clock, while under his successor, al-Mamun, two learned mathematicians were commissioned to measure a degree of the earth's circumference.

> "Choose a place in a level desert and determine its latitude. Then draw the meridian line and travel along it towards the pole-star. Measure the distance in yards. Then measure the latitude of the second place. Subtract the latitude of the first and divide the difference into the distance of the places in parasangs. The result multiplied by 360 gives the circumference of the earth in parasangs." — WIEDEMANN.

The writer just quoted describes a second method involving the measurement of the angle of depression of the horizon as seen from the top of a high mountain.

It is not improbable that western Europe acquired from eastern Asia, through Arab channels, the mariner's compass and gunpowder.

"When the night is so dark that the captains can perceive no star to orient themselves, they fill a vessel with water and place it in the interior of the ship, protected from wind; then they take a needle and stick it into a straw, forming a cross. They throw this upon the water in the vessel mentioned and let it swim on the surface. Hereupon they take a magnet, put it near the surface of the water, and turn their hands. The needle turns upon the water; then they draw their hands suddenly and rapidly back, whereupon the needle points in two directions, namely north and south." A.D. 1232 — WIEDEMANN.

The astronomical theory of the Arabs was merely that of Ptolemy. But they "were not content to consider the Ptolemaic system merely as a geometrical aid to computation; they required a real and physically true system of the world, and had therefore to assume solid crystal spheres after the manner of Aristotle." The various attempts to devise a better system all miscarried, their authors having no new guiding principle, nor superior mathematical power, and being more or less hampered by Aristotelian traditions, though Greek theories of the rotation of the earth seem not to have been unknown.

ASIATIC OBSERVATORIES

Besides the work of the Arabian astronomers themselves, it is an interesting fact that their barbarian Mongol conquerors in the East acquired a temporarily active scientific interest, founding a fine observatory at Meraga near the northwest frontier of modern Persia. The instruments used here are said to have been superior to any used in Europe until the time of Tycho Brahe in the sixteenth century. The principal achievement of this observatory was the issue of a revised set of astronomical tables for computing the motions of the planets, together with a new star catalogue. The excellence of their work may be inferred from a determination of the precession of the equinoxes within 1″. This development lasted only a few years in the latter half of the thirteenth century. A similar brief outburst of astronomical activity occurred among the Tartars at Samarkand (Russian Turkestan) nearly 200 years later, that is, a little before the time of Copernicus, and here the first new star catalogue since that of Ptolemy was compiled. It is noteworthy that there was no hostility between science and the Mohammedan church. One of the uses of astronomy indeed was to determine the direction of Mecca.

No great original idea can be attributed to any of the Arab . . . astronomers . . . They had, however, a remarkable aptitude for absorbing foreign ideas, and carrying them slightly further. They were patient and accurate observers, and skilful calculators. We owe to them a long series of observations, and the invention or introduction of several important improvements in mathematical methods. . . . More important than the actual original contributions of the Arabs to astronomy was the service that they performed in keeping alive interest in the science and preserving the discoveries of their Greek predecessors. — BERRY, pp. 82, 83.

ARABIC ALCHEMY

Turning to chemical problems, Arabian scholars concentrated attention on two aims — the transmutation of other metals into gold and the preparation of an elixir of life to cure all human disease. This work, derived from Persian, or perhaps Chinese, and Alexandrian sources (p. 151), continued for seven centuries, notably in Spain. In spite of the impossibility of attaining the direct aims, much real knowledge was gained for utilization in the beginnings of modern chemistry. The most notable among Arabic alchemical writings are those bearing the name of Jabir, a physician of the latter half of the eighth century, "the father of alchemy." But recent research indicates that they were produced by a secret society two centuries after his death. The authors seem to have separated antimony and arsenic from their sulphides and to have described the preparation of steel, the dyeing of cloth and leather, the distillation of acetic acid from vinegar, and the preparation of many chemical substances, including sulphuric and nitric acids. The idea that sulphur (fire) and mercury (liquid) are primary elements, to which salt (solid) was later added, became important — this theory surviving along with that of the Four Elements of the Greeks until the seventeenth century.

MOSLEM MEDICINE IN THE EAST

The Moslems were much interested in medicine. About A.D. 800 the Caliph Harun al-Rashid founded a hospital at Baghdad, modeled probably after the celebrated one at Jundishapur. Other hospitals followed. Exact records exist of thirty-four, with details of administration and teaching. The period of translation was followed by an enormous production of medical works in the Arabic speaking countries. What was new, resulted from keen clinical observation and from therapeutic experience. Theory was untouched and experiment impossible. Outstanding were several authors of dominating influence.

Al-Razi, or Rhazes (*c.* 850–923/4), a Persian Moslem educated at Baghdad, was the greatest physician of the Islamic world. In his monograph on measles and smallpox, he made the first clear distinction between these diseases. His chief work is a huge compilation from Greek and other authors interspersed with his own experiences. His writings include philosophy, mathematics, astronomy, and a youthful work on alchemy. As a physicist he measured specific gravities by the hydrostatic balance.

"The most famous scientist of Islam" was Abu ibn Sina, known as Avicenna (980–1037), a Persian who wrote on all the sciences then known. His great *Canon of Medicine* is the "concentrated legacy of Greek medical knowledge with the Arabs' contribution" and "the masterpiece of Arabic systematization." It was the main textbook, both in the East and the West, until the middle of the seventeenth century. After Avicenna the center of Islamic science passed to Spain.

THE MOORS IN SPAIN

A small part of the Moslems who conquered the Spanish peninsula were of pure Arab stock, more were from that part of northern Africa once known as Mauretania — whence the term Moors, generally applied to the conquering Mohammedans of the West. Later generations were largely of mixed descent through Christian mothers. In the ninth and tenth centuries a remarkable civilization arose in Spain — the highest that the Arabic speaking peoples have ever reached. Although separated politically in 756, when the only surviving Umayyad prince established an independent Caliphate at Cordova, the Spanish Moslems remained united by strong ties of common religion and culture with the vast empire, including Sicily, "of which Mecca was the religious and Baghdad the cultural and political centre." Science and trade were promoted by the pilgrimages to Mecca. In such cities as Cordova, Toledo, and Seville a type of civilization and a stage of learning were reached higher in many respects than existed at the same time and even for centuries afterward anywhere in Christian Europe.

In Spain and North Africa, as in the East, philosophers were often physicians. Ishaq al-Israili, also called Isaac Judaeus or Isaac Israeli the Elder, who was born in Egypt and flourished at Qairawan, Tunis, where he died about 932(?), a centenarian, was the earliest Jewish philosopher of the West. He wrote several widely used books on medical science, including the best work on fevers that was known during the Middle Ages and another, the most elaborate one on urine. In his book *On the Elements* he expounded Aristotelian physics. His classification of the sciences, probably the first by a Jewish philosopher, was essentially Aristotelian as modified by the Moslems. Early in the eleventh century Abu-l-Qasim, or Albucasis, of Cordova (died *c.* 1013) produced a great medical encyclopedia which included an important work on surgery illustrated by figures of instruments. Ibn Zuhr, or Avenzoar, of Seville (1091/94–1161/62) was by far the greatest among the many distinguished physicians of the Moslem West, and the most famous one of his time. His works are devoted exclusively to medicine. One of them, the *Taisir*, is one of the most important in the history of medicine because of many new descriptions of disease, including accounts of cancer of the œsophagus and of the stomach. His discovery of the Itch-Mite was anticipated by an author of the tenth century.

Ibn Rushd (1126–98) of Cordova, better known as Averroës, practiced medicine and was "the greatest of Western Muslim philosophers," famous for his commentaries on Arabic translations of the works of Aristotle, "whom he exalted above all men." His doctrine of creation as a continuous process in an ever-changing world shows him to be "an evolutionist in the true sense of the word." Opposed to Averroës was

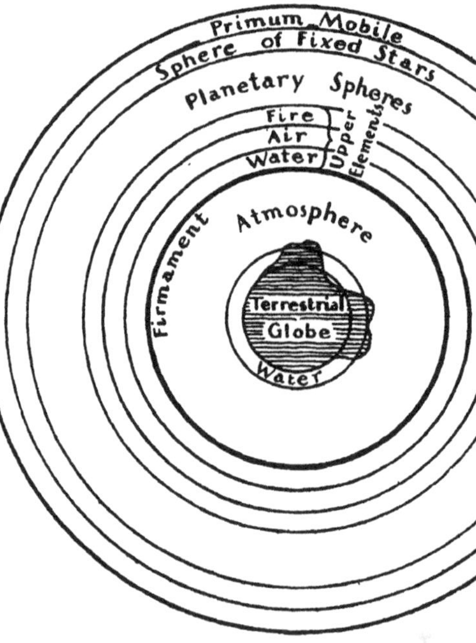

Fig. 27. — The Cosmology of Maimonides. After Singer (*Legacy of Israel*, p. 194) by permission of The Clarendon Press, Oxford, publishers.

Moses ben Maimon, called Maimonides (1135–1204). Born at Cordova and after 1153 a successful physician in Cairo, he set forth a complete system that was the culmination of Jewish philosophy. His cosmology is reflected in the great poem of Dante (p. 211). He developed the Aristotelian philosophy to include a God external to the universe, who created it out of nothing, with intelligent purpose and design. "Design . . . as Maimonides develops it, . . . is but another name for Natural Law." Averroës and Maimonides exercised influence almost comparable with that of Aristotle on the Scholastic thought of Western Europe.

In the arts and industries the Moors deserve special mention. Cordovan and Morocco leather are well known. Toledo and Damascus blades (swords) were long famous. The first paper factories in Europe were established by Moslems in Spain and Sicily. From there the art passed to Italy and made printing practicable (p. 220). The making of paper of silk rags probably was learned from Chinese workmen in Samarkand, on the ancient "silk road" between China and Byzantium, captured by the Arabs in 704; and the manufacture of paper in Baghdad began in 794.

. . . Under the caliphs, Moslem Spain became the richest, most populous, and most enlightened country in Europe. The palaces, the mosques, bridges, aqueducts, and private dwellings reached a luxury and beauty of which a shadow still remains in the great mosque of Cordova. New industries, particularly silk weaving, flourished exceedingly, 13,000 looms existing in Cordova alone. Agriculture, aided by perfect systems of irrigation for the first time in Europe, was carried to a high degree of perfection, many fruits, trees and vegetables hitherto unknown being introduced from the East. Mining and metallurgy, glass making, enamelling, and damascening kept whole populations busy and prosperous. From Malaga, Seville, and Almeria went ships to all parts of the Mediterranean loaded with the rich produce of Spanish Moslem taste and industry, and of the natural and cultivated wealth of the land. Caravans bore to farthest India and darkest Africa the precious tissues, the marvels of metal work, the enamels, and precious stones of Spain. All the luxury, culture, and beauty that the Orient could provide in return, found its way to the Moslem cities of the Peninsula. The schools and libraries of Spain were famous throughout the world; science and learning were cultivated and taught as they never had been before. Jew and Moslem, in the friendly rivalry of letters, made their country illustrious for all time by the productions of their study. . . . The schools of Cordova, Toledo, Seville, and Saragossa attained a celebrity which subsequently attracted to them students from all parts of the world. — HUME, pp. 102, 109.

FIG. 28*a*. — THE FACE OF AN ASTROLABE DATED A.H. 1010 (A.D. 1601/2). Hitti, *History of the Arabs*, p. 374. From a photograph lent by Professor Philip K. Hitti. (See also Fig. 28*b*.)

Fig. 28*b*. — The Back of an Astrolabe dated A.H. 1010 (A.D. 1601/2). Hitti, *History of the Arabs*, p. 374. From a photograph lent by Professor Philip K. Hitti. (See also Fig. 28*a*.)

The tenth century was the golden age of Moorish science in Spain. The astronomers kept alive the idea that the earth is a globe and furnished mariners with tables of latitude and longitude, while artisans produced improved astrolabes. The most brilliant geographical work was done in the twelfth century by al-Idrisi (1099/1100–1166) for the Norman King, Roger II of Sicily. It is a description of the then known world, with 70 maps.

Toledo, having been taken in 1085 by Alfonso VII, became toward the end of the century a resort for northern scholars

FIG. 29. — THE RECESSION OF ISLAM IN SPAIN. Dates under the names of places indicate when they fell to Christian forces. After Singer (*Legacy of Israel*, p. 183) by permission of The Clarendon Press, Oxford, publishers.

eager to learn of Moslem culture and science. As they could not read Arabic, the demand arose for translation into Latin. The earliest Latin manuscript to show the form of Arabic numerals was written in Spain in 976. But in the translation by John of Seville (Avendeath, *c.* 1090–1165) of an *Arithmetic* attributed to al-Khwarizmi (p. 188), elaborated by a later writer, "our so-called Arabic numeral notation, in which the digits depend on their position for their value, is used in Latin for the first time." John and other Jews translated many works from Arabic into Latin. The curiosity aroused by the "wandering Jew" with his Latin translations of Arabic learn-

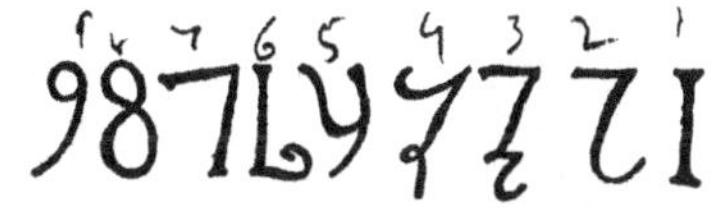

FIG. 30. — ARABIC NUMERALS IN THE EARLIEST FORM TO BE FOUND IN A LATIN MANUSCRIPT. Note the reverse order, 9–1. After Smith and Ginsburg, *loc. cit.*, by the courtesy of Professor David Eugene Smith and of Professor W. D. Reeve.

ing is a factor in the revival of science that is the subject of the next chapter.

REFERENCES FOR READING

ANDRAE, TOR, *Mohammed* (Trans. T. Menzel), 1936.

ARNOLD, T. W., "Muslim Civilization" (*Cambridge Medieval Hist.*, IV, Ch. 4).

DATTA, B., AND A. N. SINGH, *History of Hindu Mathematics*, 1935, 1938, 2 vols.

DAVIS, W. S., *Short History of the Near East*, 1922, pp. 100–159.

ELGOOD, CYRIL, *Medicine in Persia* (*Clio Medica*, vol. 14), 1934.

HUME, MARTIN, *The Spanish People*, 1901.

Legacy of Islam, Ed. by Sir Thomas Arnold and A. Guillaume, 1931.

Legacy of Israel, Ed. by C. R. Bevan and Charles Singer, 1927.

RIESMAN, DAVID, *Story of Medicine in the Middle Ages*, 1935.

SMITH, D. E., AND L. C. KARPINSKI, *Hindu-Arabic Numerals*, 1911.

Books from previous chapters: Berry, *Astronomy*, Ch. III; Dampier, *Hist. of Sci.*, 65–105; Dreyer, *Plan. Sys.*, Ch. II; Sarton, *Introduction*, II, 1–104; Singer, *Magic to Science*.

NOTE: Finger reckoning is illustrated in Byzantine style in the figure on the opposite page from a Spanish codex of the 13th century. Above are finger numerals from 100 to 900 and from 1,000 to 9,000, below are the signs for 10,000 and 20,000. After Smith and Ginsburg, *loc. cit.*, p. 23, by courtesy of Professor David Eugene Smith and Professor W. D. Reeve.

Chapter IX

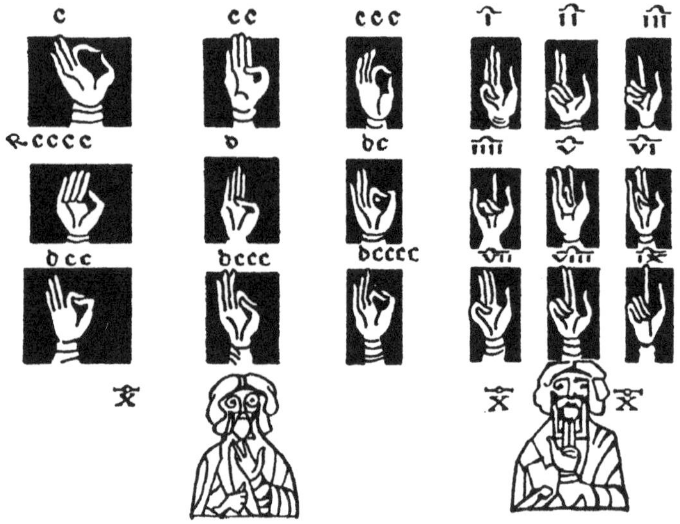

Progress of Science to A.D. 1450

It cannot be too emphatically stated that there is no historical evidence for the theory which connects the new birth of Europe with the passing away of the fateful millennial year and with it of the awful dread of a coming end of all things. Yet, although there was no breach of historical continuity at the year 1000, the date will serve as well as any other that could be assigned to represent the turning-point of European history, separating an age of religious terror and theological pessimism from an age of hope and vigor and active religious enthusiasm. . . . The change which began to pass over the schools of France in the eleventh century and culminated in the great intellectual Renaissance of the following age was but one effect of that general revivification of the human spirit which should be recognized as constituting an epoch in the history of European civilization not less momentous than the Reformation or the French Revolution. . . . The schools of Christendom became thronged as they were never thronged before. A passion for inquiry took the place of the old routine. The Crusades brought different parts of Europe into contact with one another and into contact with the new world of the East, — with a new Religion and a new Philosophy, with the Arabic Aristotle, with the Arabic commentators on Aristotle, and eventually even with Aristotle in the original Greek. . . . Whatever the causes of the change, the beginning of the eleventh century represents, as nearly as it is possible to fix it, the turning-point in the intellectual history of Europe. — H. RASHDALL, *Universities of Europe*, I, pp. 30–32.

THE CRUSADES

From the time of Mohammed's hegira from Mecca to Medina in A.D. 622 to the siege of Vienna by his followers in 1683 — a period of more than 1,000 years — Europe stood in more or less constant dread of Mohammedan conquest. Fifteen years after the hegira, Jerusalem was captured by Omar, and remained under Mohammedan control till the end of the first Crusade, since which time it has been sometimes in Christian, sometimes in Mohammedan, possession, and even today (1937) is in bitter dispute. Toleration of Christians in the Holy Land was, however, the rule until the eleventh century, and between A.D. 700 and 1000 pilgrimages to Jerusalem were frequently undertaken by Christians in the West. But after 1010 such pilgrimages began to be seriously interfered with, and matters steadily grew worse, until in 1071 Seljukian Turks displaced Arabian Mohammedans as rulers of Jerusalem. These Turks, though more rough than intolerant, eventually interfered with both trade and pilgrimages, until for this and other reasons the conquest of the Holy Land became a passion with the Christian nations.

In the spring of 1097, after several years of widespread preparation, a great host of western Christians, variously estimated at 150,000 to 600,000, gathered at Constantinople charged with war-like and religious zeal and bent on wresting Jerusalem and the Holy Land from the possession of the Mohammedan "Infidel." This was the beginning of those expeditions under the banner of the Cross — hence known as the Crusades — which may be regarded as intermittent reactions of the Christian West against the pressure of the Mohammedan East. The spirit of Christian Europe in the Middle Ages being essentially religious and ecclesiastical, it was natural that its more bold and adventurous youth should regard with jealousy and indignation the wide extent of the Mohammedan empire and especially its possession of Jerusalem and other holy places. In all, eight such Crusades are recognized by historians, with their immense influence upon Christian Europe. The expansion of the intellectual outlook due to the mere experiences of travel, for men born and bred under the parochial limitations of feudalism and monasticism, must have been great. The arts and appliances observed abroad, the different standards of all sorts, the wealth and luxury of the East, doubtless had a powerful effect upon Europe when reported or introduced by the Crusaders upon their return. When we reflect upon the ages of darkness which had rested upon Christian Europe from the fall of Rome into the hands of the barbarians to the fall of Jerusalem into the hands of the Turks — a period of almost exactly six hundred years — we may agree with those who are disposed to look upon the Crusades as an age of discovery comparable with that of the new world by Columbus and his followers — but a discovery of the East instead of the West.

The period of the Crusades extends over about two centuries, viz.: — from 1090 to 1290, and thus immediately precedes the Renaissance, of which it was apparently one of the most important preparatory factors.

SCHOLASTICISM

Before A.D. 1100 in the Latin West fear of barbarian invasion had ceased. During the following centuries trade and commerce, largely augmented by the Crusades, developed rapidly, and men acquired leisure to examine the foundations of their faith. Also in Church and State a demand arose later for learned lawyers and administrators. These influences caused students to flock to the schools, established in abbeys and cathedrals since the mandate of Charlemagne (p. 177). In Paris the great teacher Abelard (1079–1142) attracted large numbers from all parts of the Christian world to its three important schools out of which grew the university of Paris. The teaching of the schoolmen was based on the famous curriculum of "the seven liberal arts" of Capella (p. 170). These consisted of a *quadrivium* — geometry, astronomy, music, and arithmetic — and a *trivium* — grammar, logic, and rhetoric (p. 175). For success in the courts of law and the Church, the most important of these studies was logic — the "Old Logic," two books of Aristotle handed down through the Dark Ages in the Latin translation of Boethius (p. 174), with his translation of the introduction to the subject, the *Isagoge* of Porphyry. In the latter work a question was explicitly raised in a very distinct and emphatic manner, and this was to become the main subject for debate by the schoolmen.

The words in which this writer states, without resolving, the central problem of the Scholastic Philosophy, have played perhaps a more momentous part in the history of thought, than any other passage of equal length in all literature outside the canonical Scriptures. They are worth quoting at length:

"Next, concerning genera and species, the question indeed whether they have a substantial existence, or whether they consist in bare intellectual concepts only, or whether if they have a substantial existence they are corporeal or incorporeal, and whether they are separable from the sensible properties of the things (or particulars of sense), or are only in those properties and subsisting about them, I shall forbear to determine. For a question of this kind is a very deep one and one that requires a longer investigation." — RASHDALL, I, 39, 40.

It should be noted that the words, "genera and species," are used here, not in the modern sense relating to kinds of animals and plants, but as used by Aristotle in his *Categories;* and the question is one on which Plato and Aristotle differed — the old problem of the *universals:* do general ideas represent reality, God's thoughts; or are they merely man's attempt to classify his own perceptions?

The aim of the schoolmen was to establish Christian dogma on a logical basis. From their efforts gradually arose a method, an attitude of mind, and a system of philosophy that is called scholasticism. Scholasticism, accepting a philosophical system on traditional authority, argued, rather than observed, what the corresponding facts ought to be. It, nevertheless, lingered long after the Crusades were ended, and abundant survivals of it exist even today.

MEDIEVAL UNIVERSITIES

"Universities, like cathedrals and parliaments, are a product of the Middle Ages." In the eleventh and twelfth centuries there was a gradual development from the previous cathedral and monastic schools to the beginnings of the modern universities at Paris, Bologna, Oxford, and Cambridge; some of the schools, however, continued along their previous lines, especially at Chartres. From that time onward to our own, the universities have played the chief part in the advancement of learning in general and of science in particular. In this development theological influences were naturally dominant. A major event of the second decade of the twelfth century was the rediscovery of the "New Logic" of Aristotle in Latin translations from the Greek — the long lost translations of Boethius and new ones by James of Venice. But when about 1200 the *Physics* and the *Metaphysics* appeared in Latin with the comments of Averroës, difficulties began. In 1210 these works of Aristotle and Averroës were forbidden in Paris. In 1231, the pope forbade any work of Aristotle, except the "Old Logic," until expurgated. However, in 1245, no doubt through the immense influence of Albertus Magnus (p. 217), these forbidden books were required for the Master of Arts degree in the University of Paris.

In southern Italy, where traces of Greek culture remained, the medical school at Salerno seems to have been known in the ninth century, and it is sometimes called the oldest university in Europe. The law school at Bologna, in northern Italy, became well known about 1100 through the teaching of Irnerius, although the date of the University of Bologna is usually given as near the end of the twelfth century. The Masters of Arts in Paris issued regulations for students about 1200, and that year the king granted them special privileges. A few years later they were formally recognized as a corporation by the pope, and they received a university charter in 1231. These early universities began as associations or guilds of masters (Paris) or of students (Bologna) formed for self-protection. Commencement in Paris was the initiation into the guild of Masters of Arts. Such an institution of learning, having a definite curriculum, holding examinations, and conferring degrees, when recognized as such by pope or king became a university with certain privileges and immunities for its members. Not until the fifteenth century did the universities generally acquire permanent buildings, instead of having professors provide their own quarters. Laboratories for instruction were of course unknown.

> The world will never see again so brilliant a throng of ingenuous youth as gathered together in the great university towns in those years of vivid and impassioned greed for letters that followed the revival of learning. The romance of Oxford or Heidelberg or Harvard is tame compared with that electric life of a new-born world that wrought and flourished in Padua, Paris, and Alcala. — JOHN HAY, *Castilian Days*, p. 289.

Thus were the medieval universities born of the Renaissance of the Twelfth Century. In addition to the Seven Arts, the curriculum included the Three Philosophies — Natural (science), Moral, and Mental. Practically this amounted to the study of the works of Aristotle. To this might be added a "superior study" — Law, Theology, or Medicine. Teaching was by reading and commenting on a specified book. The aim was to find the meaning of the authority, not to study nature.

TRANSMISSION OF SCIENCE THROUGH MOORISH SPAIN

The meagre rivulet of classical science derived directly from Greek and Roman sources is now mingled with the current which found its way through northern Africa and Spain under the Moors. Boethius's rudimentary work was supplanted, and before 1400, the first five books of Euclid were taught at many universities. The current language of science was still Arabic, but Ptolemy's *Almagest* was translated into Latin in the second half of the twelfth century, probably with the use of Arabic numerals. Near the close of the Moorish domination of Spain, King Alfonso X of Castile (1223–84) collected at Toledo a body of Christian and Jewish scholars who under his direction prepared the celebrated Alfonsine Tables. These enjoyed a high reputation for three centuries, though first printed in 1483.

While we thus owe to the Arabs a considerable debt for preserving for the use of later ages the precious heritage of Greek learning, the revival of learning in the fourteenth century came chiefly from other quarters and would probably have come in due time even if Arabic influences had not been at work Yet it is noteworthy that early in the twelfth century re-translations of the Greek classics began to be made from the Arabic, and these may well have supplied the very limited demand for them tolerated by the church for the next hundred years. In spite of jealous exclusiveness the learning of the great schools of Granada, Cordova, and Seville gradually found its way to Paris, Oxford, and Cambridge.

During the course of the twelfth century a struggle had been going on in the bosom of Islam between the Philosophers and the Theologians. It was just at the moment when, through the favor of the Caliph Almansor, the Theologians had succeeded in crushing the Philosophers, that the torch of Aristotelian thought was handed on to Christendom. . . . — RASHDALL, I, 352.

It was from this time and from this time only (though the change had been prepared in the region of pure Theology by Peter the Lombard) that the Scholastic Philosophy became distinguished by that servile deference to authority with which it has been in modern times too indiscriminately reproached. And the discovery of the new Aristotle was by itself calculated to check the originality and speculative freedom which, in the paucity of books, had characterized the active minds of the twelfth century. The tendency of the sceptics was to transfer to Aristotle or Averroës the authority which the orthodox had attributed to the Bible and the Fathers of the Church. — RASHDALL, I, 264–366.

DAWN OF THE RENAISSANCE

In the thirteenth century it becomes plain that a new spirit is arising in Europe. We cannot fail to detect at this time the existence, even at places as far apart as Oxford and Bologna —

infinitely further apart then than now — of a widespread desire for knowledge and a zeal for learning such as had not been known for centuries. Arabic mathematical science is introduced from northern Africa by Leonardo of Pisa. A notable philosopher — the Dominican, Albertus Magnus — appears and interprets Aristotle with "all contemporary knowledge of astronomy, geography, botany, zoology, and medicine." His famous pupil St. Thomas Aquinas (1225–74) systematized the scholastic philosophy of the universe with a logical rigor quite hostile to any development of observation or experiment. He accepted the Ptolemaic system as a working hypothesis. Deeply influenced by Maimonides, his aim was to reconcile Aristotelian and Moslem knowledge with Christian theology. He succeeded so well that his philosophy soon became, and still is, the official intellectual guide of the Roman Catholic Church. Great Gothic cathedrals arise, more universities are founded, and, most noteworthy of all for the history of science, an original student of nature appears, in Roger Bacon.

By the beginning of the thirteenth century, in consequence of the opening up of communications with the East — through intercourse with the Moors in Spain, through the conquest of Constantinople, through the Crusades, through the travels of enterprising scholars — the whole of the works of Aristotle were gradually making their way into the Western world. Some became known in translations direct from the Greek; more in Latin versions of older Syriac or Arabic translations. And now the authority which Aristotle had long enjoyed as a logician — nay, it may almost be said the authority of Logic itself — communicated itself in a manner to all that he wrote. Aristotle was accepted as a well-nigh final authority upon Metaphysics, upon Moral Philosophy, and with far more disastrous results upon Natural Science. The awakened intellect of Europe busied itself with expounding, analysing and debating the new treasures unfolded before its eye. . . . — RASHDALL, I, 68.

MATHEMATICAL SCIENCE IN THE THIRTEENTH CENTURY

Increasing activity in mathematical science was due largely to Leonardo Fibonacci of Pisa, Jordanus Nemorarius in Saxony, and Roger Bacon in England.

Leonardo of Pisa, or Fibonacci (born *c.* 1170), was educated in Barbary, where his father was Pisan commercial agent, and thus became familiar with al-Khwarizmi's *Algebra*, and the Arabic numeral system. He appreciated their advantages and on his return to Italy published in his *Liber Abaci*, 1202, revised 1228, an account which gave them currency in Europe "in order that the Latin race might no longer be deficient in that knowledge." As the mathematical masterpiece of the Middle Ages, it remained a standard for more than two centuries. His algebra is rhetorical, but gains by the employment of geometrical methods. He discusses the fundamental operations with whole numbers and fractions, using the present line for division. Fractions are decomposed into parts with unit numerators as in early Egypt. Through the Arabs Leonardo inherits Egyptian as well as Greek traditions, for example, the type of fraction just mentioned, square and cube root, progressions, the method of false assumption. It would appear that when the Arabs conquered Alexandria some of the old Egyptian culture was preserved. The rule of three, partnership, powers and roots, and the solution of equations are also included.

Leonardo wrote two other books, a *Liber Quadratorum* (1220) and a *Practica Geometrica* (1225). The first is an original work on indeterminate analysis, and gives Leonardo a place between Diophantus and Fermat. The second book brings a large amount of material on geometry and trigonometry, perhaps partly from Greek sources now lost. In the *Liber Abaci* we also find the so-called Fibonacci series:

$$1, 1, 2, 3, 5, 8, 13, 21, \ldots \text{etc.},$$

in which each term is the sum of the two preceding. It is the answer to the question: How many pairs of rabbits can be produced from a single pair if each pair begets a new pair each month and every new pair becomes productive from the second month on, supposing that no pair dies.

In 1225 the emperor, impressed by the accounts of Leonardo's mathematical power, arranged a mathematical tournament of which the challenge questions are preserved:

"To find a number of which the square, when either increased or decreased by 5, would remain a square.

"To find by the methods used in the tenth book of Euclid a line whose length x should satisfy the equation $x^3 + 2x^2 + 10x = 20$.

"Three men, A, B, C, possess a sum of money u, their shares being in the ratio 3 : 2 : 1. A takes away x, keeps half of it, and deposits the remainder with D; B takes away y, keeps $\frac{2}{3}$ of it, and deposits the remainder with D; C takes away all that is left, namely z, keeps $\frac{5}{6}$ of it, and deposits the remainder with D. This deposit is found to belong to A, B, and C in equal proportions. Find u, x, y and z."

Leonardo gave a correct solution of the first and third, also a root of the cubic equation correct to nine decimals. — BALL, p. 169.

Jordanus Nemorarius (*d.* 1237) wrote important Latin works on arithmetic, geometry, and astronomy. His *De Trangulis* — the most important of these — consists of four books dealing not only with triangles, but with polygons and circles. He generally uses Arabic numerals, and denotes quantities known or unknown by letters. He solves the problem of finding two numbers having a given sum and product, by a method equivalent to our elementary algebra. This is practically the first European syncopated algebra, but seems to have become too little known to have far-reaching results in a time not yet ripe for this invention. A book *On Weights* contains elements of mechanics.

Two Oxford scholars, John of Holywood (*Sacrobosco*) and Roger Bacon, have next to be mentioned. Sacrobosco lectured at Paris on arithmetic and algebra, and wrote standard books on the former with rules but no proofs, and an astronomy of which more than sixty editions were afterwards printed. His arithmetic was the most popular source of knowledge of the "Arabic" numerals.

ROGER BACON (1214–1294?)

In the history of natural science one thirteenth century name stands out before all others, viz.: that of Roger Bacon, a member of the Franciscan order, born at Ilchester, England, in 1214. He was a pupil of Robert Grosseteste "who had especially devoted himself to mathematics and experimental science," and had studied the works of the Arabian authors. Bacon also travelled abroad and studied at the University of Paris — at that time the center of European learning. Here he took the degree of Doctor of Theology and probably also here became a Franciscan friar. He taught at Oxford, where he had a kind of laboratory for alchemical experiments. Doubtless it was for this that he became reputed as a worker in "magic" and the "black arts," for in 1257 he was forbidden by the head of his order to teach, and was sent to Paris, where he underwent great privations. In 1266 he was invited by Pope Clement IV to prepare and send to him a treatise on the sciences, and within 18 months he had written and sent three important works — his *Opus Majus*, *Opus Minus*, and *Opus Tertium*. In 1268 he returned to Oxford and there composed several more works, but under a later pope his books were condemned and he was thrown into prison where he remained until about a year before his death.

In Paris, Bacon devoted himself particularly to physical science and mathematics. His *Opus Majus* (1267) contains both a summary of ancient and current physical science, and a philosophy of learning based on Greek, Roman, and Arabic authorities. He insisted that natural science must have an experimental basis, and that astronomy and the physical sciences must be founded on mathematics, "the alphabet of all philosophy."

> He was like his contemporaries a believer in astrology. As to magic, he points out that the magnet, for example, must appear magical to the ignorant. What is the explanation of the wonderful power of words? All the miracles since the world began, almost, have been wrought by words. . . . The whole question of the "magical sciences" should be investigated by competent men specially licensed by the Pope. . . . — A. G. LITTLE, *Roger Bacon*, p. 25.

Bacon enunciated the essential principles of calendar reform, recognizing that the current plan of 365¼ days led to an error of one day in 130 years. He made an acute criticism of the arbitrary assumptions and the artificial complexity of the Ptolemaic astronomy; he discussed reflection and refraction, spherical aberration, rainbows, magnifying glasses, and shooting stars; he attributed the tides to the action of the lunar rays. In a chapter on geography, assuming the rotundity of the earth, he concludes "that the ocean between the east coast of Asia and Europe is not very broad. This . . . was quoted by Columbus in 1498. . . . It is pleasant to think that the persecuted English monk, then two hundred years in his grave, was able to lend a powerful hand in widening the horizon of mankind."

Most of this remarkable work — not printed for nearly 500 years — was so far in advance of the age that it not only failed of appreciation, but exposed the author to accusations of magic, and even to imprisonment. In spite of his many attainments he believed in astrology, in the doctrine of "signatures" under which the shape and color of leaf and flower corresponded with the special purpose for which each was designed by the Creator, and in the "philosopher's stone," and "knew" that the circle had been squared. He prophesied ships propelled swiftly by mechanical means and carriages without horses. He repudiated belief in witchcraft,[1] and paid the penalty for his courage by many years in prison.

DANTE ALIGHIERI (1265–1321)

Another notable scholar of the thirteenth century is Dante, the greatest poetical genius of the Middle Ages, who requires our notice not only because of his influence in awakening and stimulating the minds of his own and later times, but also as the author of a treatise *On Water and the Earth* (*De Aqua et Terra*) which, according to himself, was delivered at Mantua in 1320 as a contribution to the question, then much discussed, "whether on any part of the earth's surface water is higher than the earth."

[1] Not merely astrology and alchemy but even magic and necromancy were at this time the subjects of university lecture courses.

In his *Divine Comedy*, the greatest of medieval poems, he embodies the geocentric astronomy and astrology of his time. "In the whole spiritual or physical world there is hardly an

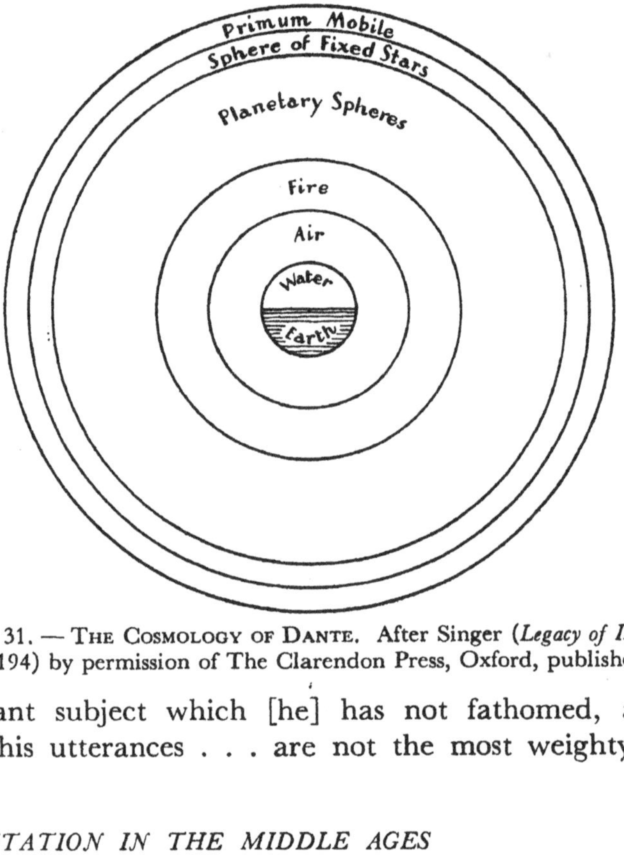

FIG. 31. — THE COSMOLOGY OF DANTE. After Singer (*Legacy of Israel*, p. 194) by permission of The Clarendon Press, Oxford, publishers.

important subject which [he] has not fathomed, and on which his utterances . . . are not the most weighty of his time."

COMPUTATION IN THE MIDDLE AGES

During the fourteenth century there was continued activity in the gradual dissemination of Arabic learning, largely through the medium of almanacs and calendars, so that Arabic computations, Euclidean geometry, and Ptolemaic astronomy became widely known. Some of these calendars emphasized the religious side and gave dates of church festivals for a series of years, others specialized in astrology, medicine, or astronomy. For ecclesiastical purposes Roman numerals were preferred, but at least an explanation of the new Arabic characters and their use was generally given.

The arithmetic of Boethius, based on Roman numerals, retained its vogue in northern Europe until about 1600. Arabic arithmetic, or algorism, based on the *Liber Abaci* of Fibonacci, employing the decimal scale and including the elements of algebra, came into general use among the Italian merchants in the thirteenth and fourteenth centuries, though not without meeting serious opposition. Outside of Italy, however, accounts were kept in Roman numerals much longer, here and there till the sixteenth century, and in the more conservative religious and educational institutions, for a hundred years longer. In such cases the actual computation was carried out on the abacus, and the result expressed in Roman numerals. Arabic numerals allowed computation without the abacus. The Florentines at the same time considerably simplified the classification of arithmetical operations, in accordance with our modern list: — numeration, addition, subtraction, multiplication, division, involution, and evolution.

Addition and subtraction were begun at the left. The multiplication table, at first little known, ended with 5×5. For further products up to 10×10, a system of finger reckoning was widely used, the rule running:

> Let the number five be represented by the open hand; the number six by the hand with one finger closed; the number seven by the hand with two fingers closed; the number eight by the hand with three fingers closed; and the number nine by the hand with four fingers closed. To multiply one number by another let the multiplier be represented by one hand, and the number multiplied by the other, according to the above convention. Then the required answer is the product of the number of fingers (counting the thumb as a finger) open in the one hand by the number of fingers open in the other together with ten times the total number of fingers closed.[1] — BALL, p. 189.

Long division naturally required the skill of a mathematical expert. The signs +, −, ÷, and the use of decimal fractions belong to a somewhat later period.

[1] In modern notation: If x is the number of fingers closed in one hand, y the number closed in the other, then

$$(5 + x)(5 + y) = (5 - x)(5 - y) + 10(x + y).$$

MATHEMATICS IN THE MEDIEVAL UNIVERSITIES

The state of mathematics in the universities toward the close of the fourteenth century may be inferred from the requirements for the master's degree at Prague (1384) and Vienna (1389). The former included Sacrobosco's *Sphere*, Euclid Books I–VI, optics, hydrostatics, theory of the lever, and astronomy. Lectures were given on arithmetic, finger-reckoning, almanacs, and Ptolemy's *Almagest*. At Vienna, Euclid I–V, perspective, proportional parts, mensuration, and a recent version of Ptolemy were required. In Leipsic, however, in 1437 and 1438 mathematical (?) lectures were confined to astrology, and conditions seem to have been much the same at the Italian universities, while Oxford and Paris probably occupied an intermediate level.

THE RENAISSANCE

With the fourteenth century we enter upon one of the most interesting and noteworthy periods of human history; viz., the Renaissance. Neither the term nor the period is, however, sharply defined, the former signifying an awakening or "new birth," the latter covering loosely the fourteenth to the sixteenth centuries. In some respects, indeed, the title might better be applied to the twelfth and thirteenth than to the later centuries. It is only necessary to recapitulate briefly some of the phenomena touched upon in the present chapter, to realize that the civilization of the later Middle Ages has been undergoing great changes. The Crusades marked the first and perhaps most important of these, while the rediscovery or recovery of the classics from Arabian and other sources in the eleventh to the thirteenth centuries, followed by the revival of (classical) learning in the fourteenth must have been powerful ferments of the medieval scholastic mind, expanded and uplifted as it was by the poetical philosophy of Dante and challenged by the naturalism and rationalism of Roger Bacon.

The great events of the fourteenth century were in part new, and in part the natural extension and development of those of the thirteenth. A strange and appalling natural phenomenon was the famous epidemic known as the "black death," a quickly fatal disease which carried off from one quarter to one half of all the inhabitants of Europe, producing social changes which are still felt.

HUMANISM

The development of better education begun in the thirteenth century was marked in the fourteenth by the founding of many now famous universities and colleges and by that revival of ancient learning which is associated especially with the name of Petrarch (1304–74). This revival, while at first chiefly literary and philosophical, brought with it translations into Latin — the current language of scholars at that time — of Aristotle and other classical writers of scientific importance, and thus aided in bringing on a new birth or renaissance in science as well as in other branches.

Precisely as there is one great name in thirteenth century literature, viz., that of Dante, which must be regarded with attention by all students of history, so in the fourteenth the name and work of Petrarch require careful consideration. Francesco Petrarca, commonly called Petrarch, a gifted Italian poet and scholar, greatly promoted the revival of ancient learning by insisting on the importance and merits of the Greek and Roman authors.

> Petrarch was less eminent as an Italian poet than as the founder of Humanism, the inaugurator of the Renaissance in Italy. . . . Standing within the kingdom of the Middle Ages, he surveyed the kingdom of the modern spirit and, by his own inexhaustible industry in the field of scholarship and study, he determined what we call the revival of learning. By bringing the men of his own generation into sympathetic contact with antiquity, he gave a decisive impulse to that European movement which restored freedom, self-consciousness and the faculty of progress to the human intellect. . . . He was the first man to collect libraries, to accumulate coins, to advocate the preservation of antique monuments, and to collate manuscripts. Though he knew no Greek, he was the first to appreciate its vast importance; and through his influence, Boccaccio laid the

earliest foundations of its study. . . . For him the authors of the Greek and Latin world were living men, — more real in fact than those with whom he corresponded; and the rhetorical epistles he addressed to Cicero, Seneca, and Varro prove that he dwelt with them on terms of sympathetic intimacy. — SYMONDS.

It is an illuminating reflection and one not without bearing on our present state that both the medieval heritage of Greek science and the Renaissance heritage of Greek literature proved barren by themselves. It was not until the one fertilized the other that there was real and vital growth. Modern thought, modern science, modern art, and modern letters are the offspring of the union. — SINGER, *From Magic to Science*, p. 99.

THE AWAKENING OF MEDICINE

The earliest reaction against the prevailing ignorance and superstition in medical practice, appeared in Southern Italy, *Magna Graecia*, where a Greek dialect was spoken until the thirteenth century or later. In the ninth century the French queen was attended by a physician from Salerno. At that ancient health-resort, not far south of Naples, a medical school had developed that was unique in not being under ecclesiastical control, and which under Norman rule (A.D. 1077–1268) became the chief medical center of Europe. From there medical knowledge spread, at first mainly through translations of Arabic works into Latin, poorly made by Constantine the African (second half eleventh century). Here appeared the first Latin directions for dissection (of a pig), *Anatomia porci*, for the use of students (*c.* 1100–10?). While at Toledo Gerard of Cremona (*c.* 1114–87) was turning into Latin the works of Rhazes, Avicenna, and other Arabic authors, at Salerno in the twelfth century great literary activity produced a number of compilations that were very popular in Western Europe, and Roger of Salerno (*c.* 1170) wrote the first Latin surgical textbook based, in part at least, on personal experience.

Meanwhile at Bologna there was, as early as 1156, a medical faculty, who taught by reading Latin translations of Arabic texts. A professor of surgery, William of Saliceto, produced an able treatise in 1275 with indications of human dissection — the first in Christendom, probably for legal evidence. Taddeo Alderotti (Thaddeus of Florence, 1223–1303) encouraged direct translation from the Greek, and his pupil, Mondino de 'Luzzi (*c.* 1275–1326), who *personally* dissected the human body in public, wrote in 1316 the first modern work on anatomy. The surgical and anatomical traditions of Bologna, largely borrowed from Avicenna, were carried in 1301 to Montpellier by Henri de Mondeville. The *Great Surgery* of Guy de Chauliac (1300–68), who studied and practiced at Montpellier, was the standard treatise on the subject during the later Middle Ages. Internal medicine lagged behind surgery. Based on the physiology of Galen, it made no progress during this period.

At Damascus, Ibn al-Nafis (1208–88/9) made an assertion of first importance when he denied the view of Galen and Avicenna that blood can pass from the right to the left side of the heart without going through the lungs. But this was not based on observation, and it remained unheeded in the East as well as in the West.

The idea of a hospital to care for the sick and needy was not new in the Middle Ages, but the first hospital in England was founded at Canterbury in 1084, and during the period of the Crusades a great number of hospitals were established throughout Europe. The eastward spread of leprosy and bubonic plague, accompanied by severe epidemics of the latter, such as the Black Death of 1347–48, induced effective, often cruel, measures to avoid contagion. Our word "quarantine" is from *Quarantina*, the forty-day period of probation required by the Republic of Ragusa in an effort to ward off the plague.

THE REVIVAL OF NATURAL HISTORY

The low state to which natural history had fallen since Roman times is shown by the popular medieval bestiaries, especially that extraordinary work called the *Physiologus*, a kind of scriptural allegory of animal life, originally Alexandrian, but surviving in mutilated forms and widely used in the early Middle Ages. It was translated from the Greek into

Latin in the fifth century, into Syriac, Arabic, and many European languages. The childish and grotesque character of this curious compendium shows how ill-adapted were the centuries of crusading to the calm pursuits of science. Interest was mainly engaged in the study of the strange and marvelous rather than the ordinary phenomena of nature.

The thirteenth century saw a great revival of natural history, chiefly in the form of huge encyclopedic compilations, rarely containing original observations. The most important authors were Frederick II and Albertus Magnus.

Frederick II (1194–1250) — King of Sicily, Holy Roman Emperor, patron of arts and sciences, sportsman, zoologist — in spite of frequent wars and incessant disputes with the pope, found time to write *On the Art of Hunting with Birds*, based in part on his own observations and experiments, including new facts in the anatomy and habits of birds. In the imperial court of Foggia and elsewhere in the Italian mainland portion of the Sicilian kingdom, men of science, like Fibonacci and Michael Scot, mingled with Provençal troubadours and Moslem singers, and it was here that Italian poetry began to blossom.

ALBERTUS MAGNUS

Albrecht von Bollstädt, the Great (1193–1280) — head of the German province of the Dominican Order, bishop, and "the greatest naturalist of the Latin Middle Ages" — wrote (*c.* 1245–60) extensive commentaries on Latin versions of nearly all the works attributed to Aristotle. In this, following the example of Maimonides, his greatest service was in reconciling Aristotelian natural philosophy with the doctrines of the Church, thus permitting Greek science to be read in the universities. Also, not only did he insist on the importance of observation and investigation, but he exemplified these principles in his greatest work, the *Opus Naturam*, the best part of which is *On Plants*, with its accurate original descriptions. The larger part, *On Animals*, is a paraphrase of Michael Scot's version of the three books of Aristotle with long commentaries, including information on marine Baltic forms and remarkable observations on the embryology of fishes. A good observer but poor theorist, Albert repeated many of Aristotle's errors and added some of his own. In astronomy he interpreted the Milky Way as an accumulation of small stars and ridiculed the current objection to antipodes. He was a capable naturalist but did no experiments in alchemy.

CHEMICAL ARTS AND ALCHEMY

Some idea of the state of chemical knowledge and arts in the later Middle Ages may be gained from medical writings and books of recipes of the time. Salernus, a physician of Salerno (*c.* 1130–60), was one of the first to mention the distillation of alcohol. Directions for preparing oil-paints, and recipes for making glues, colors, dyes, and ink, with a description of the manufacture of glass are given by Theophilus the Priest, probably a German, *c.* 1100. A formula for gunpowder appears in the work on explosives of Marc the Greek about 1300.

What astrology was to astronomy, alchemy was to chemistry; viz., the crude and often magic-working predecessor. Alchemy was introduced into the Latin world by translations from the Arabic, the earliest dated one being by Robert of Chester, A.D. 1144. A little later Gerard of Cremona translated the important work of Rhazes on alums and salts (p. 193). By 1200 the alchemical part of Avicenna's work was available in Latin, and its critical spirit was of much benefit to later Latin writers. The schoolmen were undecided between the mystical and the practical aspects of alchemy, and generally preferred to discuss rather than to perform the experiments recommended by Rhazes. Roger Bacon was one of the few experimenters. In describing the relation between sulphur and metals, Albertus Magnus spoke of an attraction which suggests the modern idea of chemical affinity, and defined flame as *ignited smoke*. While admitting the possibility of transmutation, he remarks that alchemical gold is not real gold. The best alchemical treatises of the time are from

Arabic sources and ascribed to "Geber," especially the *Summa perfectionis*, *c.* 1300, an elaborate theoretical and practical work. "We may . . . consider it as the first monument of Latin chemistry."

THE MARINER'S COMPASS

The loadstone certainly was known to antiquity as a stone having the power of attracting and carrying a load of iron. Thales described the property of certain iron ores to attract small pieces of iron, and Plato and Lucretius refer to this property. The directive power of the magnet seems to have been recognized first by the Chinese. It is mentioned in a Chinese dictionary of about A.D. 120, and it is reported to have been used by travellers on the wide plains of China since much earlier times. The first certain report of such use dates back to A.D. 215. According to Humboldt, Chinese ships navigated the Indian Ocean with the magnetic needle in the third century of our era. The Arabs also are credited with its invention and use, as stated in the preceding chapter. The first reference to it in Christian Europe is ascribed to Alexander Neckam (1157–1217) in his *De Naturis Rerum*, a work well known at the end of the twelfth century. It is mentioned also in the poem of Guiot of Provins (*c.* 1205) and in later works of the thirteenth century. One of these runs:

> No master mariner dares to use it lest he should be suspected of being a magician; nor would the sailors venture to go to sea under the command of a man using an instrument which so much appeared to be under the influence of the powers below.

There is a legend that the compass was first made commonly useful by Flavio Gioja of Amalfi, a small port on the gulf of Salerno, in 1302. However, it is known that the first technical description of a compass was written in 1269 by Peter the Stranger, a French crusader described by Roger Bacon as the greatest experimental scientist of his time. His *Epistola . . . de magnete* added considerably to current knowledge of the properties of the magnet and "was a rare and splendid exemplar of the experimental method." "Its appearance in 1269 is one of the main landmarks in the history of science" (*Sarton*). It gave details of two kinds of compasses, the floated and the pivoted, designed for astronomical use.

Columbus made the earliest definite record of the deviation of the compass while crossing the Atlantic in 1492. But it is probable that compasses in pocket sundials marked for deviation were commonly used in Germany before his time.

OPTICS

Although the burning glass was known in classic Greece and various treatises on the properties of lenses had appeared since then, spectacles seem to have been invented and made in Venice or northern Italy about 1289. A little later they were made in the Netherlands, and there is a record in China, (not earlier than the end of the century). There is no record of Moslem origin.

GEOGRAPHY

During the later Middle Ages pilgrims, Christian and Moslem, produced a massive literature on their travels, but little of scientific value. Scandinavian sailors explored the northern seas. The Genoese Lanzarote Malocello discovered the northern Canary Islands, and others voyaged along the Atlantic coast of Africa. The greatest achievements were by land into various parts of Asia. The most remarkable journey was that of Marco Polo (1254–1324), son of an experienced traveller and merchant of Venice. In company with his father and uncle, Marco left home in 1271 and proceeded overland to China. The return journey, begun in 1292, was mainly by sea and ended in Venice at the close of 1295. Three years later in a naval battle Marco was taken by the Genoese, and while in captivity he dictated his narrative to a fellow prisoner, who wrote it out in French. The story was so extraordinary that at first few believed it. Marco was the first to travel entirely across Asia to the Pacific Ocean, the first to give the West an adequate description of China and some idea of Japan, and he added to the knowledge of other eastern countries. His descriptions included natural features, minerals, plants, animals, and human activities.

The science of cartography was created about 1300 in the portolani, which were sailing-directions with charts based on direct observations of Mediterranean harbors and coasts, the first true maps in the modern sense.

CLOCKS

Clocks with wheels were known in antiquity; but they were water-clocks. The clocks with wheels that seem to have come into occasional use from the twelfth to the fourteenth centuries were no longer all water-clocks. Many of the clocks were elaborate works of art. One of the first is said to have been sent by the Sultan of Egypt in 1232 to the Emperor Frederick II.

> It resembled a celestial globe, in which the sun, moon and planets moved, being impelled by weights and wheels so that they pointed out the hour, day and night, with certainty.

Another is mentioned as in Canterbury cathedral, while still another at St. Albans, made by R. Wallingford who was abbot there in 1326, is said to have been so notable "that all Europe could not produce such another." It remained for Huygens in the seventeenth century to apply pendulums to clocks.

PAPER

The art of paper making had been introduced as early as 1189 from Moorish Spain (p. 196) into France. It entered Italy probably through Sicily, and the manufacture began in central Italy in 1276. From there the art spread to northern Italy and over the Alps. In Germany the making of paper is said to have started at Mainz about 1320, and in 1390 a paper mill was set up at Nuremberg.

Without paper there could have been no printing, and it is perhaps significant that Mainz was the source of the first dated document (A.D. 1454) that was printed from cast type.

THE INVENTION OF PRINTING

Before the middle of the fifteenth century, printing was done chiefly from fixed blocks of wood, metal, or stone, as is the case today in the printing of engravings, wood cuts, and the like. The introduction of movable types, capable of an almost infinite variety of combination was therefore a forward step of fundamental importance, since the same letter or picture could be used over and over in new combinations where previously it could be used but once. It is generally held that the invention of the art of printing from movable types was the work of Johann Gutenberg (1397–1468) of Mainz on the Rhine, aided by Johann Faust or Fust, a rich citizen of Mainz. The oldest documentary evidence of Gutenberg's activity is of 1444. The claim of Gutenberg has always been much disputed.

The enormous significance of this invention for the dissemination and preservation of knowledge cannot be overestimated. It was no longer conceivable that the scholarly production of a civilization should be so nearly obliterated as was that of Greece. No longer were libraries a monopoly of the rich. The alternative danger has become that of a surfeit and superabundance of what is unfit to survive. On the other hand, such manuscripts as had come down from ancient times were henceforth protected from loss or abuse by the ease of reproduction.

One cannot say that for the first twenty-five years of its existence printing did much, if anything, to free the profession from the shackles of mediaevalism. Not until the revival of Greek studies did men get inspiration from the true masters of science, and for at least two generations they were too busy looking for the fountains to explore for themselves the virtues of their waters. — SIR WILLIAM OSLER. (Cushing, *Life of Osler*, II, 395.)

REFERENCES FOR READING

COULTON, G. G., *The Medieval Scene*, 1930.
HASKINS, C. H., *Rise of the Universities*, 1923.
——, *Studies in History of Medieval Science*, 1924.
LITTLE, A. G., *Roger Bacon*, 1929.
MUIR, M. M. PATTISON, *Story of Alchemy and the Beginnings of Chemistry*, 1902.
RASHDALL, HASTINGS, *Universities of Europe in the Middle Ages*, 1895.
THORNDIKE, LYNN, *Science and Thought in Fifteenth Century*, 1929.

Books from previous chapters: Ball, Ch. VI, VIII, X; Cajori; Crew, Ch. IV; Dampier, pp. 84–105; Locy, Ch. VI; Singer, *Magic to Science*, pp. 59–110; *Medicine*, Ch. III; *Living Things*, pp. 67–77.

Chapter X

A New Astronomy and the Beginnings of Modern Natural Science

The breeze from the shores of Hellas cleared the heavy scholastic atmosphere. Scholasticism was succeeded by Humanism, by the acceptance of this world as a fair and goodly place given to man to enjoy and to make the best of. . . . Astronomy profited more than any other science by this revival of learning. . . . — DREYER, *Plan. Sys.*, p. 282.

The extension of the geographical field of view over the whole earth and the release of thought and feeling from the restrictions of the Middle Ages mark a division of equal importance with the fall of the ancient world a thousand years earlier. — DANNEMANN, *Naturwiss.*, III, p. 2.

THE AGE OF DISCOVERY

With the end of the fifteenth century and the beginning of the sixteenth opens one of the most marvelous chapters in all

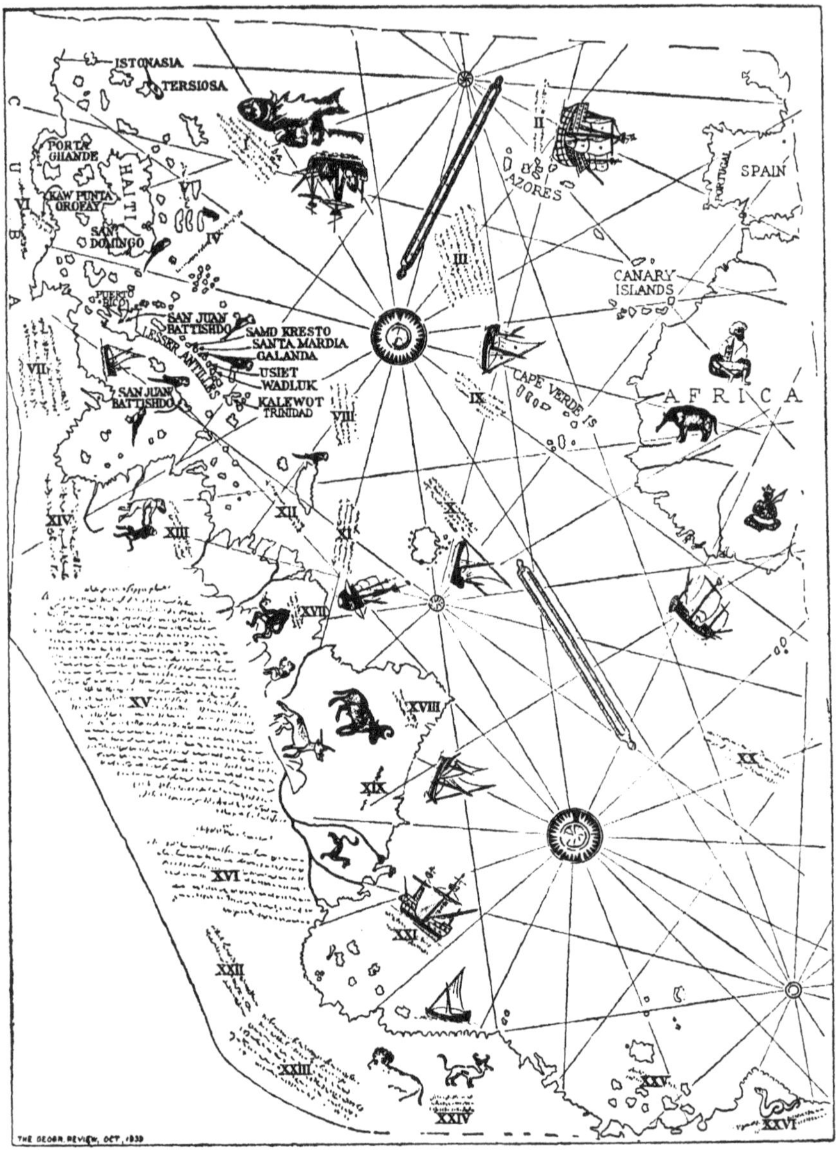

FIG. 32. — MAP OF THE ATLANTIC OCEAN IN THE TIME OF COLUMBUS. Simplified copy of Piri Re'is' map by Paul Kahle, *Geographical Review*, **23**, 637, 1933. From the drawing lent by The American Geographical Society.

NOTE: On the opposite page is a view of Tycho Brahe's observatory of Uraniborg, by Elizabeth Tyler Wolcott from the figure in a plate of Tycho Brahe, *Historia Caelestis*, 1666.

history; viz., the *Discovery of the New World.* At about the same time further explorations of the Old World attained equal extent and interest. We have referred above (p. 200) to the *Discovery of the East* by the Crusaders, and now with Columbus, Magellan, and their successors, we have an even more pregnant *Discovery of the West.* Meanwhile, Diaz and da Gama pushed the explorations of Prince Henry of Portugal, "the Navigator," to the south, and in rounding the Cape of Good Hope completed the *Discovery of the South.* To the north, explorers had already advanced to regions of perpetual snow and ice, so that in all directions there were new problems of intense interest profoundly moving the imagination of mankind.

THE REFORMATION

Another potent element was added to the already complex fermentation of medieval ideas when in 1517 a widespread insurrection began in the Christian Church, the most conservative and most powerful institution of the Middle Ages. This revolution — for such it proved to be — with which the name of Luther will always be chiefly associated, soon aroused a wave of determined opposition, naturally strongly conservative, known today as the "counter-reformation," one instrument of which was the Inquisition that had been established by Pope Innocent III early in the thirteenth century and by the sixteenth had reached large proportions. Mention should also be made of the great humanist and reformer, Erasmus (1467–1536) whose work, though not directly bearing on problems of natural science, was of immense importance in the general development of the intellectual life of his age.

The increased importance of the art of navigation reacted powerfully on the underlying sciences of mathematics and astronomy, particularly through the demand for improved astronomical instruments and tables. The Church, even, had a strong, if restricted, interest in astronomy on account of the necessity of more accurate data for its calendar.

PIONEERS OF THE NEW ASTRONOMY

Nicolaus of Cusa (1401–64), later Bishop of Brixen, wrote on *Learned Ignorance* (*De docta ignoranta*), arguing that the universe, being infinite in extent, could have no center, and that the earth has diurnal rotation. "It is now clear that the earth really moves, if we do not at once observe it, since we perceive motion only through comparison with something immovable." In mathematics he follows Euclid and Archimedes, coöperating in a translation of the latter from Greek into Latin, and dealing with the squaring of the circle. He himself tried to construct with compass and ruler a circle of area equal to a given square, and found several inaccurate values of π. These attempts at squaring the circle were attacked later by Regiomontanus.

He made a map of the known world, using central projection. He is said to have determined areas of irregular boundary by the then novel method of cutting them out from a card map and weighing, and is one of the first to emphasize the importance of measurement in all investigations. He showed independence of thinking, but his astronomical theories were too little developed — and too speculative — to constitute real progress in an age not yet quite ripe for their reception.

George Purbach, or Peurbach (1423–61), who had as a youth met Nicholas of Cusa in Rome, became professor of astronomy and mathematics in Vienna and has been called "the founder of observational and mathematical astronomy in the West." Recognizing the imperfections of the Alfonsine tables (p. 205) he published a new edition of the *Almagest* with tables of natural sines — instead of chords — computed for every ten minutes. He depended mainly, however, on imperfect Arabic translations.

His more eminent pupil and successor, Johann Müller, of Königsberg (in Bavaria), better known as Regiomontanus (1436–76), was the most distinguished scientific man of his time. After the fall of Constantinople he was among the first to avail himself of the opportunities for more direct acquaintance with the works of Archimedes, Apollonius, and Diophantus. For the defective version of the *Almagest* which had come through Arabic channels he substituted the Greek original; while his tables, published in 1475, were important both for astronomy and for the voyages of discovery of Vasco da Gama, Vespucci, and Columbus. These tables covered the period 1473 to 1560, giving sines for each minute of arc, longitudes for sun and moon, latitude for the moon, and a list of predicted eclipses from 1475 to 1530. Another work, on astrology, includes a table of natural tangents for each degree. A wealthy merchant of Nuremberg erected an elaborately equipped observatory for Regiomontanus, and the printing-press recently established there became the most important in Germany. Accepting, however, a summons to Rome to reform the calendar, he died in Rome at the age of 40.

His *De Triangulis* (1464) is the earliest modern trigonometry. Four of its five books are devoted to plane trigonometry, the other to spherical. He determines triangles from three given conditions, using sines and cosines, and employs quadratic equations successfully in some of his solutions. One of his problems is "to determine a triangle when the difference of two sides, the perpendicular on the base, and the difference between the segments into which the base is divided are given: i.e., $a - b$, $a \sin B$, $a \cos B - b \cos A$ are known; to find, a, b, c, A, B, C." Another is to construct from four given lines a quadrilateral which can be inscribed in a circle. His notation, however, is not modern. Our notation is due to Euler.

CONDITIONS NECESSARY FOR PROGRESS

The genius of Hipparchus and Ptolemy had brought Greek astronomy to its culmination. Higher it could not rise until three conditions should be fulfilled, even though here and there the heliocentric hypothesis might be adopted through an unsupported inspiration of individuals. First, there must be better astronomical instruments and more accurate observations, extended over long periods. Second, there must be improved methods of mathematical computation for the reduction and interpretation of these observations. Third, there must be substantial progress towards clear thinking as to the fundamental facts and laws of motion. These conditions were met one after another during the sixteenth and seventeenth centuries by an extraordinary series of men of genius, among whom the chief were Copernicus, Tycho Brahe, Kepler, Galileo, and Newton. Their work constitutes a great part of the history of science during these two centuries — and one of the most wonderful chapters of all time.

Of these five, Copernicus and Kepler were predominantly interested on the mathematical and theoretical side, Tycho Brahe was a great observer, Galileo combined experimental and observational skill with a new appreciation of physical laws, while Newton, building on the foundation laid by all the others, made a magnificent synthesis of their results into a rational and consistent mathematical theory of the solar system. These five represent Poland, South Germany, Denmark, Italy, and England. Scientific progress is no longer localized or dependent on princely patronage. It has now become international.

NICOLAUS COPERNICUS (1473–1543)

Nicolaus Copernicus was born in the remote little city of Thorn on the Vistula, and having relatives in the Church, prepared himself for an ecclesiastical career. This led him, after medical study at Cracow, first to the university of Vienna, then to the chief Italian universities, Bologna, Padua, Ferrara, and Rome, where he found opportunity to cultivate his mathematical talents and to master what was then known of astronomy. He became canon at Frauenburg in his native land in 1497, and from 1512 until his death thirty years later, was settled there, rendering varied public services, and practising gratuitously, as needful, the medical art he had also learned. At the same time he found it possible to devote much attention to astronomical studies.

In his study of the classical writers he came upon a statement that certain Pythagorean philosophers explained the phenomena of the daily and yearly motions of the heavenly bodies by supposing the earth itself to rotate on its axis and to have also an orbital motion. In the dedication of his great work, *The Revolutions of the Heavenly Bodies*, to the Pope, he says:

> I undertook the task of rereading the books of all the philosophers I could get access to, to see whether anyone ever was of the opinion that the motions of the celestial bodies were other than those postulated by the men who taught mathematics in the schools. And I found first, indeed, in Cicero, that Hicetas perceived that the Earth moved; and afterward in Plutarch I found that some others were of this opinion, . . .
>
> Taking this as a starting-point, I began to consider the mobility of the Earth; and although the idea seemed absurd, yet because I knew that the liberty had been granted to others before me to postulate all sorts of little circles for explaining the phenomena of the stars, I thought I also might easily be permitted to try whether by postulating some motion of the Earth, more reliable conclusions could be reached regarding the revolution of the heavenly bodies, than those of my predecessors.
>
> And so, after postulating movements, which, farther on in the book, I ascribe to the Earth, I have found by many and long observations that if the movements of the other planets are assumed for the circular motion of the Earth and are substituted for the revolution of each star, not only do their phenomena follow logically therefrom, but the relative positions and magnitudes both of the stars and all their orbits, and of the heavens themselves, become so closely related that in none of its parts can anything be changed without causing confusion in the other parts and in the whole universe. — *Harvard Classics*, **39**, 58–59.

Copernicus was not a great observational astronomer. His instruments were poor, his eyesight not keen, his location unfavorable for clear skies. His recorded observations are few, chiefly of eclipses or oppositions of planets, and of no high degree of accuracy. His interest and genius lay rather in the direction of profound analysis and careful mathematical revision of the current geocentric theory, practically unchanged since its formulation by Ptolemy thirteen centuries earlier. Unfortunately the conditions of the time were adverse to the publication of so radical an innovation as a heliocentric theory of the solar system; nor was Copernicus ever greatly interested in any publication of his results, being both indifferent to reputation and averse to controversy. In the same dedication he says:

> The contempt which I had to fear because of the novelty and apparent absurdity of my view, nearly induced me to abandon utterly the work I had begun. — *Ibid.*, 56.

Moreover, he realized the futility of publishing his revolutionary theories until he should have buttressed them with a planetary system so completely worked out that its superiority to the long-intrenched Ptolemaic system should be unquestionable — a herculean, if congenial labor. Nevertheless, he gradually formulated his astronomical system in manuscript, and about 1529 issued a *Commentariolus* giving an outline of his theory, which thus became gradually but vaguely known to scholars. Ten years later George Joachim — Rheticus — a young professor of mathematics from the Lutheran university of Wittenberg, visited Copernicus, eager to learn more of the new doctrine. The Lutheran church was not more hospitable than the Roman Catholic to scientific novelty and Luther himself called Copernicus a fool.

De Revolutionibus. — In 1540 appeared the *Prima Narratio* by Rheticus containing a considerable admixture of astrology, and in 1543 the immortal *De Revolutionibus Orbium Coelestium*, a copy reaching Copernicus, it is said, on his deathbed. He begins with certain postulates: first, that the universe is spherical; second, that the earth is spherical; third, that the motions of the heavenly bodies are uniform circular motions or compounded of such motions. The slender basis for the first and third of these may be inferred from his statement in regard to certain hypothetical causes of want of uniformity:

> Both of which things the intellect shrinks from with horror, it being unworthy to hold such a view about bodies which are constituted in the most perfect order. — Berry, p. 101.

He makes the relative character of the motions involved of fundamental importance. In his own words:

> For all change in position which is seen is due to a motion either of the observer or of the thing looked at, or to changes in the position of both, provided that these are different. For when things are moved equally relatively to the same things, no motion is perceived, as between the object seen and the observer. — BERRY, p. 101.

Thus the daily revolution of sun, moon, and stars about a stationary earth would have the same apparent effect as rotation of the earth in the opposite direction about its own axis, and the apparent yearly motion of the sun about the earth is equivalent to an orbital motion of the latter.

> "It is," he says, "more probable that the earth turns about its axis than that the planets at their various distances, the comets sweeping through space, and the endless multitude of the fixed stars, describe the same regular daily motion about the earth."

The apparent irregularities in the motions of the five known planets had been a perpetual stumbling-block to the ancient astronomers, requiring more and more complicated hypotheses for their explanations as accuracy of observations increased. The heliocentric theory of Copernicus, inaccurate as it was in some respects, afforded a simple explanation of the fact that Mercury and Venus seem merely to oscillate east and west of the sun, while Mars, Jupiter, and Saturn recede indefinitely from it, exhibiting also periodic reversals of the direction of their motion. The new explanation obviously accounted also for the variations in the brightness of these planets.

> It is certain, that Saturn, Jupiter and Mars are always nearest the earth when they rise in the evening, that is when they are in opposition to the sun, as the earth is situated between them and the sun. On the contrary, Mars and Jupiter are farthest from the earth when they set in the evening, the sun lying between them and us. This proves sufficiently that the sun is the centre of their orbits, as of those of Venus and Mercury. Since thus all planets move about one centre it is necessary that the space which remains between the circles of Venus and Mars, contain the earth and its accompanying moon. — F. DANNEMANN, *Die Naturwissenschaften*, I, p. 320.

He is, therefore, not afraid to maintain that the earth with the moon encircling it, traverses a great circle in its annual motion among the planets about the sun. The universe, however, is so vast, that the distances of the planets from the sun are insignificant in comparison with that of the sphere of the stars. He holds all this easier of comprehension, than if the

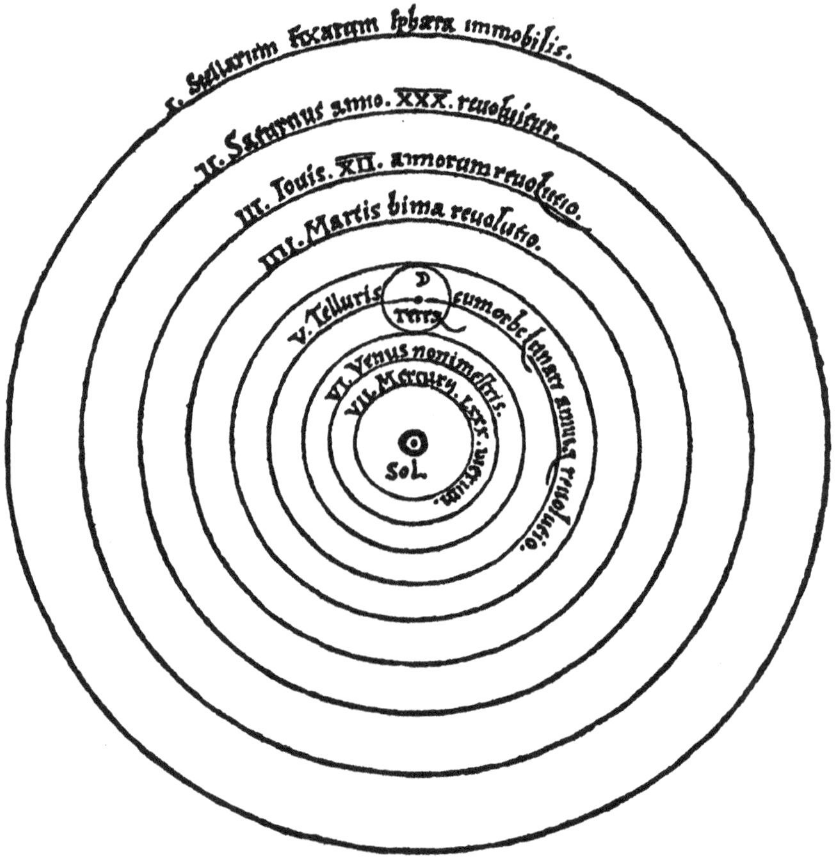

FIG. 33. — THE COPERNICAN SYSTEM. Copernicus, *De Revolutionibus Orbium Coelestium*, Lib. I, Cap. X, 1543.

mind is confused by an almost endless mass of circles, as is necessary for those who put the earth in the center of the universe.

"So in fact the sun seated on the royal throne guides the family of planets encircling it. We find thus in this arrangement a harmonious connection not otherwise realized. For here one can see why the forward and backward motions of Jupiter seem greater than those of Saturn and smaller than those of Mars." — DANNEMANN, *Naturwiss.*, I, p. 408.

His adherence to the Greek assumption of uniform circular motion leaves him still under the necessity of retaining an elaborate system of epicycles, but he rejects Ptolemy's equant.

> . . . He his fabric of the heavens
> Hath left to their disputes, perhaps to move
> His laughter at their quaint opinions wide;
> Hereafter, when they come to model heaven,
> And calculate the stars; how they will wield
> The mighty frame; how build, unbuild, contrive,
> To save appearances; how gird the sphere
> With centric and eccentric scribbled o'er,
> Cycle and epicycle, orb in orb.
> — MILTON, *Paradise Lost*, VIII.

The epicycles of Copernicus numbered however but 34 — sufficing "to explain the whole construction of the world and the whole dance of the planets" — against the 79 to which the Ptolemaic theory had gradually attained. The completeness of mathematical detail with which the whole theory is worked out cannot here be adequately described. He includes so much trigonometry as his astronomical work requires, also a revision of Ptolemy's star catalogue. He computes a very accurate value of the equinoctial precession, and interprets this correctly as due to a slow conical motion of the earth's axis, like that of a top coming to rest.

Copernicus estimates the relative sizes of moon, earth, and sun as 1 : 43 : 6937, and the distance from earth to sun — according to the method of Aristarchus — at about 1200 earth-radii, that is about $\frac{1}{20}$ of the actual.

Revolutionary as were the theories expounded by Copernicus they were not clothed in such popular form as to occasion immediate or general controversy. In dedicating his work to the Pope, Copernicus says:

Mathematics are written for mathematicians, to whom, if my opinion does not deceive me, our labors will seem to contribute something to the ecclesiastical state whose chief office Your Holiness now occupies; for when not so very long ago, under Leo X, in the Lateran Council the question of revising the ecclesiastical calendar was discussed, it then remained unsettled, simply because the length of the years and the months, and the motions of the sun and moon were held to have been not yet sufficiently determined. . . . But what I may have accomplished herein I leave to the judgment of Your Holiness in particular, and to that of all other learned mathematicians; . . . — *Harvard Classics*, **39**, 60.

Moreover criticism was in considerable measure disarmed by a fraudulent preface inserted by Osiander, a Lutheran theologian of Nuremberg, to whom the care of publication had been partially intrusted by Rheticus. In this preface, ostensibly by Copernicus himself, it is stated:

Certainly many scholars as the report of the new hypotheses of this work is already widely circulated, will have taken great offense that the mobility of the earth is taught therein, the sun on the contrary remaining stationary in the middle of the world. . . . But the proper task of the astronomer is to assemble the record of the motions in the sky by keen and careful observation. He must then ascertain the causes of these motions, or, if he is unable to discover the true causes, think out and prepare arbitrary hypotheses, by means of which these motions may be accurately computed by geometrical principles, both for the future and the past. . . . It is not necessary that his hypotheses be true; they need not even be probable. . . . — L. F. Prowe, *Nicolaus Coppernicus*, I, p. 526.

INFLUENCE OF COPERNICUS

The publication of *De Revolutionibus* was naturally a powerful stimulus to astronomical and mathematical studies. Thus Rheticus, whose relations to Copernicus had been so fruitful, calculated a new and extensive set of mathematical tables, while Erasmus Reinhold (1511–1553), who had hailed Copernicus as a new Ptolemy, in 1551 published astronomical tables — the Prutenic or Prussian — on the basis of Copernicus's work, superior to the Alfonsine, previously current.

Before the new doctrine should be completely justified or the reverse, it was necessary that certain mechanical notions should be clarified, and that more accurate observational data should be systematically collected. Copernicus had based his imposing structure on a very slender foundation of actual fact, and had professed his complete satisfaction if his theoretical results should come within ten minutes of the observed positions of the planets — a degree of accuracy which he did not, in fact, attain. On the other hand, he could indeed answer, but not rise entirely above, the traditional notions that the four elements of the ancients must have rectilinear, the heavenly bodies circular, motion; also, that if the earth rotated in twenty-four hours, loose bodies would long since have been thrown off, falling bodies would not fall, and clouds would always be left behind in the west.

Robert Recorde, in his *Castle of Knowledge* (1551), has his "Master" state to a "scholar":

that "Copernicus, a man of greate learning, of muche experience and of wondrefull diligence in obseruation, hathe renewed the opinion of Aristarchus Samius, and affirmeth that the earthe not only moueth circularlye about his own centre, but also may be, yea and is continually out of the precise centre 38 hundredth thousand miles; but bicause the vnderstanding of that controuersy dependeth of profounder knowledge than in this introduction may be vttered conueniently, I will let it passe tyll some other time." — De Morgan. (Dreyer, *Plan. Syst.*, p. 348.)

Copernicus cannot be said to have flooded with light the dark places of nature — in the way that one stupendous mind subsequently did — but still, as we look back through the long vista of the history of science, the dim Titanic figure of the old monk seems to rear itself out of the dull flats around it, pierces with its head the mists that overshadow them, and catches the first gleam of the rising sun, . . .

Like some iron peak, by the Creator
Fired with the red glow of the rushing morn.

—E. F. C. Morton, *Heroes of Science*, p. 62.

TYCHO BRAHE (1546–1601)

The first great need of the new Copernican astronomy — adequate and accurate data — was soon to be supplied by Tycho Brahe, born in 1546 of a noble Danish family. While a student at the University of Copenhagen his interest in astronomy was enlisted by an eclipse, and later, at Leipsic, he persisted in devoting to his new avocation the time and attention he was expected to give to subjects more highly esteemed for a man of birth and fortune.

Here, too, he began his life work of procuring and improving the best instruments for astronomical observations, at the same time testing and correcting their errors. Returning to Denmark from travels in Germany, his predilection for astronomy was powerfully stimulated by the appearance in the constellation Cassiopeia, in November, 1572, of a brilliant new star, which remained visible for 16 months. The great importance attached to this occurrence by Tycho and his contemporaries was due to the evidence it afforded against the truth of the Aristotelian doctrine that the heavens were immutable, since Tycho's careful observations showed that the star must certainly be more distant than the moon, and that it had no share in the planetary motions. He reluctantly published an account of the new star, expressing still his adherence to the current pre-Copernican notions of crystalline spheres for the different heavenly bodies and of atmospheric comets, all combined with astrological reflections and inferences:

This new star is neither in the region of the Element, below the Moon, nor among the orbits of the seven wandering stars, but it is in the eighth sphere, among the other fixed stars, which was what we had to prove. Hence it follows that it is not some peculiar kind of comet or other kind of fiery meteor become visible. — TYCHO BRAHE, *On a New Star*. (Shapley and Howarth, *Source Book*, p. 19.)

This may be supplemented by the following passages from Dreyer's biography:

The star was at first like Venus and Jupiter, and its effects will therefore first be pleasant; but as it then became like Mars, there will next come a period of wars, seditions, captivity, and death of princes and destruction of cities, together with dryness and fiery meteors in the air, pestilence, and venomous snakes. Lastly, the star became like Saturn, and there will therefore, finally, come a time of want, death, imprisonment, and all kinds of sad things.

As the star seen by the wise men foretold the birth of Christ, the new one was generally supposed to announce His last coming and the end of the world.

But the star had a truer mission than that of announcing the arrival of an impossible golden age. It roused to unwearied exertions a great astronomer, it caused him to renew astronomy in all its branches by showing the world how little it knew about the heavens; his work became the foundation on which Kepler and Newton built their glorious edifice, and the star of Cassiopeia started astronomical science on the brilliant career which it has pursued ever since, and swept away the mist that obscured the true system of the world. As Kepler truly said, "If that star did nothing else, at least it announced and produced a great astronomer." — DREYER, *Tycho Brahe*, pp. 50, 68, 196.

Tycho considered that the new star was formed of "celestial matter," not differing from that of which the other stars are composed, except that it was not of such perfection or solid composition as in the stars of permanent duration. It was therefore gradually dissolved and dwindled away. It became visible to us because it was illuminated by the sun, and the matter of which it was formed was taken from the Milky Way, close to the edge of which the star was situated, and in which Tycho believed he could now see a gap or hole which had not been there before.

In 1575 Tycho obtained while travelling a copy of Copernicus's *Commentariolus*, and in the following year received from King Frederick II the island of Hveen, with funds for the maintenance of an observatory upon it. As to the former his opinion is that

"The Ptolemean system was too complicated, and the new one which that great man Copernicus had proposed, following in the footsteps of Aristarchus of Samos, though there was nothing in it contrary to mathematical principles, was in opposition to those of physics, as the heavy and sluggish earth is unfit to move, and the system is even opposed to the authority of Scripture." — DREYER, *Tycho Brahe*, p. 167.

URANIBORG

The observatory of Uraniborg — the castle of the heavens — at Hveen was an extraordinary establishment.

In a large square inclosure oriented according to the points of the compass, were several observatories, a library, labora-

FIG. 34. — TYCHO BRAHE'S QUADRANT.
From Tycho Brahe, *Epistolae astronomicae*, p. 254, 1648.

tory, living-rooms and, later, workshops, a paper-mill and printing-press, and even underground observatories. The whole establishment was administered with lavish extravagance, while Tycho was neither careful of his obligations nor free from arbitrary arrogance in his personal and administrative relations. In spite of these difficulties "a magnificent series of observations, far transcending in accuracy and extent anything that had been accomplished by his predecessors" was carried on for not less than 21 years. At the same time medicine and alchemy were also cultivated.

Concerned as he was to secure the greatest possible accuracy, Tycho constructed instruments of great size; for example, a wooden quadrant for outdoor use with a brass scale of some ten feet radius, permitting readings to fractions of a minute.

A smaller but more serviceable azimuth quadrant of brass gave angles to the nearest minute. He had a copper globe constructed at great expense with the positions of some 1,000 stars carefully marked upon it.

The very precision of his observations tended to confirm his scepticism of the Copernican hypothesis, as it seemed incredible that the earth's supposed orbital motion should cause no change which he could detect in the position and brightness of the stars. He was also misled by supposing that the stars had measurable angular magnitude. He was not successful in making any fundamental improvement in the relatively crude methods of time measurement, depending himself on a wheel-mechanism without the regulating pendulum, and an apparatus of the sand-glass or clepsydra type.

In 1577 Tycho made observations on a brilliant comet, and drew from them important theoretical inferences; namely, that instead of being an atmospheric phenomenon, the comet was at least three times as remote as the moon, and that it was revolving about the sun at a greater distance than Venus — unimpeded by the familiar crystalline spheres. He was even led, in discussing apparent irregularities of its motion, to suggest that its orbit might be oval — foreshadowing one of Kepler's great discoveries.

According to the current view of his time, comets

were formed by the ascending from the earth of human sins and wickedness, formed into a kind of gas, and ignited by the anger of God. This poisonous stuff falls down again on people's heads, and causes all kinds of mischief, such as pestilence, Frenchmen (!), sudden death, bad weather, etc. — DREYER, *Tycho Brahe*, p. 64.

Eleven years later Tycho published a volume on the comet as a part of a comprehensive astronomical treatise which was, however, never completed. About the same time his royal patron died, and the new administration proved less sympathetic with the great astronomer's work and less indulgent with his extravagance and personal eccentricities.

Following a series of disagreements, Tycho withdrew from his observatory in 1597, spent the winter in Hamburg, and after negotiations with different sovereigns, accepted the invitation of the Emperor Rudolph to settle in Prague in 1599. Here he again organized a staff of assistants, including, to the great advantage of himself and of his science, the young Kepler, but his further progress was prematurely terminated by death in 1601, at the age of 55.

Tycho's chief services to the progress of astronomy consisted first, in the superior accuracy of his instruments and observations, heightened by repetition and the systematic correction of errors; second, in the extension of these observations over a long series of years. In both respects he departed from current practice, and anticipated the modern. In point of accuracy his errors of star-places seem rarely to have exceeded 1′ to 2′, and he even determined the length of the year within one second. While he recomputed almost every important astronomical constant, he accepted the traditional distance of the sun.

Kepler gave striking evidence later of his confidence in Tycho's accuracy by writing:

"Since the divine goodness has given to us in Tycho Brahe a most careful observer, from whose observations the error of 8′ is shewn in this calculation, . . . it is right that we should with gratitude recognize and make use of this gift of God. . . . For if I could have treated 8′ of longitude as negligible I should have already corrected sufficiently the hypothesis . . . discovered in chapter xvi. But as they could not be neglected, these 8′ alone have led the way towards the complete reformation of astronomy, and have made the subject-matter of a great part of this work." — BERRY, p. 184.

On the other hand, Tycho was not strong on the theoretical side. He was never willing to accept the Copernican hypothesis of rotation and orbital motion of the earth — maintaining, for example, that if the earth moved, a stone dropped from the top of a tower must fall at a distance from the foot. Again with reference to the apparent displacement of the stars which would be expected to result from orbital motion of the earth, he says:

A yearly motion would relegate the sphere of the fixed stars to such a distance that the path described by the earth must be insignificant in comparison. Dost thou hold it possible that the space between the sun, the alleged centre of the universe, and Saturn amounts to not even $\frac{1}{700}$ of that distance? — DANNEMANN, II, p. 123.

Sensible, however, of the weakness of the Ptolemaic theory, he devised an ingenious compromise in which the planets revolved about the Sun in their respective periods, and the entire heavens about the earth daily — all of which is not mathematically different from the Copernican theory.

Pierre de la Ramée, or Petrus Ramus, a French mathematician and philosopher, impatient with the cumbrous astronomical hypotheses of the ancients, and unsatisfied with Copernicus's proposed simplification, published a work in 1569 expressing the hope

"that some distinguished German philosopher would arise and found a new astronomy on careful observations by means of logic and mathematics, discarding all the notions of the ancients."

Within a few months he discussed the matter at length with Tycho Brahe at Augsburg. Without accepting Ramus's views, the young astronomer did make it his life work to lay the necessary foundation for such a new astronomy. Thirty years later, Möstlin, professor at Tübingen, wrote his former student Kepler — then aged 28 —

that Tycho "had hardly left a shadow of what had hitherto been taken for astronomical science, and that only one thing was certain, which was that mankind knew nothing of astronomical matters."

KEPLER

Born late in 1571 in Würtemberg, of Protestant parents in very straitened circumstances, Johann Kepler's whole life was a struggle against poverty, ill-health, and adverse conditions. In 1594, abandoning with some hesitation theological studies, for which his acceptance of the new Copernican hypothesis disqualified him, he was appointed lecturer on mathematics at Gratz. Students were few, and his duties included the preparation of a yearly almanac, containing, besides what its name implies, a variety of weather predictions and astrological information. "Mother Astronomy," he says, "would surely have to suffer hunger if the daughter Astrology did not earn their bread."

Becoming thus more interested in astronomy, "there were," he says, "three things in particular: viz., the number, the size, and the motion of the heavenly bodies, as to which I searched zealously for reasons why they were as they were and not otherwise." The first result which seemed to him important, though somewhat fantastic from our standpoint, was a crude correspondence between the planetary orbits and the five regular solids, published in 1596 under a title which may be abridged to *Cosmographic Mystery.*

The Earth is the circle, the measurer of all. Round it describe a dodecahedron; the circle including this will be Mars. Round Mars describe a tetrahedron; the circle including this will be Jupiter. Describe a cube round Jupiter; the circle including this will be Saturn. Then inscribe in the Earth an icosahedron; the circle inscribed in it will be Venus. Inscribe an octahedron in Venus; the circle inscribed in it will be Mercury. — BREWSTER, *Martyrs of Science*, p. 195.

Kepler declared that he would not renounce the glory of this discovery "for the whole Electorate of Saxony." The cor-

FIG. 35. — JOHANN KEPLER (1571–1630).
From *Opera omnia*.

respondence of the dimensions of this fantastic geometrical construction with the distances of members of our solar system is in reality far from close, but both Tycho Brahe and Galileo seem to have been favorably impressed by the book.

The difficulties of Kepler's position as a Protestant in Gratz led him, after a preliminary visit, to accept an engagement as Tycho's assistant at Prague.

In 1602 Kepler succeeded Tycho as imperial mathematician. Most fortunately, also, he secured possession of his chief's great collection of observations, though not of the instruments — a matter of less consequence, since Kepler like Copernicus was a mathematician rather than an observer. To the study of these records he devoted the next 25 years. Among all the planetary observations of Tycho Brahe those of Mars presented the irregularities most difficult of explanation, and it was these which, having been originally assigned to Kepler, engrossed his attention for many years, and in the end led to some of his finest discoveries.

The Copernican theory like the Ptolemaic involved the resolution of the motion of each planet into a main circular motion, modified by superimposing other circular motions — epicycles — successively upon it, each circle being the path of the center of the next. Even after disentangling the essential irregularities of Mars's orbit from those merely due to irregular motion of the earth, Kepler could still obtain no satisfactory agreement with Tycho's records, of which, as has been said, he refused to doubt the accuracy. Abandoning the restriction of circular motions, he experimented with other closed curves, of which the ellipse is simplest. Taking the sun at a focus, the problem was at last solved, theory and observation reconciled within due limits of error. At the same time uniform motion was also abandoned, Kepler's scientific imagination leading him to the great discovery that the planet traverses its orbit in such a manner that a line joining it to the sun would describe sectors of equal area in equal times, the planet thus moving fastest when nearest the sun.

Of Kepler's celebrated three laws, the first two are:

The planet describes an ellipse, the sun being in one focus.

The straight line joining the planet to the sun sweeps out equal areas in equal intervals of time.

These results were published in 1609 as part of extended *Commentaries on the Motions of Mars.*

The great problem was solved at last, the problem which had baffled the genius of Eudoxus and had been a stumbling-block to the Alexandrian astronomers, . . . The numerous observations made by Tycho Brahe, with a degree of accuracy never before attained, had in the skilful hand of Kepler revealed the unexpected fact that Mars describes an ellipse, in one of the foci of which the sun is situated, and that the radius vector of the planet sweeps over equal areas in equal times. . . . Kepler had therefore, unlike all his predecessors, not merely put forward a new hypothesis which might do as well as another to enable a computer to construct tables of the planet's motion; he had found the actual orbit in which the planet travels through space. . . .

In the history of astronomy there are only two other works of equal importance, the book *De Revolutionibus* of Copernicus and the *Principia* of Newton. — DREYER, *Plan. Syst.*, 392, 401.

Resuming later the tendency of his *Cosmographic Mystery*, he published in 1619 his *Harmony of the World*, containing his third law:

The squares of the times of revolution of any two planets (including the earth) about the sun are proportional to the cubes of their mean distances from the sun.

In his delight he exclaims, "What sixteen years ago I regarded as a thing to be sought, that for which I joined Tycho Brahe, for which I settled in Prague, for which I have devoted the best part of my life to astronomical contemplations; at length I have brought to light, and have recognized its truth beyond my most sanguine expectations. . . ." — WILLIAM WHEWELL, *Hist. of the Inductive Sciences*, I, p. 294.

Archimedes of old had said "Give me a place to stand on, and I shall move the world." Tycho Brahe had given Kepler the place to stand on, and Kepler did move the world.

It should be borne in mind that Kepler's results depend not on *a priori* theory for their confirmation, but upon actual observations supporting them and interpreted by them. The great further step of showing that the three laws are not independent and empirical, but mathematical consequences of a single mechanical law still awaited the genius of Newton.

Kepler's notions in regard to force and motion are still crude. Thus, for example, having in mind an analogy with magnetism, Kepler says in his *Epitome of the Copernican Astronomy* (1618–21):

"There is therefore a conflict between the carrying power of the sun and the impotence or material sluggishness (inertia) of the planet; each enjoys some measure of victory, for the former moves the planet from its position and the latter frees the planet's body to some extent from the bonds in which it is thus held . . . but only to be captured again by another portion of this rotary virtue." — BERRY, p. 196.

Elsewhere he says:

"We must suppose one of two things; either that the moving spirits, in proportion as they are more removed from the sun, are more feeble; or that there is one moving spirit in the centre of all the orbits, namely, in the sun, which urges each body the more vehemently in proportion as it is nearer; but in more distant spaces languishes in consequence of the remoteness and attenuation of its virtue." — WHEWELL, I, p. 289.

He recognized the necessity of a force exercised by the sun, but believed it inversely proportional to the distance instead of to the square of the distance. His notions of gravity are expressed in his book on Mars:

"Every bodily substance will rest in any place in which it is placed isolated, outside the reach of the power of a body of the same kind. . . . Gravity is the mutual tendency of cognate bodies to join each other (of which kind the magnetic force is), so that the earth draws a stone much more than the stone draws the earth. Supposing that the earth were in the centre of the world, heavy bodies would not seek the centre of the world as such, but the centre of a round, cognate body, the earth; . . . And if the earth and the moon were not kept in their orbits by their animal force, the earth would ascend towards the moon one fifty-fourth part of the distance, while the moon would descend the rest of the way and join the earth, provided that the two bodies are of the same density. If the earth ceased to attract the water all the seas would rise and flow over the moon." — DREYER, p. 399.

Kepler's last important published work was his *Rudolphine Tables* (1627), embodying the accumulated results of Tycho's work and his own, and remaining a standard for a century. It is noteworthy that during Kepler's work on these tables, mathematical computation was peacefully revolutionized by the introduction of logarithms, newly discovered by Napier and Bürgi.

In 1628, after vain attempts to collect arrears of his salary as imperial mathematician, he even joined Wallenstein as astrologer, but died soon after at Regensburg in 1630.

Kepler also wrote an important work, the *Dioptrice* (1611), with a mathematical discussion of refraction and the different forms of the newly invented telescope, the whole constituting the foundation of modern optics. In it he develops the first correct theory of vision, "Seeing amounts to feeling the stimulation of the retina, which is painted with the colored rays of the visible world. The picture must then be transmitted to the brain by a mental current, and delivered at the seat of the visual faculty." He supposes that color depends on density and transparency, and that refraction is due to greater resistance of a dense medium. He enunciates the law that intensity of light varies inversely as the square of the distance. "In proportion as the spherical surface from whose centre the light proceeds is greater or smaller, so is the strength or density of the light-rays which fall on the smaller sphere to the strength of those rays which fall on the larger sphere." He explains the estimation of distance by binocular vision. He supposes the velocity of light to be infinite. His more purely mathematical work will be mentioned in a later chapter.

If Kepler had burnt three-quarters of what he printed, we should in all probability have formed a higher opinion of his intellectual grasp and sobriety of judgment, but we should have lost to a great extent the impression of extraordinary enthusiasm and industry, and of almost unequalled intellectual honesty, which we now get from a study of his works. — BERRY, p. 197.

Kepler says: "If Christopher Columbus, if Magellan, if the Portuguese, when they narrate their wanderings, are not only excused, but if we do not wish these passages omitted, and should lose much pleasure if they were, let no one blame me for doing the same." — WHEWELL, I, 293.

GALILEO

In the same year, 1521, in which Magellan established the rotundity of the earth by sailing around it, Luther appeared before the Diet of Worms and thus launched an organized opposition, since known as Protestantism, to the Roman Catholic church. The later years of Kepler and Galileo fell within the period of the Thirty Years' War, of which neither was to witness the close. Permanent English settlements in America had just begun. Galileo (1564–1642) was born only three days before the death of Michael Angelo, "nature seeming to signify thereby the passing of the sceptre from art to science," and in the same year with Shakespeare. He exerted a mighty influence on the development of science in many fields, and in particular laid the foundations of modern dynamics.

It is a remarkable circumstance in the history of science, that astronomy should have been cultivated at the same time by three such distinguished men as Tycho, Kepler, and Galileo. While Tycho, in the 54th year of his age, was observing the heavens at Prague, Kepler, only 30 years old, was applying his wild genius to the determination of the orbit of Mars, and Galileo, at the age of 36, was about to direct the telescope to the unexplored regions of space. . . . Tycho was destined to lay the foundation of modern astronomy, by a vast series of accurate observations made with the largest and the finest instruments. It was the proud lot of Kepler to deduce the laws of the planetary orbits from the observations of his predecessors; while Galileo enjoyed the more dazzling honor of discovering by the telescope new celestial bodies, and new systems of worlds. — BREWSTER, *Martyrs of Science*, p. 189.

Coming into a world still dominated by the Aristotelian tradition, Galileo is puzzled by the conflict between his own observations and the accepted theories, but firm and fearless in his convictions, he eagerly and powerfully controverts the older notions, incidentally gaining enemies as well as disciples.

In his whole point of view and habit of mind Galileo embodied the attitude and spirit of modern science. He was keenly alert in observing, analyzing, and reflecting on natural phenomena, eager and convincing in his expositions, sceptical and intolerant of mere authority, whether in science, philosophy, or theology. It was a true instinct of the conservatives to recognize in him the champion of a principle fatally hostile to their own. Between these antagonistic principles no permanent peace was possible.

While still a mere youth, he discovered the regularity of pendulum vibrations by observing the slow swinging of the cathedral lamp of Pisa (1582). Before he was 25 he published work on the hydrostatic balance (1586), and on the center of gravity of solids.

Many years later (1638) he showed that the hypothesis of uniform acceleration accounted correctly for the observed relations between space, time, and velocity, and that the path of a projectile is a parabola. In the words of a recent writer, when Galileo

> deduced by experiment, and described with mathematical precision, the acceleration of a falling body, he probably contributed more to the physical sciences than all the philosophers who had preceded him.

Hearing of the telescope invented in Holland, he constructed one for himself, by means of which he discovered sun spots, the mountains of the moon, the satellites of Jupiter, the rings of Saturn, and the phases of Venus. Galileo's discoveries on the surface of the moon were naturally offensive to contemporary Aristotelians, to whom that body was perfectly smooth and spherical. He appealed in vain to ocular evidence, considering himself that such a moon as they imagined would have been "but a vast unblessed desert, void of animals, of plants, of cities, and men — the abode of silence and inaction —

senseless, lifeless, soulless, and stripped of all those ornaments which now render it so varied and so beautiful."

He noted that while the fixed stars differed in brightness, it was only the planets which presented a disk-like appearance. A careful examination of Jupiter led to a more sensational discovery. January 7, 1610, he observed three small stars near the planet Jupiter, two to the east and one to the west, all in a line parallel to the ecliptic. The next day all three were on the west of Jupiter and nearer to each other. It seemed that Jupiter must be moving in the wrong direction. The next night was cloudy, but on the 10th the two stars visible were again on the east. He was forced to infer that these were moving with respect to Jupiter, and he drew the correct conclusion "that there were in the heavens three stars which revolve around Jupiter, in the same manner as Venus and Mercury revolve around the sun." January 13th he discovered the fourth satellite.[1]

His results were published in *The Sidereal Messenger*, announcing "great and very wonderful spectacles, and offering them to the consideration of every one, but especially of philosophers and astronomers; which have been observed by Galileo Galilei . . . by the assistance of a perspective glass lately invented by him; namely in the face of the moon, in innumerable fixed stars in the Milky Way, in nebulous stars, but especially in four planets which revolve around Jupiter at different intervals and periods with a wonderful celerity; which, hitherto not known to any one, the author has recently been the first to detect, and has decreed to call the Medicean stars." — Whewell, p. 277.

The reception which these discoveries met with from Kepler is highly interesting, and characteristic of the genius of that great man. . . . "I immediately began to think how there could be any addition to the number of the planets without overturning my Cosmographic Mystery, according to which Euclid's five regular solids do not allow more than six planets round the sun . . . I am so far from disbelieving the existence of the four circumjovial planets, that I long for a telescope, to anticipate you, if possible, in discovering two round Mars, as the proportion seems to require, six or eight

[1] It appears that Simon Mayer also discovered the four great satellites of Jupiter some days earlier. (J. H. Johnson, *Brit. Astr. Assoc. Jour.*, Jan., 1931.)

round Saturn, and perhaps one each round Mercury and Venus." — FAHIE, *Galileo*, p. 104.

In a very different spirit did the Aristotelians receive the *Sidereal Messenger* of Galileo. The principal professor of philosophy at Padua resisted Galileo's repeated and urgent entreaties to look at the moon and planets through his telescope; and he even labored to convince the Grand Duke that the satellites of Jupiter could not possibly exist. — BREWSTER, p. 33.

"There are seven windows given to animals in the domicile of the head, through which the air is admitted to the tabernacle of the body, to enlighten, to warm, and to nourish it. What are these parts of the microcosmos? Two nostrils, two eyes, two ears, and a mouth. So in the heavens, as in a macrocosmos, there are two favorable stars, two unpropitious, two luminaries, and Mercury undecided and indifferent. From this and many other similarities in nature, such as the seven metals, etc. . . ., we gather that the number of planets is necessarily seven. Moreover, these satellites of Jupiter are invisible to the naked eye, and therefore can exercise no influence on the earth, and therefore would be useless, and therefore do not exist." — FAHIE, p. 103.

It was inevitable that such a man as Galileo should accept the Copernican hypothesis. He writes to Kepler in 1597:

"I esteem myself fortunate to have found so great an ally in the search for truth. It is truly lamentable, that there are so few who strive for the true and are ready to turn away from wrong ways of philosophizing. But here is no place for bewailing the pitifulness of our times, instead of wishing you success in your splendid investigations. I do this the more gladly, since I have been for many years an adherent of the Copernican theory. It explains to me the cause of many phenomena which under the generally accepted theory are quite unintelligible. I have collected many arguments for refuting the latter, but I do not venture to bring them to publication.

"That the moon is a body like the earth I have long been assured. I have also discovered a multitude of previously invisible fixed stars, outnumbering more than ten times those which can be seen by the naked eye, — forming the Milky Way. Further I have discovered that Saturn consists of three spheres which almost touch each other." — DANNEMANN, II, pp. 23, 24.

While none of Galileo's astronomical discoveries were either necessary or sufficient to confirm the Copernican theory, their support was exceedingly important. Thus the slow motion of

FIG. 36. — GALILEO GALILEI (1564–1642).
From *Opera*, 1744.

FIG. 37. — TITLE-PAGE OF GALILEO'S DIALOGUE ON THE TWO CHIEF SYSTEMS OF THE WORLD, 1632.

the sun spots across the disk and their subsequent reappearance showed rotation of that body, the satellites of Jupiter and particularly the phases of Venus, analogous to those shown by the moon, obviously harmonized with the Copernican theory. This implied at least that the planets shone by reflected sunlight, and it had indeed been insisted against that theory that Venus and Mercury under it must show phases till then undiscovered.

In 1632 Galileo published his celebrated *Dialogue on the Two Chief Systems of the World, Ptolemaic and Copernican*, a work comparable in magnitude and importance with Copernicus's *Revolutions*. In the curious preface he says:

"Judicious reader, there was published some years since in Rome a salutiferous Edict, that, for the obviating of the dangerous Scandals of the present Age, imposed a reasonable Silence upon the Pythagorean Opinion of the Mobility of the Earth. There want not such as unadvisedly affirm, that the Decree was not the production of a sober Scrutiny, but of an ill-formed passion; and one may hear some mutter that Consultors altogether ignorant of Astronomical observations ought not to clipp the wings of speculative wits with rash prohibitions. My zeale cannot keep silence when I hear these inconsiderate complaints. I thought fit, as being thoroughly acquainted with that prudent Determination, to appear openly upon the Theatre of the World as a Witness of the naked Truth. . . . I hope that by these considerations the world will know that if other Nations have Navigated more than we, we have not studied less than they; and that our returning to assert the Earth's stability, and to take the contrary only for a Mathematical *Capriccio*, proceeds not from inadvertency of what others have thought thereof, but (had one no other inducements), from these reasons that Piety, Religion, the Knowledge of the Divine Omnipotency, and a consciousness of the incapacity of man's understanding dictate unto us." — Fahie, p. 246.

In the first of the four conversations into which the work is divided, the Aristotelian theory of the peculiar character of the heavenly bodies is subjected to destructive criticism, with emphasis on such phenomena as the appearance of new stars, of comets and of sun spots, the irregularities of the moon's surface, the phases of Venus, the satellites of Jupiter, etc.

"When we consider merely the vast dimensions of the celestial sphere in comparison with the littleness of our earth . . . and then think of the speed of the motion by which a whole revolution of the heavens must be accomplished in one day, I cannot persuade myself that the heavens turn while the earth stands fast." — DANNEMANN, II, p. 31.

Adducing not merely the sun spots themselves, but their rapid variation, he insists that the universe is not rigid and permanent, but constantly changing or, as science has more and more emphasized since his day, passing through consecutive, related phases or *evolving*.

"I can listen only with the greatest repugnance when the quality of unchangeability is held up as something preëminent and complete in contrast to variability. I hold the earth for most distinguished exactly on account of the transformations which take place upon it." — DANNEMANN, II, p. 34.

He begins to see the fallacy of the objections that if the earth rotated, a body dropped from a masthead would be left behind by the ship and that movable objects could be thrown off centrifugally at the equator. As positive arguments in support of the Copernican system, he urges particularly the retrogressions and other irregularities of the planets, and also the tides.

Of the famous controversy of Galileo with the Inquisition, it may here suffice to quote the judgment of the court:

"The proposition that the sun is in the centre of the world and immovable from its place is absurd, philosophically false, and formally heretical; because it is expressly contrary to the Holy Scriptures," etc.

and another passage from the biographer already cited:

For over fifty years he was the knight militant of science, and almost alone did successful battle with the hosts of Churchmen and Aristotelians who attacked him on all sides — one man against a world of bigotry and ignorance. If then, . . . once, and only once, when face to face with the terrors of the Inquisition, he, like Peter, denied his Master, no honest man, knowing all the circumstances, will be in a hurry to blame him. — FAHIE, pp. 313, 326.

Of Galileo's still more remarkable services to physics and dynamics, something will be added in a later chapter.

THE INQUISITION: BRUNO (1548–1600)

The judgment and sentence of the Inquisition upon Galileo, forms one of the darkest pages in the history of Science. Another victim of the Inquisition was Giordano Bruno, an Italian philosopher, who, having joined the Dominican order at the age of fifteen, was later accused of impiety and subjected to persecution. Bruno fled from Rome to France, and later to England, where at Oxford he disputed on the rival merits of the Copernican and the so-called Aristotelian systems of the universe. In 1584 he published an exposition of the Copernican theory. Bruno, moreover, attacked the established religion, the monks, the Jewish records, miracles, etc., and after revisiting Paris, and residing for a time in Wittenberg, rashly returned to Italy, where he was apprehended by the Inquisition and thrown into prison. After seven years of confinement he was excommunicated and, on Feb. 17, 1600, burnt at the stake. Thus was the end of the sixteenth century illuminated by the flames of martyrdom.

MEDICAL AND CHEMICAL SCIENCES

The discovery and publication in 1478 of the *De Medicina* of Celsus made the best of ancient medicine available to the physicians of the Renaissance. It gained enormous influence and authority. Probably for this reason the name Paracelsus was assumed by that erratic and radical Swiss physician and astrologer, Theophrast von Hohenheim (1493–1541), whose great merit is his courage in opposing scholasticism, authority in science and medicine, and in substituting his own experience. His science had a strong religious motive. For him alchemy was the search for the invisible forces of nature, the universal science of life and motion. Life being considered a chemical process, and disease a disharmony between man and his environment, he prescribed inorganic salts in place of the herbs and extracts then commonly used. "The true use of chemistry," he said, "is not to make gold but to prepare medicines." He believed that in nature there must be remedies specific for each disease, and he anticipated some aspects of the homœopathic doctrine of Hahnemann. While he did a service in altering the direction of chemical effort, he mixed much mystical nonsense with his reforms. Very different was his Veronese contemporary, Girolamo Fracastoro (1483–1553), whose work *De contagione*, 1546, entitles him to be called the father of epidemiology. He recognized three forms of infection: (1) by contact, (2) by "fomites," i.e., soiled clothing, etc., (3) by transmission from a distance. He compared contagion with fermentation and imagined it to be due to specific germs (*seminaria*) inhaled with the breath and capable of rapid multiplication — a clever guess verified three centuries later. He described and named syphilis and was the first to distinguish typhus fever from typhoid and from bubonic plague. In France, Ambroise Paré (1510–90), beginning as a barber, became the leading surgeon. He initiated important reforms in practice, especially in using milder measures than boiling oil for gun-shot wounds and the ligature in place of the hot iron to check bleeding, and he interpreted the anatomy of Vesalius to French surgeons.

Another physician of the sixteenth century, Georg Bauer, or in Latin *Agricola* (1490–1555) in his great work *De re metallica*, 1556, gave a complete description of the mining and metallurgy of his time. In two earlier works describing all the minerals then known, he established the method that has been the model for all subsequent descriptions of minerals.

ANATOMY AND PHYSIOLOGY: VESALIUS

Hardly less important, meantime, than the studies of Copernicus, Tycho Brahe, Galileo, and Kepler upon the heavenly bodies were those of the Belgian anatomist, Andreas Vesalius (1514–64), and his successors upon the human body. The medical reformers of the previous generation, called Galenists, had rejected Arabic medicine, expecting to find all knowledge of the medical sciences in the Greek texts of Galen and of Dioscorides. Teaching was still in the scholastic style, the professor reading from the book, while barbers dissected to illustrate Galen. Disgusted with this sort of instruction under Jacobus Sylvius at Paris, Vesalius went to Padua, where he could dissect for himself, and at the age of twenty-four was given his M.D. and the chair of surgery. His great work, *De humani corporis fabrica*, illustrated by beautiful wood engravings, was printed in 1543, the same year as the *De revolutionibus* of Copernicus.

This date has been taken to mark the end of the Middle Ages and the beginning of the Renaissance in Science, when men of science, influenced by the humanistic Renaissance with its revival of classical literature and art, fostered by the princes and princely merchants of Italy, turned from medieval subservience to authority and from Arabic science, and, guided by the classical Greek authors, revived their spirit of inquiry into the facts of nature. Henceforth there was steady progress "by the method of observation and experiment and a willingness [to forego speculation and] to accept the fact *that* one event follows another without seeking to know *why* it does so." This movement is strongly marked in the fields of anatomy and physiology, zoology, and botany, as well as in astronomy and physics. Surveying the Renaissance naturalists as a whole, one sees that they were mostly physicians learned in Greek, who took Galen, Aristotle, or Dioscorides as a guide and added their own observations, frequently in the form of accurate and artistic engravings. Anatomy and physiology were not separate, and one author might write in several of the biological fields. The innovations of these pioneers of the sixteenth century are too numerous to be fully treated here.

In his great biography Roth says of Vesalius:

"My investigations have shown that Vesalius is really the founder of modern anatomy, the founder in the truest sense of the word. . . . Vesalius [is] the first who actually knew the human body in an exact and comprehensive way, the first to break down and overthrow the omnipotent belief in bookish tradition. . . . This achievement was based on his own labors. . . . Anatomy, and, at the same time, anatomical investigation, is Vesalius' creation." — F. H. GARRISON, *Bull. Soc. Med. Hist.*, **4**, 48, 1916.

The *Fabrica* is a book on human anatomy so complete that later works may be regarded as commentaries upon it. It deals also with comparative anatomy and incidentally with the functions of almost every organ. The discovery in a monkey of a structure described by Galen as human gave Vesalius the clue to the source of Galen's errors. Henceforth his aim

FIG. 38. — THE HEART IN CROSS SECTION AS FIGURED BY VESALIUS. A,B, vena cava; C, aorta; D, pulmonary artery; E,F, coronary vessels; G, right ventricle; H, left ventricle; I, interventricular septum. Reproduced in the original size from the *Fabrica*, 1543. Courtesy of the Boston Medical Library.

was to describe and figure all parts of the human body as they would appear in a living subject. His dissections show great skill, notably that of the brain, and are recorded in his text and in detailed drawings by his fellow-countryman, Jan Kalkar, a pupil of Titian. Vesalius, however, was not free from anatomical errors and never was able to get away from the physiology of Galen, although he expresses wonder that blood could pass from the right to the left side of the heart through the septum, in which he could find no openings. The *Fabrica* ends with a brief chapter describing his experiments in vivisection, some like Galen's (p. 167), others new.

The book met with a storm of opposition from the Galenists, and Vesalius, after burning his notes and drawings, departed in 1544 for Madrid to become physician to the Emperor, Charles V. His scientific career was ended, but his book had such success that a new edition was printed in 1555.

This enabled him to make considerable improvements, especially to strengthen his statement of disbelief in the passage of blood through the cardiac septum.

Under the Republic of Venice, Padua enjoyed a measure of academic freedom not to be found elsewhere in Europe. It was an intellectual center of the first rank, and the tradition established there by Vesalius remained through a long line of brilliant successors. The first of these, Realdus Columbus, 1559, as well as Servetus, 1553, and, later, Andrea Cesalpinus anticipated Harvey (p. 293) by postulating a circulation of blood through the lungs — Cesalpinus included the systemic circulation. But these statements without evidence had no lasting effect. The statement of Servetus was made incidentally in a theological work that gave such offense to the Pope, as also to Calvin, that he escaped the one only to be burned by the other with the whole edition, except three copies, of his book.

Gabriele Falloppius (1523–62), who succeeded Columbus at Padua, made important contributions to anatomy, including a description of the uterine tubes that bear his name. He was followed by his pupil, Hieronymus Fabricius of Aquapendente (1537–1619). The anatomical theater still at Padua is a perpetual monument to the fame of this great anatomist and surgeon. He adopted the comparative method, especially in his works on embryology, the earliest illustrated books on the subject. The second of these is the more important. It gives the first description, since Aristotle, of the early stages of the chick. In it Fabricius figures structures as present at an earlier stage than they really exist, giving ground for the idea of "preformation," so prominent in later discussions. His most fruitful work, however, was *De venarum ostiolis* (*On the Valves of the Veins*), 1603, in which the valves are correctly described and figured. Nevertheless, he was unable to see that they prevent blood from flowing in the manner described by Galen. That was left to his pupil, William Harvey, as we shall see in a later chapter. Meanwhile in Rome, Bartolomeo Eustachius (1524–74) was making the first studies in anatomical variations of individuals. His work on the ear is remembered in the name of the Eustachian tube.

NATURAL HISTORY AND NATURAL PHILOSOPHY

One of the first to exhibit the scientific spirit of the Renaissance in the north, was Nicholas of Cusa, Nicolaus Cusanus, a pioneer in astronomy (p. 227), who also recorded "the first biological experiment of modern times," in which he showed that a plant gained something of weight from the air. His *De staticis experimentis*, in which many investigations are suggested, shows appreciation of how the experimental method should be applied.

A typical figure of the Italian Renaissance is the Florentine, Leonardo da Vinci (1452–1519), whose achievements in physical science will be mentioned later. A great artist, he left few paintings, but his extraordinary constructive imagination is shown in a great mass of papers, of which some 5,000 leaves are extant. Turning his attention from art to natural science, he studied human anatomy, the structure of the eye, and the flight of birds. In his notebooks he anticipated modern points of view in regard to fossils and changes in the crust of the earth. In anatomy he makes human dissections and accurate and beautiful anatomical drawings. He even compares the flow of the blood with the circulation of water from the sea to the clouds and back to the hills as rain, a century in advance of Harvey's great discovery. Unfortunately, his work, being unpublished, had no such influence as its originality deserved.

A more serious student of fossils was Bernard Palissy (*c.* 1510–89), famous as the maker in Paris of pottery decorated with raised and colored figures of animal and plant forms. In 1580 he published a book in which he discussed petrified wood, and fossil fishes and shells, and took the then advanced ground (as Xenophanes and Pythagoras had done, however, some two thousand years earlier) that these are in reality what they appear to be, i.e., petrified remains of plant and animal life. He identified some of the shells with living forms and concluded that the places where they were found had once been covered by the sea or fresh water. His bold stand, marking one of the first steps in modern times toward a rational geology, added to his Protestant religion in rousing an opposition that threw him into the Bastile, where he died.

The spirit of inquiry that blossomed in the sixteenth century is illustrated by three outstanding zoologists who wrote books based almost entirely on their own observations and illustrated by figures of things they had seen. The chief work of Pierre Belon (1517–64) is on birds (1555) and contains a remarkable plate, the first of its kind, showing the homologies in the skeletons of man and bird. His book on fishes and the dolphin (1551) is the first printed book on marine animals, and his short treatise on cone-bearing trees is the first monograph of a plant group. A more accurate observer was Guillaume Rondelet (1507–66) of Montpellier, who described a large number of marine and fresh-water fishes and other aquatic animals (1554–55) illustrated with good wood cuts. The work on fishes by Hippolyto Salviani (1514–72) of Rome is famous for its beautiful copper plates drawn from nature (1554–58).

Conrad Gesner (1516–65) was the most famous naturalist of his time — on account of his vast erudition called by Cuvier "the German Pliny." After a short term as professor of Greek at Lausanne, he took the M.D. at Basel, and settled in his native town, Zurich, where he was appointed lecturer on natural science and, later, town physician. He was a prolific writer on language, theology, botany, zoology, medicine, milk, and mountaineering. His catalogue of all known writings in Latin, Greek, and Hebrew (1545) is famous; his book on fossils is the first one illustrated; and his great work, *Historia Animalium*, is said to mark the beginning of modern zoology. In five huge volumes (1551–87) the animal kingdom is divided according to the system of Aristotle. The aim is to make easily accessible all the trustworthy knowledge of his time. The work is a compilation combined with much that is original. Each species is illustrated by a wood cut, those of familiar forms being especially full of life and spirit, exhibiting skill of high degree in both artist and engraver. The fifth volume appeared after his heroic death from plague during an epidemic he was striving to check in Zurich. What remained of over 1,500 drawings and cuts, that he had prepared for a greater work on plants, were gathered and published in 1753. Ulysis Aldrovandi (1522 or 27–1605 or 07), professor at Bologna and founder of its botanic garden, prepared the largest collective work on animals ever published. Five of the thirteen volumes appeared during his life, the publication finally being completed at state expense. He took Gesner for his model, and while excelling him in bulk did not equal him in quality. The engravings are coarse and the text contains much that is useless or untrue. The best part is the treatise on Insects, 1602.

Herbalists of the thirteenth century had begun to get away from the crude drawings of their predecessors and to turn to nature. But "the German Fathers of Botany" produced the first printed figures drawn from observation. Otto Brunfels (1489–1534) of Mainz, in his *Herbarium vivae eicones*, 1530–36, gave clear line drawings that marked an epoch in natural illustration. He was followed by Leonard Fuchs (1501–66), an excellent observer, whose *Historia stirpium*, 1542, is a landmark in the history of science, with its wood cuts of extraordinary beauty and truth. But these authors were interested in plants merely as drugs, and they wasted much time in trying to identify German plants with the descriptions of Dioscorides; for no drug was regarded as genuine unless it met that test. Jerome Bock (Hieronymus Tragus, 1494–1554), on the other hand was the first to attempt original description so complete that illustrations would not be needed in his *New. Kreutter Buch*, 1539, intended for popular use. In the meantime, knowledge of American plants was coming to Europe through the Spaniards. The first of them was Gonzalo de Oviedo (1478–1557) whose rude engravings in his history of the West Indies, *c.* 1536, include the earliest figures of Indian corn and the pineapple. "The one botanical genius of the German Renaissance" was Valerius Cordus (1515–44), M.D. and

Docent at Marburg, who was "the first to teach men to cease from dependence on the poor descriptions of the ancients and to describe plants anew from Nature." He collected in Germany, Switzerland, and Italy. At the time of his death from fever in Rome he had completed his *Historia plantarum*, which was left for Gesner to edit and publish, 1561. He made very considerable contributions to plant morphology in addition to his clear, straightforward systematic descriptions, each one from the living plant in flower and in fruit. These included more new forms than all previous work in Germany, as well as many plants well known to the ancients, with the briefest possible mention of their medical qualities. The most popular of all books of plants at this time was the *Commentaries on Dioscorides* by Pietro Andrea Mattioli of Siena (1500–77), based in part on the famous Julia Anicia manuscript from Constantinople. He added his own observations and original illustrations, making a new book of the *Materia medica*, which went through many editions. Of the first, 1544, it is said that more than 33,000 copies were sold, and it remained the standard textbook for two hundred years. The botanical work of Andrea Cesalpinus (1519–1603) attracted little attention during his life, but it was highly regarded by Linnaeus for the philosophical and artificial system of classification.

While Galileo was beginning the study of mechanics (p. 279), another branch of physics was being cultivated in England by William Gilbert of Colchester (1544–1603), President of the Royal College of Physicians, whose famous work, *De Magnete*, 1600, presented the first rational treatment of electrical and magnetic phenomena since the *Epistola* of Peter the Stranger (see p. 219). He regarded the earth as a great magnet, and, though not convinced of the truth of the Copernican theory, attributed the earth's rotation to its magnetic character. He even extended this idea to the heavenly bodies, with an animistic tendency. His great discovery was that the loadstone is merely iron which has been subjected to magnetic force.

The most prolific writer on natural philosophy and physical science of the sixteenth century was G. della Porta[1] (1541–1615), a native of Naples and a resident of Rome, founder of an early scientific academy there, and afterwards of the famous Accademia dei Lincei of Rome.

In his *Natural Magic*, della Porta is first to describe a camera obscura. He compared it to the eye, noting the correspondence of its parts. He touched also on many interesting properties of lenses, and referred to spectacles, some forms of which had long been known (see p. 220). His work *On Refraction* deals largely with binocular vision, and is a criticism of the work of Euclid and Galen on that subject. The author hints also at a crude telescope, and may have known some form of stereoscope. Della Porta was a tireless and fruitful, if not especially original, thinker and worker.

REFERENCES FOR READING

Brewster, David, *Martyrs of Science*, Ed. 2, 1846.
Castiglioni, A., *Renaissance of Medicine in Italy*, 1934.
Fahie, J. J., *Galileo, His Life and Work*, 1903.
Foster, Michael, *Lectures on the History of Physiology*, 1901, pp. 1–24.
Dreyer, J. L. E., *Tycho Brahe*, 1890.
Galileo Galilei, *Dialogues on Two New Sciences* (tr. by Crew and De Salvio), 1914.
Gilbert, William, *On the Magnet* (tr. by P. F. Mottelay), 1893.
Hawks, E., *Pioneers of Plant Study*, 1928.
Lodge, Oliver, *Pioneers of Science*, 1893.
Miall, L. C., *Early Naturalists*, 1912, pp. 1–98.
Noyes, Alfred, *Watchers of the Sky*, 1922.
Smith, Preserved, *History of Modern Culture*, I, 1930.

From previous Lists: Berry, *Hist. of Astron.*, Ch. IV–VI; Dreyer, *Plan. Systems*, Ch. XII–XVI; Locy, *Biol.*, Ch. VIII, IX, XIX; Osler, *Mod. Med.*, pp. 119–163.

[1] Not Giovanni Battista della Porta (1542–97), a sculptor.

Note: The figure on the opposite page represents Stevin's triangle, redrawn by E. T. W. For description see page 287.

Chapter XI

Mathematics and Mechanics in the Sixteenth and Early Seventeenth Centuries

It was not alone the striving for universal culture which attracted the great masters of the Renaissance, such as Brunelleschi, Leonardo da Vinci, Raphael, Michael Angelo and especially Albrecht Dürer, with irresistible power to the mathematical sciences. They were conscious that, with all the freedom of the individual phantasy, art is subject to necessary laws and, conversely, with all its rigor of logical structure, mathematics follows esthetic laws. — F. Rudio. (Moritz, *Memorabilia Math.*, p. 183.)

The miraculous powers of modern calculation are due to three inventions: the Arabic Notation, Decimal Fractions and Logarithms. — Cajori, *History of Mathematics*, p. 161.

The invention of logarithms and the calculation of the earlier tables form a very striking episode in the history of exact science, and, with the exception of the *Principia* of Newton, there is no mathematical work published in the country which has produced such important consequences, or to which so much interest attaches as to Napier's *Descriptio*. — Glaisher, "Logarithms," *Encycl. Brit.*, Ed. 9.

THE RENAISSANCE OF MATHEMATICS AND PHYSICS

The period from the invention of printing, about 1450, to the middle of the seventeenth century was one of very great importance for mathematics and mechanics as well as for astronomy. At the beginning, Arabic numerals were known, but the mathematics even of the universities hardly extended beyond the early books of Euclid and the solution of simple cases of quadratic equations in rhetorical form. At the end of the period the foundations of modern mathematics and mechanics were securely laid.

AIMS AND TENDENCIES OF MATHEMATICAL PROGRESS

In the centuries just preceding, the chief applications of mathematics had connected themselves with the relatively simple needs of trade, accounts and the calendar, with the graphical constructions of the architect and the military engineer, and with the sines and tangents of the astronomer and the navigator. During the period in question some of these applications became increasingly important, and at the same time mathematics was more and more cultivated for its own sake. Mathematicians became gradually a more and more distinctly differentiated class of scholars; mathematical textbooks took shape. The beginnings of this evolution have been dealt with already; its further progress is now to be traced.

The larger achievements and tendencies of the period in mathematical science were the following:

In Arithmetic, decimal fractions and logarithms were introduced, regulating and immensely simplifying computation; a general theory of numbers was developed; in Algebra, a compact and adequate symbolism was worked out, including the use of the signs $+$, $\div$, $\times$, $-$, $=$, $()$, $\sqrt{\ }$, and of exponents; equations of the third and fourth degree were solved, negative and imaginary roots accepted, and many theorems of our modern theory of equations discovered.

In Geometry, the computation of π (the ratio of the circumference of every circle to its diameter) was carried to many decimals, the beginnings of projective geometry were made, and a so-called method of indivisibles developed, foreshadowing the integral calculus; in plane and spherical Trigonometry, the theorems and processes now in use were worked out, and extensive tables computed.

In Mechanics, ideas about force and motion, equilibrium and center of gravity, were gradually clarified.

Underlying some of these new developments are the dawning fundamental concepts; function, continuity, limit, derivative, infinitesimal, on which our modern mathematics has been built up. Descartes, Newton and Leibniz are soon to make their revolutionary discoveries in analytic geometry and the calculus.

We have seen that up to about 1500 the chief stages in the development of mathematics have been the introduction and improvement of Arabic arithmetic for commercial purposes (though accounts were kept in Roman numerals until 1550 to 1650), the rediscovery of Greek geometry, and the improvement of trigonometry in connection with its increasing use in astronomy, navigation, and military engineering. The development of science has been powerfully promoted by the general intellectual emancipation of the period of the Renaissance, while mathematical progress, beginning earlier, has been both a consequence and a method of the general advance. The diffusion and the preservation of scientific knowledge have derived immense advantage from the new art of printing and from expanding commercial intercourse. Algebra, almost helpless in Greek times because, for lack of proper symbolism, expressed only in geometrical or rhetorical form, has been converted by a process of abbreviation, at first into a syncopated form, intermediate between the rhetorical and our modern purely symbolic notation. Important scientific instruments for observation and measurement have been invented.

PACIOLI

The earliest printed book on the whole of arithmetic and algebra was published at Venice in 1494 by Luca Pacioli (1445–*c.* 1514), a Franciscan monk born in Tuscany. Rules are here given for the fundamental operations of arithmetic, and for extracting square roots. Commercial arithmetic is treated at considerable length by the newer algoristic or Arabic methods. The method of arbitrary assumption corrected by proportion is used effectively, for example:

> To find the original capital of a merchant who spent a quarter of it in Pisa and a fifth of it in Venice, who received on these transactions 180 ducats, and who has in hand 224 ducats.
>
> Assume that his original capital was 100 ducats; then the surplus would be $100 - 25 - 20 = 55$, but this is $\frac{5}{4}$ of his actual surplus $224 - 180$, therefore his original capital was $\frac{4}{5}$ of $100 = 80$ ducats.

Some of Pacioli's commercial problems are exceedingly complicated. He solves numerical equations of the first and second degree, but admits only positive roots and considers the solution of cubic equations, as well as the squaring of the circle, impossible. Addition is denoted by *p* or $\bar{p}$, abbreviating the Latin *plus*, equality sometimes by *ae*, a beginning of syncopated algebra. The introduction of the radical sign with indices $\sqrt{2}$, $\sqrt{3}$ and of the signs $+$ and $-$ date from about this time.

In geometry Pacioli, like Regiomontanus, employs algebraic methods. In the spirit of the Renaissance he brings the feeble mathematics of the universities into fruitful relations with the practical mathematics of the artist and the architect. The inscribed hexagon and the equilateral triangle play their part as guild secrets in the development of Gothic architecture. The question is not "How to prove," but "How to do."

On the other hand, the current scholastic tendency to drift into mystical interpretation is exemplified by the following (condensed) extract from Pacioli:

> There are three principal sins, avarice, luxury, and pride; three sorts of satisfaction for sin, fasting, almsgiving, and prayer; three persons offended by sin, God, the sinner himself, and his neighbour; . . . three degrees of penitence, contrition, confession, and satisfaction, which Dante has represented as the three steps of the ladder that lead to purgatory, the first marble, the second black and rugged stone, and the third red porphyry. . . . There are three enemies

of the soul: the Devil, the world, and the flesh. There are three things which are of no esteem: the strength of a porter, the advice of a poor man, and the beauty of a beautiful woman. There are three vows of the Minorite Friars: poverty, obedience, and chastity. There are three terms in a continued proportion. . . . Three principal things in Paradise: glory, riches, and justice. There are three things which are especially displeasing to God: an avaricious man, a proud poor man, and a luxurious old man. And all things in short, are founded in three; that is, in number, in weight, and in measure. — MORITZ, 2145.

GEOMETRY IN ART

Brunelleschi (1377–1446), the famous architect of the early Renaissance, made a perspective view of the Signoria in Florence in a sort of box with clouds. The famous doors of the Baptistery by his contemporary Ghiberti show the development of perspective in the marked contrast between the earlier and the later panels. Raphael in his School of Athens includes himself and Bramante in a group of mathematicians.

Leonardo da Vinci (1452–1519), one of the intellectual giants of the Renaissance, eminent alike in art, science and engineering, gave the first correct explanation of the partial illumination of the darker part of the moon's disk by reflection from the earth; and his notebooks reveal accurate observations on the optics of a narrow beam of light in a dark room, a camera obscura. He calls mechanics the paradise of the mathematical sciences, because through it one first gains the fruit of these sciences. He denies the possibility of perpetual motion, saying "Force is the cause of motion and motion the cause of force." He discusses the lever, the wheel and axle, bodies falling freely or on inclined planes, foreshadowing Galileo. Contrary to the Aristotelian tradition he asserts that everything tends to continue in its given state, and he even enunciates the fundamental principle that for simple machines forces in equilibrium are inversely as the virtual velocities.

"Whoever," he says, "appeals to authority applies not his intellect but his memory." "While Nature begins with the cause and ends with the experiment, we must nevertheless pursue the opposite plan, beginning with the experiment and by means of it investigating

> the cause." "No human investigation can call itself true science, unless it comes through mathematical demonstration." "He who scorns the certainty of mathematics will not be able to silence sophistical theories which end only in a war of words."

Unfortunately, his power of expression being far inferior to his real genius, his work in this field remained unpublished, and therefore relatively unfruitful.

Leonardo and other great artists of his time — notably Albrecht Dürer of Nuremberg (1471–1528) — developed the geometrical theory of perspective. For the purpose of accurately representing the human head Dürer made both plans and elevation. "Intelligent painters and accurate artists," he says, "at the sight of works painted without regard to true perspective must laugh at the blindness of these people, because to a right understanding nothing impresses more disagreeably than falsehood in a painting, regardless of the diligence with which it has been made. That such painters, however, are pleased with their own mistakes is due to the fact that they have not learned the art of measurement, without which no one can become a true workman." All this had importance both for modern art and modern geometry.

Characteristic of this period is the so-called *Margarita Philosophica* published in many editions from 1503 to 1600. It was the first modern encyclopedia printed, and gives in its twelve books "a compendium of the trivium, the quadrivium, and the natural and moral sciences."

Michael Stifel (1487–1567), a German Lutheran minister who started as an Augustine monk, wrote an important arithmetical treatise, *Arithmetica Integra* (1544). This book, which appeared with a preface of Melanchthon, contains many original results. Stifel gives an extensive study of the binomial coefficients (no longer in the semi-mystical character of the Neo-Pythagorean triangular numbers, etc.), and has already the so-called "triangle of Pascal." He approaches closely the notion of logarithms (p. 276) and is one of the first to introduce negative numbers. He has also some improvements of current notation.

As a curiosity and sign of the times we may add that Stifel also developed a fantastic arithmetical interpretation of the Bible, identifying Pope Leo X with the beast in *Revelation* and

Fig. 39. — Two Sixteenth Century Modes of Reckoning. From the *Margarita Philosophica*, courtesy of Professor David Eugene Smith.

predicting the immediate end of the world — with results disastrous to his person as well as his reputation.

The low state of computation at this time is illustrated with startling clearness by a bulletin on the blackboard at Wittenberg, in which Melanchthon urgently invited the academic youth to attend a course on arithmetic, adding that the beginnings of the science are very easy, and even division can with some diligence be comprehended.

ROBERT RECORDE (1510–58)

Robert Recorde, the "morning star of English mathematical literature" (Cajori), studied at Oxford and graduated in medicine at Cambridge in 1545, later becoming "royal physician." His work entitled *The Grounde of Artes*, or arithmetic, one of the earliest mathematical books printed in English (1540), ran through more than 27 editions and exerted a great influence on English education. In the "Preface to the Loving Reader" he says:

Sore ofttimes have I lamented with myself the unfortunate condition of England, seeing so many great Clerks to arise in sundry other parts of the World, and so few to appear in this our Nation; whereas for pregnancy of natural wit (I think) few Nations do excell English-men. But I cannot impute the cause to any other thing, then to the contempt or misregard of Learning. For as English-men are inferiour to no men in mother Wit, so they pass all men in vain Pleasures, to which they may attain with great pain and labour; and are slack to any never so great commodity, if there hang of it any painfull study or travelsome labour.

The book itself is in the form of a dialogue or catechism beginning:

The Scholar speaketh.

"Sir, such is your authority in mine estimation, that I am content to consent to your saying, and to receive it as truth, though I see none other reason that doth lead me thereunto; whereas else in mine own conceit it appeareth but vain, to bestow any time privately in learning of that thing that every Child may and doth learn at all times and hours, when he doth any thing himself alone, and much more when he talketh or reasoneth with others."

He employs the symbol + "whyche betokeneth too muche, as this line – plaine without a crosse line betokeneth too little."

In 1557 he published an algebra under the alluring title *Whetstone of Witte*, using the sign = for equality, which he says he selected because "noe 2 thynges can be moare equalle" than two parallel straight lines.

ALGEBRAIC EQUATIONS OF HIGHER DEGREE

The main achievement of Italian mathematicians of the sixteenth century was the solution of the equations of the third, later of the fourth degree. Much of the early work was done at the university of Bologna, where Scipio del Ferro solved, about 1515, the equation $x^3 + mx = n$, without publishing the result. About 1535 Tartaglia discovered Ferro's solution and also a solution of the equation $x^3 + px^2 = n$. There was considerable public interest in these mathematical speculations, and public contests on the solution of the third degree equations were held. In one of them Tartaglia triumphed over one of Ferro's pupils who had doubted his achievement. Later Tartaglia confided his method under a pledge of secrecy to Cardan, who published the solution in his *Ars Magna* of 1545. The result was a considerable amount of ill-will between Tartaglia and Cardan, and a public contest between the former and Cardan's pupil, Ferrari (p. 273), who also solved the equation of the fourth degree.

NICCOLO FONTANA, OR TARTAGLIA (1500–57)

Tartaglia, a man of the humblest origin, lectured at Verona and Venice, and first won fame by successfully meeting a challenge to solve mathematical problems, all of which proved, as he had anticipated, to involve cubic equations.

His *Nova Scienza* (1537) discusses falling bodies, and many problems of military engineering and fortification, the range of projectiles, the raising of sunken galleys, etc.

The title-page is chiefly occupied by a large plate, which represents the courts of Philosophy, to which Euclid is doorkeeper, Aristotle and Plato being masters of an inmost court, in which Philosophy sits throned, Plato declaring by a label that he will let nobody in who does not understand Geometry. In the great court there is a cannon being fired, all the sciences looking on in a crowd — such as Arithmetic, Geometry, Music, Astronomy, Cheiromancy, Cosmography, Necromancy, Astrology, Perspective, and Prestidigitation! A wonderfully modest-looking gentleman, with his hand upon his heart, stands among the number, with a you-do-me-too-much-honour look upon his countenance; Arithmetic and Geometry are pointing to him, and under his feet his name is written — Nicolo Tartalea. — HENRY MORLEY, *Jerome Cardan*, I, p. 221.

The *Invenzioni* (1546) gives his solution of the cubic equation. A *Treatise on Numbers and Measures* (1556, 1560) gives a method for finding the coefficients in the expansion of $(1 + x)^n$ for $n = 2, \dots 6$. It contains also a wide range of problems from commercial arithmetic and a collection of mathematical puzzles. The following examples may illustrate these:

"Three beautiful ladies have for husbands three men, who are young, handsome, and gallant, but also jealous. The party are travelling, and find on the bank of a river, over which they have to pass, a small boat which can hold no more than two persons. How can they pass, it being agreed that, in order to avoid scandal, no woman shall be left in the society of a man unless her husband is present?"

"A ship carrying as passengers 15 Turks and 15 Christians encounters a storm, and the pilot declares that in order to save the ship and crew one half of the passengers must be thrown into the sea. To choose the victims, the passengers are placed in a circle, and it is agreed that every 9th man shall be cast overboard, reckoning from a certain point. In what manner must they be arranged so that the lot may fall exclusively upon the Turks?"

"Three men robbed a gentleman of a vase containing 24 ounces of balsam. Whilst running away they met in a wood with a glass-seller of whom in a great hurry they purchased three vessels. On reaching a place of safety they wish to divide the booty, but they find that their vessels contain 5, 11, and 13 ounces respectively. How can they divide the balsam into equal portions?" — BALL.

There is no other treatise that gives so much information concerning the arithmetic of the sixteenth century, either as to theory or application. The life of the people, the customs of the merchants, the struggles to improve arithmetic, are all set forth here by Tartaglia in an extended but interesting fashion.

Tartaglia, anticipating Galileo, taught that falling bodies

of different weight traverse equal distances in equal times, and that a body swung in a circle if released flies off tangentially.

JEROME CARDAN (*1501–76*)

Cardan led a life of wild and more or less disgraceful adventure, strangely combined with various forms of scientific or semi-scientific activity — particularly the practice of medicine. He studied at Pavia and Padua, travelled in France and England, and became professor at Milan and Pavia.

His *Ars Magna* (1545) contains the solution of the cubic equation fraudulently obtained from his rival Tartaglia. After its publication the aggrieved Tartaglia challenged Cardan to meet him in a mathematical duel. This took place in Milan, August 10, 1548, but Cardan sent his pupil Ferrari in his place. Cardan had left for parts unknown. As Tartaglia began to explain to the crowd the origin of the strife and to criticise Ferrari's 31 solutions, he was interrupted by a demand that judges be chosen. Knowing no one present he declines to choose; all shall be judges. Being finally allowed to proceed he convicts his opponent of an erroneous solution, but is then overwhelmed by tumultuous clamor with demands that Ferrari must have the floor to criticise his solution. In vain he insists that he be allowed to finish, after which Ferrari may talk to his heart's content. Ferrari's friends are vehement; he gains the floor and chatters about a problem which he claims Tartaglia has not been able to solve till the dinner hour arrives and Tartaglia, apprehending still worse treatment, withdraws in disgust.

Ferrari (1522–65), this disciple of Cardan, even succeeded in giving a general solution of the equation of the fourth degree, beyond which, as has been shown only in the last century, the solution can in general no longer be similarly expressed.

In the history of mathematics, one of Cardan's most important contributions was his work on probability. He is believed to have been the first to discuss this subject. Of his scientific inventions, one was an improved suspension of the compass needle. He was also eminent as an astrologer.

SYMBOLIC ALGEBRA: VIETA

Of still greater importance in the history of algebra is F. Vieta (1540–1603) a lawyer of the French court. He won the interest of Henry IV by solving a complicated problem proposed by an eminent mathematician, as was the custom of the time, as a challenge to the learned world. This involved an equation of the 45th degree which he succeeded in solving by a trigonometric method. Later he was employed to interpret the cipher despatches of the hostile Spaniards. His *In Artem Analyticam Isagoge* is the earliest work on *symbolic* algebra. In it known quantities are denoted by consonants, unknown by vowels, the use of homogeneous equations is recommended, the first six powers of a binomial given, and a special exponential notation introduced. He shows that the celebrated classical problems of trisecting a given angle and duplicating a cube involve the solution of the cubic equation, and makes important discoveries in the general theory of equations — for example resolving polynomials into linear factors and deriving from a given equation other equations having roots which differ from those of the first by a constant or by a given factor. He solves Apollonius's famous problem of determining the circle tangent to three given circles, and expresses π by an infinite series. He devises systematic methods for the solution of spherical triangles.

DEVELOPMENT OF TRIGONOMETRY

Many circumstances combined to promote the development of trigonometry at this period. It was needed by the military engineer, the builder of roads, the astronomer, the navigator, and the map-maker, whose work was tributary to all of these.

Rheticus (George Joachim, 1514–76) — "the great computer whose work has never been superseded," known as collaborator with Copernicus — worked out a table of natural sines for every 10 seconds to fifteen places of decimals. We owe to him our familiar formulas for sin $2x$ and sin $3x$. The notation sin, tan, etc., and the determination of the area of a spherical triangle date from about this time. To this period belong also the very important work of Mercator on map-making and the reform of the calendar by Pope Gregory XIII.

MAP-MAKING

Mercator (Gerhard Kremer, 1512–94) devoted himself, first in Louvain, later in Duisburg, to mathematical geography, and gained his livelihood by making maps, globes, and astronomical instruments, combined in later life with teaching. His great world map, completed in 1569, marks an epoch in cartography. The first "Atlas" was published by his son in 1595. He gives a mathematical analysis of the principles underlying the projection of a spherical surface on a plane.

> "If," he says, "of the four relations subsisting between any two places in respect to their mutual position, namely difference of latitude, difference of longitude, direction and distance, only two are regarded, the others also correspond exactly, and no error can be committed as must so often be the case with the ordinary marine charts and so much the more the higher the latitude."

Mercator's projection amounts to a projection of a sphere on a plane by which the meridians and the parallels pass into two sets of parallel straight lines perpendicular to each other. Angles are preserved in magnitude, but areas remote from the equator are disproportionately expanded. A straight line on the chart corresponds with the course of a ship steering a constant course. This Mercator projection belongs, together with the stereographic projection known to Ptolemy, among the most important methods of map-making.

Map-making flourished, especially in the Netherlands, under the influence of Mercator and his pupils. The maps of the Blaeu family in Amsterdam (17th century) became famous.

THE GREGORIAN CALENDAR

Until 1582 the Julian calendar (p. 160) remained in force with $365\frac{1}{4}$ days each year and a gradually increasing error amounting at this time to ten days. Under the auspices of Pope Gregory XIII the days from October 5 to 15, 1582, were dropped and the number of leap-years in 400 reduced from 100 to 97. Religious jealousies prevented the adoption of this reform in Protestant Germany for a century, while England postponed it until 1752.

LOGARITHMS, A NEW INVENTION FOR COMPUTATION

The invention of logarithms would appear to have been a natural sequel of any adequate theory and notation for exponents. Thus Stifel in his arithmetic (1544) had tabulated small integral powers of 2 — from $\frac{1}{4}$ to 64 — and shown the correspondence between multiplication of these powers and addition of the indices or exponents, but his use of exponents was too limited, he lacked the apparatus of decimal fractions necessary for the practical application of the method and probably had no conception of the vast labor-saving possibilities so near at hand.

In 1614 John Napier published at Edinburgh his *Mirifici Logarithmorum Canonis Descriptio*, for which the time was so fully ripe that an enthusiastic reception was at once assured. Napier as a devout Protestant, stimulated by fear of an impending Spanish invasion, busied himself with inventions "proffitabill & necessary in theis dayes for the defence of this Iland & withstanding of strangers enemies of God's truth & relegion." Among these were a mirror for burning distant ships, and a sort of armored chariot. Impressed by the tremendous calculations then in progress by Rheticus, Kepler, and others in connection with the development of the new astronomy, Napier made a vastly more important invention. His definition of a logarithm rests on the following kinetic basis:

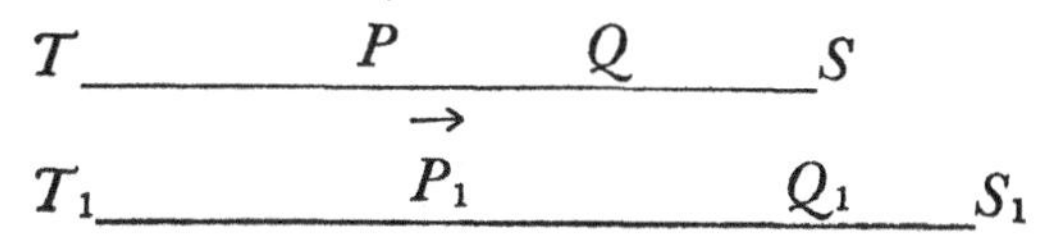

TS is a straight line of definite length; T_1S_1 extends to the right indefinitely. Moving points P and P_1 start from T and T_1 with equal initial speeds; the latter continues at the same rate, the former is retarded so that its speed is always proportional to its distance from S. If equal intervals are taken on T_1S_1 the corresponding intervals in TS will grow smaller to the right. When P is at any position Q, the logarithm of QS is represented by the corresponding length T_1Q_1 on the other line. It may be shown in fact that if in our notation $PS = x$, $T_1P_1 = y$, $TS = l$, $\frac{dx}{dy} = -\frac{x}{l}$. This conception involving a functional relation between two variables went much deeper than the comparison of discrete numbers by Stifel.

Napier's conception of a logarithm involved a perfectly clear apprehension of the nature and consequences of a certain functional relationship, at a time when no general conception of such a relationship had been formulated, or existed in the minds of mathematicians, and before the intuitional aspect of that relationship had been clarified by means of the great invention of coördinate geometry made later in the century by René Descartes. A modern mathematician regards the logarithmic function as the inverse of an exponential function; and it may seem to us, familiar as we all are with the use of operations involving indices, that the conception of a logarithm would present itself in that connection as a fairly obvious one. We must however remember that, at the time of Napier, the notion of an index, in its generality, was no part of the stock of ideas of a mathematician, and that the exponential notation was not yet in use. — HOBSON.

Independent tables were computed by the astronomer Bürgi and published at Prague in 1620. Both Napier and Bürgi, while not basing their work directly on the relation which we should express by the equivalent equations $x = a^y$ and $y = \log_a x$, were in effect avoiding fractional values of y by taking values of a near 1, their actual values being $a =$ .9999999 and $a = 1.0001$, respectively. Napier was also influenced by his desire that sines and cosines (which are by definition proper fractions) have positive logarithms. It should be noted that Napier's logarithms are not the natural logarithms with base e, which are commonly used in the calculus. The relation is

$$\text{Nap.}\log x = 10^7 \log_{\frac{1}{e}} \frac{x}{10^7}.$$

If we introduce our modern graphical interpretation of $y = \log_a x$, Bürgi is concerned with the determination of abscissas

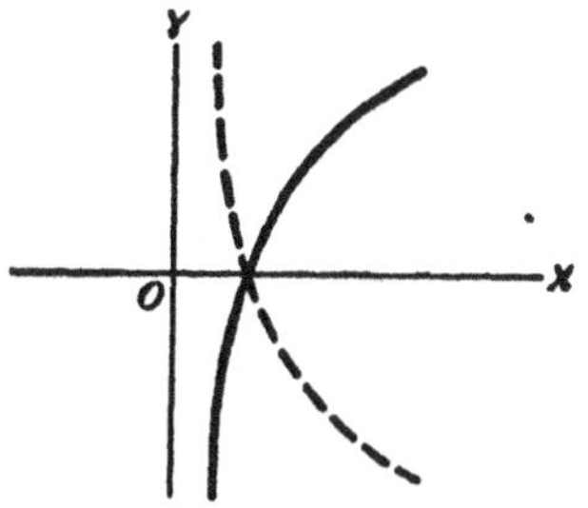

FIG. 40. — LOGARITHMIC CURVES.

of points where the exponential curve is met by the horizontal straight lines $y = c$, where c takes successive integral values. A base a near 1 naturally gives values of x near each other, the full line corresponding to a base a, greater than 1, the dotted curve to a base a less than 1 where $a_1a_2 = 1$.

In 1615 Henry Briggs, afterwards Savilian Professor of Geometry at Oxford, wrote of Napier "I hope to see him this summer, if it please God, for I never saw book which pleased me better, or made me more wonder." In connection with this and later visits it was soon discovered that great simplification in the practical use of logarithms would result from taking log 1 = 0 and log 10 = 1 and giving up the restriction of logarithms to integral values, thus making the decimal parts of all logarithms depend wholly on the sequence of digits. Napier had been so predominantly interested in trigonometric applications that his table consisted not of logarithms of abstract numbers, but of 7-place logarithms of the trigonometric functions for each minute. In connection with his change of the base, Briggs developed interesting methods of interpolating and testing the accuracy of logarithms. He gives the logarithms from 1 to 20,000 and from 90,000 to

100,000 to 14 places, computing also 10-place trigonometric tables with an angular interval of 10 seconds.

Kepler recognized immediately the enormous significance of the new logarithmic method and addressed an enthusiastic panegyric to Napier in 1620, not knowing that he had died in 1617. What if logarithms had been invented in time to save Kepler his vast computations? — GUTZMER.

Ezechiel De Decker, a Dutch surveyor, assisted by Adriaen Vlacq, a Dutch bookseller residing in Gouda, published in 1626–27 the logarithms from 1 to 100,000 and thus filled the gap in Briggs's table, and this is the basis for the tables since published. In more recent times methods of interpolation have been employed which are more powerful and less laborious, while ordinary computation has been simplified by avoiding the use of too many decimal places, and by the mechanical device of the slide-rule. The modern computing machine naturally tends to supersede the logarithmic method. Among the remarkable computations characteristic of the sixteenth century may be mentioned Ludolph van Ceulen's achievement in computing π to 35 decimal places, using regular polygons of 96 and 192 sides. German writers in consequence have sometimes attached his name to this important constant.

In England Thomas Harriott (1560–1621) and William Oughtred (1575–1660) rendered important services in introducing the most recent advances in arithmetic, algebra, and trigonometry. The former rejected negative and imaginary roots indeed, but used the signs $>$ and $<$, denoted a^2 by aa, etc. Oughtred, the inventor of the slide-rule, used the symbols $\times$ and : :, also the contractions for sine, cosine, etc.

TWO NEW SCIENCES

Even after Galileo's condemnation by the Inquisition, though old, infirm, and nearly blind, his scientific ardor was unquenched, and in 1638 he published (at Leyden) a work on mechanics under the title, *Conversations and Mathematical Demonstrations on Two New Branches of Science*, which constituted the most notable progress in mechanics since Archimedes. He says:

> My purpose is to set forth a very new science dealing with a very ancient subject. There is, in nature, perhaps nothing older than motion, concerning which the books written by philosophers are neither few nor small; nevertheless I have discovered by experiment some properties of it which are worth knowing and which have not hitherto been either observed or demonstrated. Some superficial observations have been made, as, for instance, that the free motion (*naturalem motum*) of a heavy falling body is continuously accelerated; but to just what extent this acceleration occurs has not yet been announced; for so far as I know, no one has yet pointed out that the distances traversed, during equal intervals of time, by a body falling from rest, stand to one another in the same ratio as the odd numbers beginning with unity.
>
> It has been observed that missiles and projectiles describe a curved path of some sort; however no one has pointed out the fact that this path is a parabola. But this and other facts, not few in number or less worth knowing, I have succeeded in proving; and what I consider more important, there have been opened up to this vast and most excellent science, of which my work is merely the beginning, ways and means by which other minds more acute than mine will explore its remote corners.
>
> This discussion is divided into three parts; the first part deals with motion which is steady or uniform; the second treats of motion as we find it accelerated in nature; the third deals with the so-called violent motions and with projectiles. . . . — *Two New Sciences*, p. 153.

Throughout this work Galileo depends on the results of experiments, rather than mere speculation, on measurements made as exactly as his instruments permitted. He doubted the ancient and scholastic ideas of the "levity" of air and that "nature abhors a vacuum" and devised apparatus by which he weighed air and determined its specific gravity with reference to water. His estimate was 1/400; the true value is 1/773. When confronted with the fact that a pump [1] would

[1] "This pump worked perfectly so long as the water in the cistern stood above a certain level; but below this level the pump failed to work. When I first noticed this phenomenon I thought the machine was out of order; but the workman whom I called in to repair it told me the defect was not in the pump but in the water which had fallen too low to be raised through such a height; and he

not draw water to a height of more than "18 cubits" (about 33 feet) he failed to recognize that this represented the pressure of the atmosphere, but by means of a cylinder with a

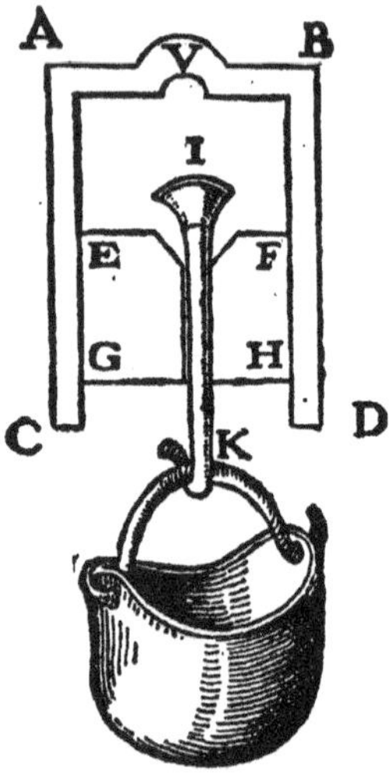

FIG. 41. — DEVICE FOR MEASURING THE FORCE OF A VACUUM. ABDC, a cylinder of metal or glass; EFHG, wooden stopper perforated to receive the iron wire IK. When the cylinder is filled with water and inverted with a weighted vessel on the hook K, the valve I prevents air from entering, and a vacuum forms at V. From Galileo, *Dialogues Concerning Two New Sciences*, trans. Crew and de Salvio, p. 14. Permission of The Macmillan Company.

closely fitted piston and valve for admission of water he weighed the "force of a vacuum." He shows experimentally that a body descends an inclined plane with uniformly accelerated motion.

In a board 12 ells in length a groove half an inch wide was made. It was drawn straight and lined with very smooth parchment. The

added that it was not possible, either by a pump or by any other machine working on the principle of attraction, to lift water a hair's breadth above eighteen cubits; whether the pump be large or small this is the extreme limit of the lift. Up to this time I had been so thoughtless that, although I knew a rope, or rod of wood, or of iron, if sufficiently long, would break by its own weight when held by the upper end, it never occurred to me that the same thing would happen, only much more easily, to a column of water. And really is not that thing which is attracted in the pump a column of water attached at the upper end and stretched more and more until finally a point is reached where it breaks, like a rope, on account of its excessive weight? . . ." *Loc. cit.*, p. 16.

board was then raised at one end, first one ell, then two. Then Galileo let a polished brass ball roll through the groove and determined the time of descent for the whole length of the groove. If on the other hand he let the ball roll through only one quarter of the length, this required just half the time. . . . The distances were to each other as the squares of the times, — *Ibid.*, p. 178,

a law verified by hundredfold repetitions for all sorts of distances and slopes. The time was still determined by weighing water escaping through a small orifice. He shows by ingenious experiments the dependence of velocity on height alone, and that a freely falling body has the necessary energy to reach its original level. The whole theory of the falling body is now easily deduced.

When, therefore, I observe a stone initially at rest falling from an elevated position and continually acquiring new increments of speed, why should I not believe that such increases take place in a manner which is exceedingly simple and rather obvious to everybody? If now we examine the matter carefully we find no addition or increment more simple than that which repeats itself always in the same manner. This we readily understand when we consider the intimate relationship between time and motion; for just as uniformity of motion is defined by and conceived through equal times and equal spaces (thus we call a motion uniform when equal distances are traversed during equal time-intervals), so also we may, in a similar manner, through equal time-intervals, conceive additions of speed as taking place without complication; thus we may picture to our mind a motion as uniformly and continuously accelerated when, during any equal intervals of time whatever, equal increments of speed are given to it. . . .

Hence the definition of motion which we are about to discuss may be stated as follows: A motion is said to be uniformly accelerated, when starting from rest, it acquires, during equal time-intervals, equal increments of speed. . . .

The time in which any space is traversed by a body starting from rest and uniformly accelerated is equal to the time in which that same space would be traversed by the same body moving at a uniform speed whose value is the mean of the highest speed and the speed just before acceleration began. . . .

The spaces described by a body falling from rest with a uniformly accelerated motion are to each other as the squares of the time-intervals employed in traversing these distances. . . . — *Ibid.*, pp. 161, 173, 174.

Galileo passes from falling bodies to pendulums, in which the friction of the inclined plane is absent and air resistance negligible. He appreciates the possibility of utilizing the pendulum for time measurement, and devises a simple apparatus for the purpose, foreshadowing the invention of the clock. He discovers that the time of vibration of the pendulum varies as the square root of the length.

He analyzes correctly the component motions of a projectile, recognizing the law of the parallelogram of motion, as distinguished from the parallelogram of forces discovered by Newton. He shows that whether the initial direction of aim is horizontal or not, the path described is a parabola with axis vertical, explicitly neglecting air resistance and change of direction of vertical force.

I now propose to set forth those properties which belong to a body whose motion is compounded of two other motions, namely, one uniform and one naturally accelerated; these properties, well worth knowing, I propose to demonstrate in a rigid manner. This is the kind of motion seen in a moving projectile; its origin I conceive to be as follows: . . .

A projectile which is carried by a uniform horizontal motion compounded with a naturally accelerated vertical motion describes a path which is a semi-parabola. — *Ibid.*, pp. 244, 245.

All this Dynamics was practically pioneer work of enormous importance for the future of mechanics. Heretofore motion had been held to imply a continuous cause; with Galileo this was true only of change in motion. For him space and time were fundamental concepts, but the nature of force was unknown. The mathematics of the age was inadequate for dealing with motion due to a variable force.

In Statics Galileo had somewhat more from the ancients to build upon. To him we owe the formulation of the law of virtual velocities — applying dynamical ideas to problems of statics. If two forces are in equilibrium they are proportional to the corresponding paths, or: What one by any machine gains in power is lost in distance. The parallelogram or triangle of forces in equilibrium however escapes him, and his ideas about impulse though remarkably in advance of his time were not fully worked out.

He investigates strength of materials under tension and fracture, with reference to practical applications in construction. He draws just inferences in regard to the relation between strength and size of plants and animals as well as machines, comparing for example hollow bones and straws with solid bodies of similar mass. He derives an important formula for the stiffness of a horizontal beam supported at one end and regarded as a lever. He discusses the curve formed by a cord suspended between two points, recognizing that it is not a parabola.

In Hydrostatics he reviews the known work of Archimedes and corrects the error of the Aristotelians in regard to the dependence of floating on specific gravity. He develops the modern theory that the fundamental factor in the mechanics of fluids is that they consist of freely moving particles yielding to the slightest force. He makes effective application of the principle of virtual velocities to fluids. At the close of his third conversation he expresses his modest confidence in the great future of his new ideas.

> The theorems set forth in this brief discussion will, when they come into the hands of other investigators, continually lead to wonderful new knowledge. It is conceivable that in such a manner a worthy treatment may be gradually extended to all the realms of nature. — *Ibid.*, p. 154.

— a prediction magnificently fulfilled in succeeding generations.

Among other branches of physics in which Galileo accomplished work of value may be mentioned the expansion by heat — the beginnings of thermometry, experiments on the acoustics of vibrating cords and plates, discovering the dependence of harmony on the ratio of the rates of vibration, and the relations of length, thickness, and tension of cords.

He explains resonance and dissonance. He assumes light to have a finite velocity, but does not succeed in measuring it.

Let each of two persons take a light contained in a lantern, or other receptacle, such that by the interposition of the hand, the one can shut off or admit the light to the vision of the other. Next let them stand opposite each other at a distance of a few cubits and practice until they acquire such skill in uncovering and occulting their lights that the instant one sees the light of his companion he will uncover his own. After a few trials the response will be so prompt that without sensible error the uncovering of one light is immediately followed by the uncovering of the other, so that as soon as one exposes his light he will instantly see that of the other. Having acquired skill, at this short distance, let the two experimenters, equipped as before, take up positions separated by a distance of two or three miles and let them perform the same experiment at night, noting carefully whether the exposures and occultations occur in the same manner as at short distances; if they do, we may safely conclude that the propagation of light is instantaneous; but if time is required at a distance of three miles which, considering the going of one light and the coming of the other, really amounts to six, then the delay ought to be easily observable. If the experiment is to be made at still greater distances, say eight or ten miles, telescopes may be employed, each observer adjusting one for himself at the place where he is to make the experiment at night; then although the lights are not large and are therefore invisible to the naked eye at so great a distance, they can readily be covered and uncovered since by aid of the telescopes, once adjusted and fixed, they will become easily visible. . . . — *Ibid.*, p. 43.

He seeks to apply to astronomical phenomena the new discoveries in magnetism.

Galileo was not chiefly interested in mathematics, but he emphasizes the dependence of other sciences upon it. He gives an acute discussion of infinite, infinitesimal and continuous quantities leading up to the conclusion "that the attributes 'larger,' 'smaller,' and 'equal' have no place either in comparing infinite quantities with each other or in comparing infinite with finite quantities." Again "the finite parts of a *continuum* are neither finite nor infinite but correspond to every assigned number."

In commenting on Galileo's achievements, Lagrange the great mathematician of the eighteenth century says:

> These discoveries did not bring to him while living as much celebrity as those which he had made in the heavens; but to-day his work in mechanics forms the most solid and the most real part of the glory of this great man. The discovery of Jupiter's satellites, of the phases of Venus, and the Sun-spots, etc., required only a telescope and assiduity; but it required an extraordinary genius to unravel the laws of nature in phenomena which one has always under the eye, but the explanation of which, nevertheless, had always baffled the researches of philosophers. — FAHIE, *Galileo*, p. 348.
>
> Leonardo da Vinci likens a scientific conquest to a military victory in which theory is the field marshal, experimental facts the soldiers. The philosophers who preceded Galileo had, in the main, been trying to fight battles without soldiers. — CREW.

A PIONEER IN MECHANICS: STEVIN

Even before Galileo, Simon Stevin of Bruges (1548–1620), a man who thought independently on mechanical problems, made the first really important advances since Archimedes, eighteen centuries earlier. First active as a bookkeeper in Flanders, he came to the Northern Netherlands, where he taught mathematics to Prince Maurice of Nassau, who appointed him quartermaster-general of the Dutch army. He became an authority on civil and military engineering. He was influential in improving methods of public statistics and accounting, and advocated decimal weights and measures. Appreciating the possibilities of the decimal fraction he asserted (1585) that fractions are quite superfluous, and every computation can be made with whole numbers, but he did not realize the simplest notation. The honor of this great invention he shares with Bürgi of Cassel. Another of his inventions was a sailing carriage carrying 28 people and outstripping horses.

In a treatise on statics and hydrostatics (1586) he introduced comparatively new and powerful geometrical methods for dealing with mechanical problems. Among the most interesting is his discussion of the inclined plane by means of an endless chain hanging freely over a triangle with unequal sides. Excluding the inadmissible hypothesis of perpetual motion, the uniform chain must be in equilibrium in any position. The hanging portion is by itself in equilibrium, therefore the two inclined sections must balance each other, and either would be balanced by a vertical force corresponding to the altitude of the triangle. Arriving thus at the parallelogram of forces in equilibrium, he expresses his astonishment by exclaiming "Here is a wonder and yet no wonder."

In studying pulleys and their combinations he arrives at the far-reaching result that in a system of pulleys in equilibrium "the products of the weights into the displacements they sustain are respectively equal" — a remark containing the principle of virtual displacement. He reaches correct results in regard to basal and lateral pressure by reasoning analogous to that about the chain, and by assuming on occasion that a definite portion of the liquid is temporarily solidified. By ingenious experiments he proves the dependence of fluid pressure on area and depth, and takes proper account of upward and lateral pressure. He studies the conditions of equilibrium for floating bodies, showing that the center of gravity of the body in question must lie in a perpendicular with that of the water displaced by it, and that the deeper the center of gravity of the floating body the more stable is the equilibrium.

In analyzing the lateral pressure of a fluid Stevin anticipates the calculus point of view by dividing the surface into elements on each of which the pressure lies between ascertainable values. Increasing the number of divisions, he says it is manifest that one could carry this process so far that the difference between the containing values should be made less than any given quantity however small — all quite in harmony with our present definitions of a *limit*.

Stevin's work and that of Galileo seem to have been quite independent of each other, the former confining his theory to statics, the latter laying a solid foundation for the new science of dynamics. Torricelli, a disciple of Galileo best known for his invention of the mercurial barometer (p. 302), but writing much on geometry, extended dynamics to liquids, studying the character of a jet issuing from the side of a vessel.

Throughout this period the universities lagged. In Italy Galileo lectured to medical students who were supposed to need astronomy for medical purposes — i.e., astrology. At Wittenberg there was a professor for arithmetic and the sphere, and one for Euclid, Purbach's planetary theory and the *Almagest*, but their students were few. . . . So, says a German writer, we face the extraordinary fact that the most educated of the nation were as helpless in the problems of daily life as a marketwoman of today. The university lectures in mathematics were mainly confined to the most elementary computation — matters taught more thoroughly in the commercial schools, particularly after the invention of printing.

REFERENCES FOR READING

GALILEO GALILEI, *Dialogues Concerning Two New Sciences* (Trans. Crew and de Salvio), 1914.

HOBSON, E. W., *John Napier and the Invention of Logarithms*, 1914.

MACH, ERNST, *Science of Mechanics*, Ed. 4 (for Galileo and Stevin), 1919.

MORLEY, HENRY, *Jerome Cardan*, 1854.

From previous lists: Ball, *Hist. Math.*, Ch. XII, XIII; Fahie, *Galileo;* Lodge.

NOTE: The diagrams of muscular action facing this page are from Borelli, *De motu animalium*, 1685, Plate III. Courtesy of the Boston Medical Library.

Chapter XII

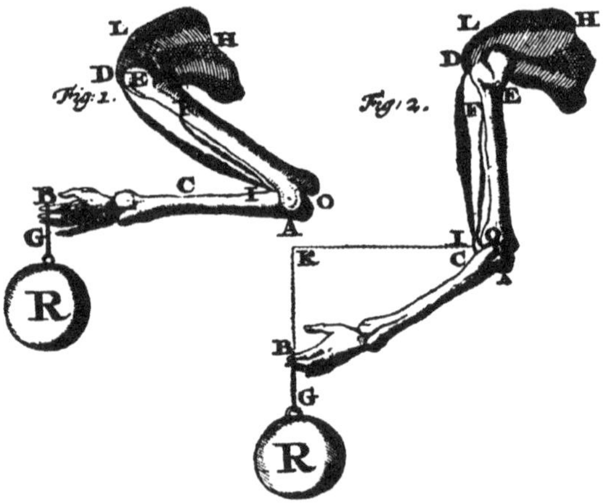

Natural and Physical Science in the Seventeenth Century

The genius of seventeenth-century Europe brought forth many imperishable masterpieces in literature and art, . . . and gave birth to liberty and popular government . . . But the supreme glory of the seventeenth century — or, more accurately, of the hundred and fifty years beginning with Copernicus, Cardan, and Vesalius, and ending with Newton and Huygens — and the chief importance of that period in history, lies in its scientific achievements. Among all the brilliant discoveries of that age, none was more dazzling or ultimately more momentous than that of science itself. — P. SMITH, *History of Modern Culture*, p. 144.

Within a century the method of constructing Nature by reasoning, more or less logically, from accepted philosophical ideas, was replaced by the method of appealing to Nature by experiment, and adopting only such laws as were in quantitative agreement with the results of experiment. . . . The search in the mind for first causes gave way to a search in Nature for observable order and regularity, coupled with a disinclination to speculate beyond the facts. — E. N. DA C. ANDRADE, "Science in the Seventeenth Century," *Nature*, **142,** 19–30, 1938.

THE PROGRESS OF SCIENCE IN THE SEVENTEENTH CENTURY

In the seventeenth century were seen most, if not all, of the achievements that mark the transition from the Renaissance to the Period of Modern Science. Of foremost importance were the formulation of the mathematical laws of gravitation and the development of the infinitesimal calculus, and fundamental for the development of modern physics were the discovery of the laws of motion, the physics of the pendulum, and the pressure of the atmosphere. These are discussed in Chapters XI and XIII. The demonstration of the circulation of the blood is the beginning of modern physiology. Almost equally important to chemistry and to biology is the discovery that combustion and life depend upon one and the same substance in the air. Fundamental in biology are the discovery of unit structures, "cells," in the tissues of plants and animals, and the experimental proof of sex in plants, as in animals; and that the ancient belief in the "spontaneous generation" of certain kinds of animals is untrue. The foundation for the systematic description of animals and plants was laid by the definition of "species."

The progress of observational and experimental science was accelerated by the invention of "philosophical instruments," such as the air-pump, reflecting telescope, microscopes — simple and compound, the barometer, the thermometer, and the pendulum-clock.

THE NEW PHILOSOPHY

Sir Francis Bacon (1560–1626), Baron of Verulam, Lord Chancellor of England, was the pioneer philosopher of the seventeenth century. He made no scientific discovery, and Dr. Harvey, who worked while Bacon wrote, said of him, "the Lord Chancellor writes of science like — a Lord Chancellor." Nevertheless, because of his official position and immense philosophical and literary ability, he was able to draw universal attention to the methods of science and especially to the method of investigation by induction, so that his indirect service to science was great. Some idea of Bacon's method may be gained from the following passages in his great work, the *Novum Organum*, 1620:

For man is but the servant and interpreter of nature: what he does and what he knows is only what he has observed of nature's order in fact or in thought; beyond this he knows nothing and can do nothing. For the chain of causes cannot by any force be loosed or broken, nor can nature be commanded except by being obeyed. . . . And all depends on keeping the eye steadily fixed upon the facts of nature and so receiving their images simply as they are. — (*Works*, I (2), p. 58.)

After discussing the current errors, "idols," of philosophy, he goes on to say:

Those who have handled sciences have been either men of experiment or men of dogmas. The men of experiment are like the ant; they only collect and use: the reasoners resemble spiders, who make cobwebs out of their own substance. But the bee takes a middle course; it gathers its material from the flowers of the garden and of the field, but transforms and digests it by a power of its own. Not unlike this is the true business of philosophy; . . . — *Novum Organum*, Aphorism XCV (*Works*, I (2), p. 131).

René Descartes (1596–1650) (Cartesius), whose life and fundamental contributions to mathematics are described in the next chapter, was another pioneer of modern philosophy. In the hope of placing science on firm ground, he attempted in his *Discourse on Method* (1637) and later works to explain the whole material universe as a machine. This included the phenomena of corporeal life, which he believed to occur of mathematical necessity without intervention of any kind of spiritual force; although the idea of God had the central place in his philosophy and he affirmed the immortality of the soul of man, carefully avoiding any offense to the Church.

ORGANIZATION OF THE FIRST SCIENTIFIC SOCIETIES

Scientific societies have played a very important part in the advancement of science by exchange and discussion of ideas and by publication. With the possible exception of an academy formed by Leonardo da Vinci in the fifteenth century — the first devoted chiefly to science was probably that founded by della Porta at Naples in 1560 and named Academia Secretorum Naturae. The requirement for membership was to have made some discovery in natural science. Della Porta fell under ecclesiastical suspicion as a practitioner of the black arts, and although acquitted was ordered to close his "Academy." The Accademia dei Lincei (of the Lynx), founded at Rome in 1603, included both della Porta and Galileo among its early members, and still flourishes. The famous Accademia del Cimento (Experiment) was founded in 1657 by pupils of Galileo in Florence, including its patrons Grand Duke Ferdinand II and his brother, Prince Leopold de Medici.

The first scientific society in northern Europe is the Royal Society of London, which had its origin about 1645 in secret weekly meetings of "divers worthy persons, inquisitive into natural philosophy and other parts of human learning" at Gresham College and elsewhere in London, also for a time in Oxford. Dr. Wallis, one of the first members, thus describes (1696) these early meetings:

"Our business was (precluding matters of theology and state affairs) to discourse and consider of philosophical enquiries, and such as related thereunto: — as Physick, Anatomy, Geometry, Astronomy, Navigation, Staticks, Magneticks, Chymicks, Mechanicks, and Natural Experiments; with the state of these studies and their cultivation at home and abroad. We then discoursed of the circulation of the blood, the valves of the veins, the *venae lacteae*, the lymphatic vessels, the Copernican hypothesis, the nature of comets and new stars, the satellites of Jupiter, the oval shape (as it then appeared) of Saturn, spots on the sun and its turning on its own axis, the inequalities and selenography of the moon, the several phases of Venus and Mercury, the improvement of telescopes and grinding of glasses for that purpose, the weight of air, the possibility or impossibility of vacuities and nature's abhorrence thereof, the Torricellian experiment in quick silver, the descent of heavy bodies and the degree of acceleration therein, with divers other things of like nature, some of which were then but new discoveries, and others not so generally known and embraced as now they are; with other things appertaining to what hath been called the New Philosophy, which from the time of Galileo at Florence and Sir Francis Bacon (Lord Verulam) in England, hath been cultivated in Italy, France, Germany, and other parts abroad, as well as with us in England." — (Huxley.)

After the Restoration the meetings could be held openly, and in 1662 the Royal Society was chartered by Charles II. The *Philosophical Transactions* have been published continuously to this day, and the F.R.S. (Fellow of the Royal Society) has remained the highest distinction of British men of science.

The French Academie des Sciences had a similar origin before 1638 in weekly meetings of savants to make experiments and discuss their discoveries. The Academy was chartered by Louis XIV in 1666 with twenty-one members, appointed and granted salaries as government officials. The corresponding Berlin Academy began in 1700.

THE CIRCULATION OF THE BLOOD

In 1598 Padua stood highest among the universities, with Galileo in the chair of mathematics expounding his quantitative experimental methods and Fabricius in the chair of anatomy preparing his work on the values of the veins to be published in 1603. It was to Padua that William Harvey (1578–1657), after graduating at Cambridge, went in 1598 to study medicine, taking the M.D. degree four years later. Upon his return he received the M.D. in Cambridge and settled in London. It is evident that his medical practice did not prevent him from making observations and experiments on the movements of the heart and blood, for his notes show that he described the circulation of the blood in the first course of lectures after his appointment in 1615 as Lumleian Lecturer to the College of Physicians. His great work, *De Motu Cordis*, was published in 1628. Familiar with the works of Galen and Vesalius, with the description of the pulmonary circulation by R. Columbus (1559), and with the views of Fabricius on the valves, Harvey approached the problem from an experimental, mechanical point of view. First from the study of the *action* of the heart in a great variety of living animals, including invertebrates, he obtained a correct idea of its mechanism and of the relation of its contraction to the pulse. He saw that in mammals *all* the blood in the right ventricle must pass through the lungs to reach the left ventricle. Then he extended this view to the rest of the body, and showed that the *volume* of blood pumped from the left ventricle into the aorta could be accounted for only by the assumption that it returned to the right ventricle through the veins, thus flowing in a continuous *circuit.* Finally, this assumption explained the arrangement of the *valves* in the veins, permitting flow in one direction only toward the heart, not away from it as Fabricius had thought. Harvey's theory at first met strong opposition from the Galenists, but he lived to see it generally accepted without any dispute as to priority.

But Harvey never saw the peripheral connections between arteries and veins. The first to demonstrate this connection was Marcello Malpighi (1628–94) professor at Bologna, in his first work, two letters to his friend, Borelli, published in 1661, which gave the first anatomical basis for a true conception of the respiratory process. Finding the lung of the dog too difficult, he turned to the simpler lung of the frog and with "a microscope of two lenses" saw in the living frog the passage of blood through the capillary vessels from arteries to veins. In the mesentery of the frog this was even clearer. He showed that the blood does not escape into the air spaces, but always is contained in tubules. He saw also the corpuscles but mistook them for fat globules.

It remained, however, for Antony van Leeuwenhoek (1632–1723) of Delft, to complete the work of Harvey on the circulation of the blood by observations on a large number of animals. The most striking of these were observations on the flow of blood through the capillaries in the tails of young fishes and tadpoles. He also described and figured for the first time the corpuscles.

Knowledge of the circulation of the body-fluids of man and other mammals is incomplete without the lymph. The first view of the lymphatic vessels was obtained by Gasparo Aselli (1581–1626) before Harvey's work was published. In 1622 he discovered the "lacteals" in the mesentery of a dog and recognized that they conveyed material from the intestine, but thought that they discharged into the liver. This error was corrected by Jean Pecquet, 1651, when he discovered the thoracic duct connecting the lacteals with the veins at the base of the neck; and the description of the lymph vessels of other parts of the body was made two years later by Thomas Bartholin and Olof Rudbeck.

> The new theory of the circulation made for the first time possible true conceptions of the nutrition of the body, it cleared the way for the chemical appreciation of the uses of the blood, it afforded a basis which had not existed before for an understanding of how the life of any part, its continued existence and its power to do what it has to do in the body, is carried on by the help of the blood. — FOSTER, *Lectures*, p. 48.

ADVANCES IN HUMAN ANATOMY

During the seventeenth century important advances in human anatomy were made in England, especially with regard to the structure of the liver and other glands and the brain. Knowledge of the glands was made more precise by the work of Niels Stensen (Nicolaus Steno) of Copenhagen (1662). He clarified the distinction previously made by F. Sylvius, between secreting glands with ducts and the ductless (lymphatic) glands, and presented a rational theory of secretion. Knowledge of the reproductive organs was greatly advanced by Reinier de Graaf in 1672.

EMBRYOLOGY

Modern embryology may be said to have begun with the work on the development of the chick which Harvey described in his *Treatise on Generation* in 1651. By means of a hand lens he was able to trace the development of the chick in a general way back to the beginning of the second day. He found that the embryo arises from the light-colored spot in the yolk, called by Fabricius the *cicatricula*. Harvey saw this dilate and become divided into circles, forming what we now call the *area embryonalis*, and this he compared to the iris of the eye and called the *oculum ovi*. At the end of the third day he discovered a fine line of red around the edge of the spot, "and nearly in its center there appears," he says, "a leaping point of the color of blood so small that when it contracts it almost entirely escapes the eye, when it dilates it shows like the smallest spark of fire. Such is the outset of animal life which the plastic force of nature puts in motion from the most insignificant beginnings." He was unable to distinguish any other parts of the embryo until the fifth day, and he describes in a very interesting way the gradual appearance of the various organs of the embryo as he saw them from that time until the perfect chick is formed.

Harvey stimulated a long series of researches by his assertion that "all animals whatever, even viviparones also, nay Man himself, to be made of an egge: and that the first conception of all living creatures which bring forth young are certain egges . . ." By "egg" he meant any structure or substance endowed with the *generative principle*, and hence he was unable to reject entirely the idea of spontaneous generation. He accepted Aristotle's view that the function of the male is to supply a sort of contagion, an immaterial *aura seminalis*.

Another great generalization made by Harvey is his theory that "the more perfect animals with red blood are made by epigenesis, or the superaddition of parts."

Malpighi, working with a better microscope (1672 and 1673) found that the blood is not "the first engendered part," as Harvey had asserted, for he could see an outline of the embryo before the appearance of the "*punctum sanguineum*." This small detail of observation led Malpighi to reject Harvey's whole theory of epigenesis, and to assert instead that the embryo as a whole is *predelineated* in the unincubated egg. Thus arose a great discussion that lasted for more than a century.

SPONTANEOUS GENERATION

The influence of membership in the Accademia del Cimento is seen in the attack on the question of spontaneous generation by Francesco Redi (1668). By well-conceived experiments he proved that maggots do not arise spontaneously in putrid flesh but are deposited by or hatched from the eggs of flies. He attempted to make this principle general, even dissecting out the eggs of a Mantis. But he failed to explain the presence of grubs in living plants. This gap was filled by Malpighi in his work on galls and by Antonio Vallisnieri in his *Observations on the Fly of Rosebushes*, 1713. Meanwhile, Leeuwenhoek had demonstrated the development of mussels, flies, and aphids from eggs; and the idea of spontaneous generation would have been dead had not the microscope revealed minute forms of life hitherto unsuspected.

MINUTE STRUCTURE OF ANIMALS AND PLANTS

Galileo is said to have been the effective inventor of the compound microscope — a Galilean telescope reversed; and to have made the first biological observation, on the eye of an insect, with such an instrument. Its optical properties were worked out by Kepler and by Huygens, the latter inventing an improved eyepiece. By 1625 many opticians were making compound microscopes, but working naturalists soon found simple lenses more serviceable. The first systematic investigation with the microscope was made by the Accademia del Linci and a member, Francesco Stelluti, produced in 1630 the first printed figure made with the aid of the microscope. Robert Hooke (1635–1703) of Oxford examined many objects with his compound microscope, publishing the results in his *Micrographia* (1665). Of special importance is his method of making *thin sections* and his discovery in a plant tissue, cork, of cavities he called *cells*.

Leeuwenhoek, whose only language was Dutch, made simple microscopes with which he examined all sorts of things, and reported his discoveries to the Royal Society in letters that were translated and published in the *Philosophical Transactions*, with his sketches. Besides his important observations on the circulation and on the reproduction of certain invertebrates, he was the first to describe yeast cells, bacteria, various protozoa, and the like. In a letter dated 1677, translated into

Latin and published in 1678, he announced a major event in the history of biology — the discovery by a medical student, Johan Ham, of "animalcules" (spermatozoa) in human semen. Later he described the spermatozoa of many animals and concluded that the true germ is the spermatozoon, the egg

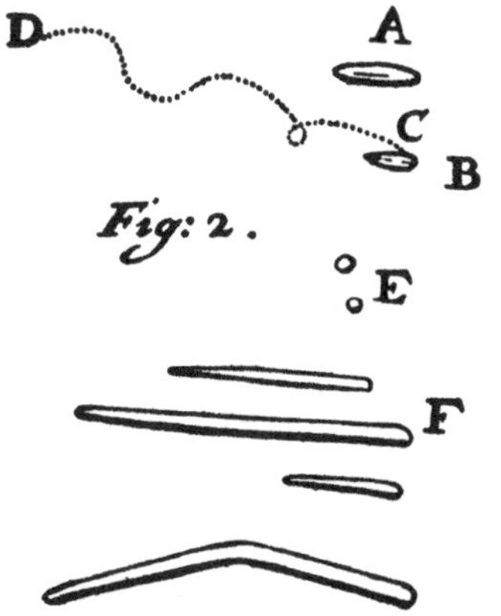

FIG. 42. — EARLIEST DRAWING OF BACTERIA, "Animalcules" from between human teeth. A, an actively motile form; B, another, more numerous; D-C, its path; E, very numerous minute forms that "moved like gnats playing in the air"; F, most abundant form, non-motile. After Leeuwenhoek, *Phil. Trans.*, **17**, No. 197, Fig. 2.

being merely nutritive. Thus started the controversy between the Ovists and the Spermists that lasted well into the nineteenth century.

Malpighi's work on the lung brought to him the F.R.S. and practically all his works after that were published by the Royal Society. His work on the *Viscera*, 1666, includes detailed descriptions of the minute anatomy of the liver, kidney, and spleen that go far beyond any previous work. His book on the *Silk Worm*, 1669, is the first detailed description of insect anatomy. His *Anatome Plantarum*, 1675–79, beautifully illustrates the structure of flowering plants; includes descriptions of cells, called by him utricules, sacculi, and globules; and gives the first description of the development of the seed and embryo. Thus he is a founder of both animal and plant histology.

It may be truly said of Malpighi that whatever part of natural knowledge he touched he left his mark; he found paths crooked and he left them straight, he found darkness and he left light. . . . Doubtless Malpighi was reaping what Harvey had sown; doubtless he was also reaping what Galileo had sown; doubtless also the microscope gave him a tool which none before him had possessed. It was just the putting these three things together which parts him from the old times, and makes him the beginning of the new. — FOSTER, *loc. cit.*, p. 120.

The *Anatomy of Plants*, 1672–82, by Nehemiah Grew (1641–1712), also well illustrated, entitles him to rank with Malpighi as a founder of plant anatomy. But he failed to understand the cellular structure of plants, attributing the appearance of sections to a lace-like inter-weaving of fibers. He extracted the green pigment from leaves without, of course, realizing the supreme importance of this substance.

The last great microscopist of the seventeenth century was Jan Swammerdam (1637–80). His *Historia Insectorum Generalis*, was printed in 1669, but the main part of his extraordinary work on the metamorphosis of insects, the *Biblia Naturae*, based on incredibly minute dissections, did not appear until published through the interest of friends long after his death and had little effect on seventeenth-century science.

THE IDEA OF "SPECIES"

The extensive writings of John Ray (1627 or 28–1705) on animals and plants with improved arrangement of the larger groups, clearly defined by anatomical characters, mark a new era in descriptive biology. He was the first to introduce a precise conception of "species" as a group of individuals derived from similar parents and capable of reproducing their kind. This new definition of the ancient Aristotelian term has great theoretical importance. The number of species is fixed because God rested on the seventh day from all his works. Ray recognized, however, the possibility of specific variability. His greatest improvement in botanical classification was the division of herbs into Monocotyledons and Dicotyledons, and he suggested a similar division of trees. His

Wisdom of God Manifested in the Creation, 1691, was very popular and supplied many examples of purposive adaptation and of design in nature.

SEX IN PLANTS

The ancient idea of sex in the date-palm and in figs had almost been forgotten when Grew published (*Phil. Trans.*, 1676) the suggestion of Sir Thomas Millington of Oxford that the anthers of a flower "doth serve as the male, for the generation of the seed." This was accepted as generally applicable by Ray, and was opposed by Malpighi. The first proof of this important link in the chain of biological theory was supplied by R. J. Camerarius of Tübingen. He showed by convincing experiments with a large number of diverse plants (1694) that perfect seeds cannot be formed unless pollen (produced in the anthers) reaches the pistil of the flower.

PHYSICS, CHEMISTRY, AND PHYSIOLOGY

One of the first (the other was Harvey) to apply the new quantitative physics to physiological problems was Santorio Santorio (or Sanctorius, 1561–1636), of Padua. He gauged body temperature with various modifications of Galileo's thermoscope, rated the pulse by comparison with the pendulum, noted relative humidity by means of a loop of gut with a weight attached, and in a chair suspended from a steelyard weighed himself under various conditions of diet, rest, etc. Hence he has been called the first student of metabolism. The "Florentine thermometer," filled with mercury and in the modern form except the scale, was developed by the Accademia del Cimento before 1666.

The mechanistic philosophy of Descartes included a mechanical explanation of human functions in his book *On Man* (*De Homine*, 1662), which has been described as the first textbook of physiology written after the modern fashion. It contains no new observation, but, accepting Harvey's theory of the circulation of the blood, it describes a machine that would be capable of performing all the acts of the human body under the command of the Rational Soul from its seat at the center of the brain. With a similar point of view but very different method, the real founder of bio-physics was G. A. Borelli (1608–79), an enthusiastic student of Galileo's works, member

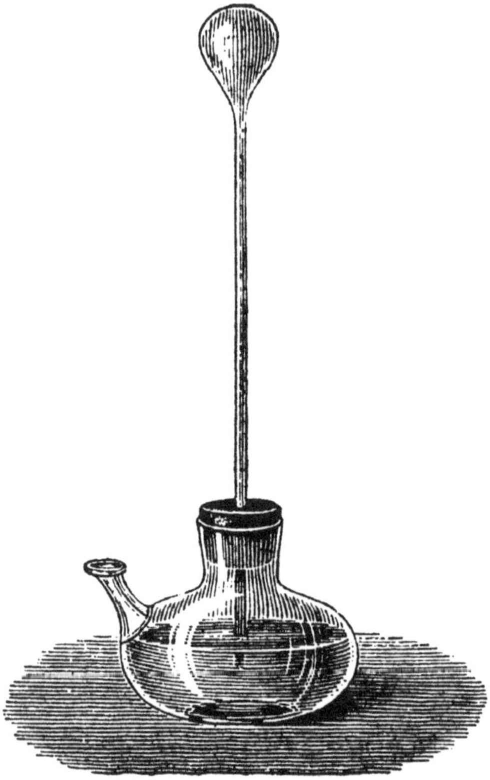

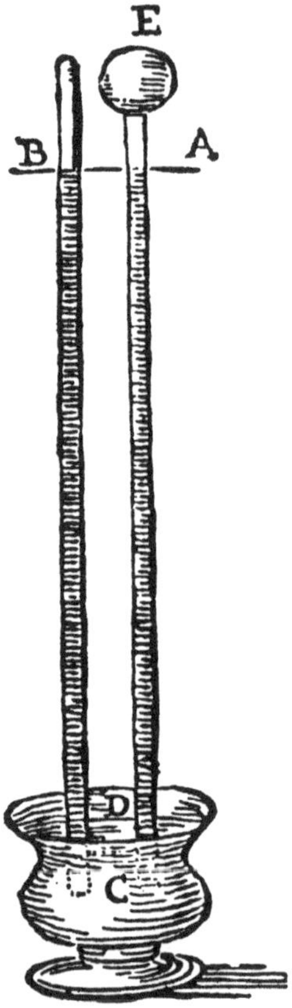

FIG. 43 (left). — GALILEO'S THERMOSCOPE. Fluid rises or falls in the tube when the air in the bulb is cooled or warmed. After Traumüller u. Gerland; from Dannemann, *Naturwiss.*, II, 65.

FIG. 44 (right). — TORRICELLI'S BAROMETER. A and B, two forms of glass tube open at the lower end; D, level of mercury in basin C; BA, height of mercury in each tube; EA, empty space. After Torricelli, *Esperienze dell'Argento Vivo,* 1644; from *Neudrucke von Schriften . . .* , No. 7, 1897.

of the Accademia del Cimento, and friend of Malpighi, who made important contributions to mathematics, astronomy, and physics.

Regarding physiology as a branch of physics, his chief interest was in animal motion, of which he distinguished two kinds — skeletal and visceral. His treatment of the first laid the foundation of muscular mechanics and that of the latter sometimes brilliantly hit the mark. His work, *De Motu Animalium* (1680–81), created a band of followers who often carried iatro- (i.e., medico-) physics to absurd lengths.

Evangelista Torricelli (1608–47), a disciple of Galileo, improved his master's experiment of weighing "the resistance of a vacuum" by substituting mercury for the water and piston. He suspected that the weight of the mercury standing about 30 inches in the tube was sustained, not by the vacuum above, but by pressure of the atmosphere on the mercury in the cup below. It remained, however, for Blaise Pascal (1623–62), "the great moralist," to plan the crucial experiment that proved the "Torricellian vacuum" to be a barometer. During an ascent of the Puy-de-Dôme in 1648 the height of the mercury was found to decrease in comparison with a similar instrument observed at the base of the mountain. A more convincing proof of atmospheric pressure was demonstrated by Otto von Guericke of Magdeburg, 1648, with two hemispheres fitted together, forming a hollow globe and evacuated by an air-pump of his invention.

The Hon. Robert Boyle (1627–91) F.R.S., "the Father of Chemistry and Brother of the Earl of Cork" after school at Eton was sent abroad with a tutor and at Florence he was occupied, he says:

> In the new paradoxes of the great star-gazer Galileo, whose ingenious books, perhaps because they could not be so otherwise, were confuted by a decree from Rome; his highness the pope, it seems, presuming, and that justly, that the infallibility of his chair extended equally to determine points in philosophy as in religion, and loathe to have the stability of that earth questioned in which he had established his kingdom. — *Works*, Vol. I. (*Penny Cyclopaedia*, Vol. 5, p. 298.)

Soon after his return he was admitted to the "Invisible College" that later was the Royal Society. He lived in Oxford from 1654 to 1668 and thereafter in London. At Oxford he established a laboratory and employed as his assistant Robert Hooke, who was afterward "Curator of Experiments" to the Royal Society. Having heard of Guericke's air-pump, they constructed a better one with a vacuum chamber for experiments. When a barometer was in the chamber the mercury in it could be caused to rise or fall at will by changing the air-pressure. Boyle was much interested in the "spring" (elasticity) of the air. With a U-tube closed at one end he reversed the barometer. The tube being full of air, mercury was poured in so that the air was compressed at the closed end; its volume, carefully measured, was inversely proportional to the height of mercury in the other limb — *Boyle's law* (1662). Working together and separately, Boyle and Hooke did a great variety of experiments with the vacuum chamber. The first of Boyle's many scientific works, *New Experiments, Physico-Mechanical* . . ., 1660, (second edition 1662), is a milestone in the history of physics and of the physiology of breathing. In this book is described the experiment of placing a mouse and a candle in the glass receiver and gradually exhausting the air. The candle went out and the mouse died at nearly the same time. This was the fundamental experiment in the physiology of respiration. Boyle also showed that fishes require air dissolved in water in which they live, and further, that sulphur will not burn in a vacuum but can be ignited by a burning-glass when air is admitted. Hooke found that gun-powder will burn in a vacuum, and that either sulphur or charcoal will burn if dropped into molten nitre (saltpetre, KNO_3). Gun-powder being a mixture of these three substances, he concluded that the nitre supplies something like that part of air which can sustain life and combustion. This he called the nitrous part. Hooke also in experiments on artificial respiration before the Royal Society (1667) extended the method of Vesalius to prove that an animal may be kept alive by air blown through the lungs and that the renewal of the air, not the movement of the chest or lungs, is the essential thing.

John Mayow (1643–79), working independently at Oxford, gave an excellent description of the mechanism of human breathing, and was the first to suggest that respiration consists essentially in the union of material in the blood with a particular part of the air that he called "nitro-aërial particles" (*On Respiration*, 1668). In a later work (1674) in which the "Nitro-Aërial Spirit" is discussed at length, he attributes animal-heat to respiration and, with citations of the works of Boyle and others, gives descriptions of his own experiments including those with a mouse or a candle in a bell-jar inverted over water, which prove that either combustion or respiration removes from the air a portion that can be measured, leaving an inactive residue. This first description of what is now called oxygen is not always clear and remained unnoticed until rediscovered a century later.

More fortunate was Boyle, whose first love was chemistry, in the immediate success of his book, *The Sceptical Chymist*, 1661, in which he ridiculed the four elements of the philosophers and the three elements (occult qualities) of the alchemists, and insisted that the term "element" should be restricted to irreducible substances, making a clear distinction between elements and compounds. Boyle was the first to give a method (probably not original) for the preparation of phosphorus, *The Aerial Noctiluca*, 1680, and to describe the properties of this interesting substance.

He experimented also with magnetism and electricity. In a pamphlet of 1675, he approves the idea that amber, wax, glass, etc., when electrified by rubbing give off "a substantial emanation," or "electrical effluvia," and as proof of this he cites the odor produced at the time.

Boyle's criticism of the mystics was applicable to J. B. van Helmont (1577–1644), who had adopted the vitalism and chemistry of Paracelsus, in modified form, while at the same time rejecting the idea of transmutation. He recognized that a metal dissolved by an acid remained in the solution, from which it could be recovered. It was he who coined the word "gas," having in mind "the chaos of the Ancients," for effluvia arising from various fermentations (i.e., ebullitions), which he regarded as the essential process in chemical actions, digestion, and all other physiological changes. Among the gases that he described, the first was "gas sylvestre," derived from burning charcoal, fermenting beer, action of vinegar on shells, and other sources. This gas is not air, he said, but a form of water, of which all things not air are made. That all vegetables are produced out of the single element water, he proved by growing a willow in a pot moistened with rain water or distilled water. At the end of the experiment, lasting five years, the weight of the soil in the pot remained the same while the tree had gained 164 pounds, "derived from water alone." A good quantitative experiment with a wrong conclusion, characteristic of much of van Helmont's work. His importance lies chiefly, perhaps, in his influence upon Franciscus Sylvius (1614–72) a chemist of a very different type, a great teacher who headed at Leyden the first university chemical laboratory, where he became professor in 1658. He rejected every idea of the occult and paid especial attention to the properties of acids and salts. His great contribution was his insistence upon the value of chemical knowledge as a means of solving vital problems, thus founding a materialistic chemical physiology.

A FALSE THEORY OF COMBUSTION: PHLOGISTON

A German contemporary of Boyle, J. J. Becher (1635–82), whose alchemistic writings had considerable vogue, adopted as elements three "earths" — the "mercurial," the "vitrifiable" and the "combustible." To the last one — "the material and principle of fire, not fire itself" — G. E. Stahl (1660–1734) gave the name *phlogiston.* This he believed to be something dispelled during combustion and the calcining of metals. Because from a metallic calx (oxide) the metal could be recovered by burning with charcoal, the metal was held to have absorbed phlogiston in that process from the charcoal, which having mostly disappeared, was regarded as almost pure phlogiston. Conversely, when the metal was calcined (oxidized) by burning without charcoal, it was held to have lost its phlogiston. This theory satisfied the chief requirement imposed on any new theory: viz., that of accounting for the facts (as then known) and was advocated by natural philosophers for the next hundred years, although it was necessary to assume that phlogiston possessed negative gravity, or "levity," since its loss increased weight.

In his physiological chemistry Stahl departed widely from the materialistic views of Sylvius and he utterly repudiated the mechanical physiology of Descartes and Borelli. No anatomist, nevertheless he recognized that there is an essential difference in texture between natural non-living bodies and living beings which disintegrate rapidly when from them has departed the sensitive soul (*anima senstitiva*) by which and for which they are created. This sensitive soul, which may be compared to the "entelechy" of Aristotle (p. 95), according to Stahl differs to a marked degree from the rational soul of Descartes and the sensitive soul of van Helmont. It controls directly all activities of the living body. Chemical changes in the body may be like those in the laboratory, but this cannot be taken for granted. Stahl was thus the founder of an "animistic," or vitalistic physiology that still has its advocates.

MEDICAL SCIENCE AND MEDICAL THEORY IN THE SEVENTEENTH CENTURY

Many superstitions that can be traced back to Babylonian times still prevailed in medical practice — such as the Royal Touch for the King's Evil (scrofula), astrology, and the notion of demonic possession of the insane. Much good work was done in collecting information about structural changes associated with disease, but there was lacking a genius to organize a science of pathological anatomy. There were various medical sects — the iatro-mathematical, or iatro-physical, who based their practice upon the mechanical physiology of Borelli; the iatro-chemical who followed the teachings of van Helmont and Sylvius; and the followers of Stahl, chiefly at

Montpellier. It was this state of medicine that was ridiculed in the plays of Molière.

In strong contrast stands Thomas Sydenham (1624–89) of London, a friend of Boyle and of Locke, the philosopher. He returned to the principles of Hippocrates — the healing power of nature and the bed-side study of diseases. His classic work, *The Method of Treating Fevers*, 1666, opens with the definition of disease as, "An effort of nature, striving with all her might, to restore the patient by the elimination of morbific matter," which embodies the Hippocratic idea and also is interesting for its implication of the modern idea of disease as a struggle for existence between pathogenic matters (such as microbes) and the inner forces of the body. Sydenham studied by observation the natural history of diseases as specific entities. His great achievement was the initiation of a new approach by the scientific description and analysis of all the cases before him. He was the founder of modern clinical medicine.

REFERENCES FOR READING

BOYLE, ROBERT, *The Sceptical Chymist*, 1661 (Everyman's Library).

HARVEY, WILLIAM, *On the Movement of the Heart and the Blood*, 1628 (Everyman's Library).

MALLOCH, ARCHIBALD, *William Harvey*, 1929.

REDI, FRANCESCO, *Experiments on the Generation of Insects*, 1688, Trans. M. Bigelow, 1909.

SHIPLEY, A. E., "The Revival of Science in the Seventeenth Century," 1913 (*Vanuxem Lectures*, Princeton, pp. 97–144; also *Camb. Hist. Engl. Lit.*, Vol. VIII).

SINGER, CHARLES, *The Discovery of the Circulation of the Blood*, 1922.

TAYLOR, C. M., *The Discovery of the Nature of the Air*, 1934.

From previous lists: Foster, *Hist. Physiol.;* Miall, *Early Nat.;* Smith, *Hist. Mod. Cult.*, Vol. I.

Chapter XIII

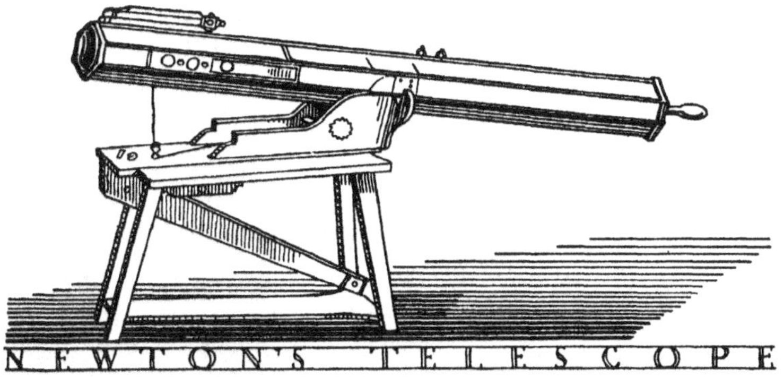

Beginnings of Modern Mathematical Science

. . . All the sciences which have for their end investigations concerning order and measure, are related to mathematics, it being of small importance whether this measure be sought in numbers, forms, stars, sounds, or any other object; . . . accordingly, there ought to exist a general science which should explain all that be known about order and measure, considered independently of any application to a particular subject, and . . . indeed, this science has its own proper name, consecrated by long usage, to wit, *mathematics*. . . . And a proof that it far surpasses in facility and importance the sciences which depend upon it is that it embraces at once all the objects to which these are devoted and a great many others besides. . . . — DESCARTES, *Direction of the Mind.* (H. A. P. Torrey, *Philosophy of Descartes*, p. 72.)

As long as algebra and geometry proceeded along separate paths, their advance was slow and their applications limited. But when these sciences joined company, they drew from each other fresh vitality and thenceforward marched on at a pace toward perfection. — LAGRANGE. (Moritz, p. 81.)

MATHEMATICAL PHILOSOPHY. ANALYTIC GEOMETRY. DESCARTES

The invention of analytic geometry by Descartes in 1637 and the almost contemporary introduction of integral calculus as the method of "indivisibles" may be regarded as the real beginning of modern mathematical science. Thanks to these fruitful ideas the science has during the three centuries that have since elapsed made extraordinary progress both in its own internal development and in its application throughout the range of the physical sciences.

René Descartes was born in Touraine in 1596, and after the education appropriate for a youth of family and some years of fashionable life in Paris, entered the army, then in Holland. His military career continued till 1621 with incidental opportunity for his favorite speculations in mathematics and philosophy. Some of his most fruitful ideas dated from dreams and his best thinking was habitually done before rising.

> It is impossible not to feel stirred at the thought of the emotions of men at certain historic moments of adventure and discovery — Columbus when he first saw the Western shore, . . . Franklin when the electric spark came from the string of his kite, Galileo when he first turned his telescope to the heavens. Such moments are also granted to students in the abstract regions of thought, and high among them must be placed the morning when Descartes lay in bed and invented the method of coördinate geometry. — WHITEHEAD, *Introduction*, p. 122.

In order to devote himself more completely to his favorite studies he settled in Holland in 1629, devoting the next four years to writing a treatise, entitled *Le Monde*, upon the universe. In 1637 he published his great *Discourse on the Method of Good Reasoning and of Seeking Truth in Science.*[1] This begins with a prefatory note:

> If this discourse appear too long to be read at once, it may be divided into six parts: and, in the first, will be found various considerations touching the Sciences; in the second, the principal rules of the Method which the Author has discovered; in the third, certain of the rules of Morals which he has deduced from this Method; in the fourth, the reasonings by which he establishes the existence

[1] *Discours de la Méthode pour bien conduire sa raison et chercher la vérité dans les sciences.*

NOTE: On the opposite page is a view of Newton's telescope, by Elizabeth Tyler Wolcott; after R. S. Ball, *Great Astronomers.* Compare it with Fig. 49.

of God and of the Human Soul, which are the foundations of his Metaphysic; in the fifth, the order of the Physical questions which he has investigated, and, in particular, the explication of the motion of the heart and of some other difficulties pertaining to Medicine, as also the difference between the soul of man and that of the brutes; and, in the last, what the Author believes to be required in order to greater advancement in the investigation of Nature than has yet been made, with the reasons that have induced him to write.

DISCOURSE ON METHOD. PART I

Good sense is, of all things among men, the most equally distributed; for every one thinks himself so abundantly provided with it, that those even who are the most difficult to satisfy in everything else, do not usually desire a larger measure of this quality than they already possess. And in this it is not likely that all are mistaken; the conviction is rather to be held as testifying that the power of judging aright and of distinguishing Truth from Error, which is properly what is called Good Sense or Reason, is by nature equal in all men; and that the diversity of our opinions, consequently, does not arise from some being endowed with a larger share of Reason than others, but solely from this, that we conduct our thoughts along different ways, and do not fix our attention on the same objects. — DESCARTES, *Discourse on the Method*, trans. John Veitch (Open Court Pub. Co., 1907), pp. vii, 1.

Of his cardinal precepts:

The *first* was never to accept anything for true which I did not clearly know to be such; that is to say, carefully to avoid precipitancy and prejudice. . . .

The *second*, to divide each of the difficulties under examination into as many parts as possible, and as might be necessary for its adequate solution.

The *third*, to conduct my thoughts in such order that, by commencing with objects the simplest and easiest to know, I might ascend by little and little, and . . . step by step, to the knowledge of the more complex; . . .

And the *last*, in every case to make enumerations so complete, and reviews so general, that I might be assured that nothing was omitted. — *Ibid.*, p. 19.

Three appendices dealt with optics, meteors, and geometry, the last containing the beginnings of analytical geometry.

The relation of his philosophy to mathematics may be indicated in the following passages.

The long chains of simple and easy reasonings by means of which geometers are accustomed to reach the conclusions of their most difficult demonstrations, had led me to imagine that all things, to the knowledge of which man is competent, are mutually connected in the same way, and that there is nothing so far removed from us as to be beyond our reach, or so hidden that we cannot discover it, provided only we abstain from accepting the false for the true, and always preserve in our thoughts the order necessary for the deduction of one truth from another.

. . . Considering that of all those who have hitherto sought truth in the Sciences, the mathematicians alone have been able to find any demonstrations, that is, any certain and evident reasons, I did not doubt but that such must have been the rule of their investigations. I resolved to commence, therefore, with the examination of the simplest objects, not anticipating, however, from this any other advantage than that to be found in accustoming my mind to the love and nourishment of truth, and to a distaste for all such reasonings as were unsound. — *Ibid.*, pp. 19, 20.

When . . . I asked myself why was it then that the earliest philosophers would admit to the study of wisdom only those who had studied mathematics, as if this science were the easiest of all and the one most necessary for preparing and disciplining the mind to comprehend the more advanced, I suspected that they had knowledge of a certain mathematical science different from that of our times. . . .

I believe I find some traces of these true mathematics in Pappus and Diophantus, who, although they are not of extreme antiquity, lived nevertheless in times long preceding ours. But I willingly believe that these writers themselves, by a culpable ruse, suppressed the knowledge of them; like some artisans who conceal their secret, they feared, perhaps, that the ease and simplicity of their method, if become popular, would diminish its importance, and they preferred to make themselves admired by leaving to us, as the product of their art, certain barren truths deduced with subtlety, rather than to teach us that art itself, the knowledge of which would end our admiration. — DESCARTES, *Rules for the Direction of the Mind*, trans. H. A. P. Torrey, pp. 70, 71.

Descartes had attempted the solution of a historic geometrical problem propounded by Pappus. From a point *P* perpendiculars are dropped on m given straight lines and also on n other given lines. The product of the m perpendiculars is in a constant ratio to the product of the n; it is required to determine the locus of P. Pappus had stated without proof that for $m = n = 2$ the locus is a conic section, Descartes showed this algebraically — Newton afterwards conquering the difficulty by unaided geometry. For the accepted definition of a tangent as a line between which and the curve no other line can be drawn, he introduced the modern notion of limiting position of a secant. In connection with this he considered a circle meeting the given curve in two consecutive points, a perpendicular to the radius of the circle being a common tangent to the circle and the given curve. The circle was not that of curvature, but was introduced because the center lies on the normal of the curve. He recognized the possibility of extending his methods to space of three dimensions, but did not work out the details. His geometry contained also a discussion of the algebra then known, and gave currency to certain important innovations, in particular the systematic use of a, b, and c, for known, x, y, and z, for unknown quantities; the introduction of exponents; the collection of all terms of an equation in one member; the free use of negative quantities; the use of undetermined coefficients in solving equations; and a rule of signs for studying the number of positive or negative roots of equations. He even fancied that he had found a method for solving an equation of any degree.

It is important to distinguish just what Descartes contributed to mathematics in his analytic geometry. Neither the combination of algebra with geometry nor the use of coördinates was new. From the time of Euclid quadratic equations had been solved geometrically, while latitude and longitude involving a system of coördinates are of similar antiquity. The great step made by Descartes was his recognition of the equivalence of an equation and the geometrical locus of a point whose coördinates satisfy that equation. The equation defines the curve, the curve depicts the equation. Each exhibits properties of the other. The systematic and powerful machinery of algebra becomes available for solving geometrical problems, while, on the other hand, the geometrical illustration makes the algebra graphic and concrete. The advantage is comparable with that conferred by the possession of two arms or eyes, or even two senses, under a common will.

Later works dealt with philosophy and physical science, in particular with a theory of vortices. He argues that all matter is in motion and that this must result in the formation of vortices in a pervading subtle fluid, or ether. The sun is the center of one great vortex, each planet of its own, thus avoiding the objectionable assumption of action at a distance and approximating vaguely the future nebular hypothesis. Newton thought it worth while to refute this theory, which was chiefly notable as a bold attempt to interpret the phenomena of the universe by means of a single mechanical principle.

Descartes's achievements in mathematics leave no doubt of his exceptional intellectual power. He had neither the data nor the scientific method for accomplishing similar results in other branches of science, and in mathematics he would doubtless have accomplished much more had he not expended his energies so widely in over-confident reliance on his logical method. He died at Stockholm in 1650.

INDIVISIBLES. KEPLER, CAVALIERI

While Descartes was thus as it were incidentally laying the foundations of modern geometrical analysis, his Italian contemporary, Bonaventura Cavalieri (1598–1647) was rendering a similar service to the integral calculus in developing his theory of indivisibles.

The problem of measuring the length of a curve or the area of a figure having a curved boundary, or the volume of a solid bounded by a curved surface goes back indeed to comparatively ancient Greek times. Most notable in this direction was the work of Archimedes. Kepler, attempting to resolve astronomical difficulties by the hypothesis of elliptical orbits, is confronted at once with the problem of determining the length of the circumference of an ellipse. He gives the approximation

$\pi(a + b)$ where a and b are the semi-axes. This is exact for the circle ($a = b$) and fairly close if a and b are nearly equal, as in most of the planetary orbits. Interesting himself in current methods of measuring the capacity of casks, Kepler published in 1615 his *Nova Stereometria Doliorum Vinariorum*, in which he determined the volumes of many solids bounded by surfaces of revolution. The Greek method had in case of the circle, etc., depended on an "exhaustion" process of inscribing and circumscribing polygons differing less and less from the curve both in boundary and in area. Kepler however divided his solid into sections, determined the area of a section and then sought the sum. He lacked an adequate system of coördinates, a clearly defined conception of a limit, and an effective method of summation. In view of the intrinsic difficulty of this important problem, however, the extent of his success is remarkable.

He also sought to determine the most economical proportions for casks, etc., expressing his view of the underlying mathematical theory by the theorem "In points where the transition from a less to the greatest and again to a less takes place, the difference is always to a certain degree imperceptible."

Kepler discriminates between 93 different solids of revolution, to several of which he gives names of fruits: apple, pear, lemon, etc. There are several integrations and summations of series in his books. Well known in astronomy is the so-called Kepler's equation:

$$x = e \sin x + M, \qquad e, M \text{ constants},$$

one of the first transcendental equations, and which Kepler solved only in a special case.

Kepler was one of the active promoters of the young art of computing and using logarithms. He was deeply impressed by Napier's work, which he got in 1619, and soon afterwards published his own theory and tables of logarithms, which were very ephemeral, but paved the ground for the work of Briggs.

Another contribution to mathematics is his work on space and plane harmonics, which led him to the study of star polygons and star polyhedra. He discovered two new star polyhedra.

Cavalieri, in 1635, adopted the form of statement that a line consists of an infinite number of points, a surface of an infinity of lines, a solid of an infinity of surfaces, but later revised this on the basis of the assumption "that any magnitude may be divided into an infinite number of small quantities which can be made to bear any required ratios one to the other." On this basis, open as it was to criticism, he solved simple area problems involving the parabola and the hyperbola.

The principle of comparing areas by comparing lengths of a system of parallel lines crossing them is easily illustrated in

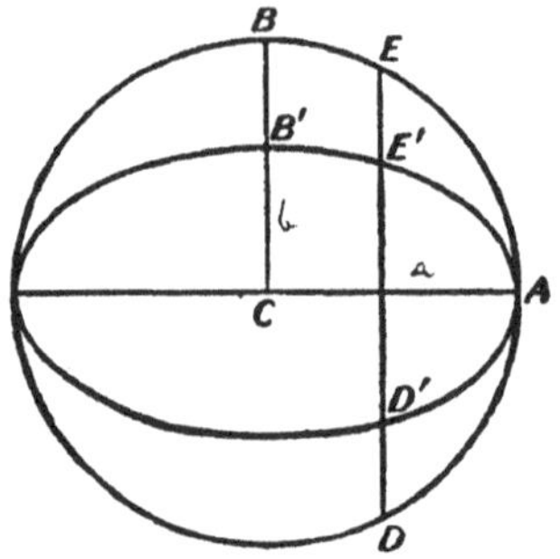

$$\frac{DE}{D'E'} = \frac{CB}{CB'} = \frac{a}{b}$$

FIG. 45. — CIRCLE AND ELLIPSE, AREAS COMPARED.

the case of the ellipse by comparing it with the circle having as its diameter the major axis of the ellipse. If CA $= a$ and $CB^1 = b$ are the semi-axes of the ellipse the two curves are known to be so related that every vertical chord of the circle is in a fixed ratio $a : b$ to the part of it lying within the ellipse. The area of the circle must bear the same relation to the area of the ellipse. The transition from length to area while not rigorously worked out by Cavalieri does not necessarily involve the false assumption that area consists of the sum of parallel lines. A similar method is evidently applicable to volumes. Thus was anticipated one of the most interesting and important processes of modern mathematics — integration as a summation.

Similarly Cavalieri determined volumes by a consideration of the thin sections or elements into which they may be resolved by parallel planes. The principle that "two bodies have the same volume if sections at the same level have the same area" is still known by his name.

Descartes's work with tangents seems not to have led him to develop the fundamental ideas of the differential calculus (because he worked with normals), and it appeared that the integral calculus would be evolved first from the work of Cavalieri.

PROJECTIVE GEOMETRY: DESARGUES

Hardly less interesting than the new ideas of Descartes and Cavalieri are those of their contemporary Desargues (1593–1662), an engineer and architect of Lyons, who made important researches in geometry. But for the still more brilliant geometrical achievements of Descartes, these might have led to the immediate development of projective geometry, the elements of which are contained in Desargues's work. In general this geometry instead of dealing with definite triangles, polygons, circles, etc., in the Euclidean manner, is based on a consideration of all points of a straight line, of all lines through a common point and of the possible effects of setting up an orderly one-to-one correspondence between them. In particular, Desargues makes a comparative study of the different plane sections of a given cone, deducing from known properties of the circle analogous results for the other conic sections.

In his chief work Desargues enunciates the propositions:

1. A straight line can be considered as produced to infinity and then the two opposite extremities are united.
2. Parallel lines are lines meeting at infinity and conversely.
3. A straight line and a circle are two varieties of the same species.

On these he bases a general theory of the plane sections of a cone.

Desargues contented himself with enunciating general principles, remarking: — "He who shall wish to disentangle this proposition will easily be able to compose a volume." He met Descartes while employed by Cardinal Richelieu at the siege of Rochelle, and they with others met regularly in Paris for the discussion of the new Copernican theory and other scientific problems.

He says "I freely confess that I never had taste for study or research either in physics or geometry except in so far as they could serve as a means of arriving at some sort of knowledge of the proximate causes. . . . for the good and convenience of life, in maintaining health, in the practice of some art, . . . having observed that a good part of the arts is based on geometry, among others the cutting of stones in architecture, that of sun-dials, that of perspective in particular."

Perceiving that the practitioners of these arts had to burden themselves with the laborious acquisition of many special facts

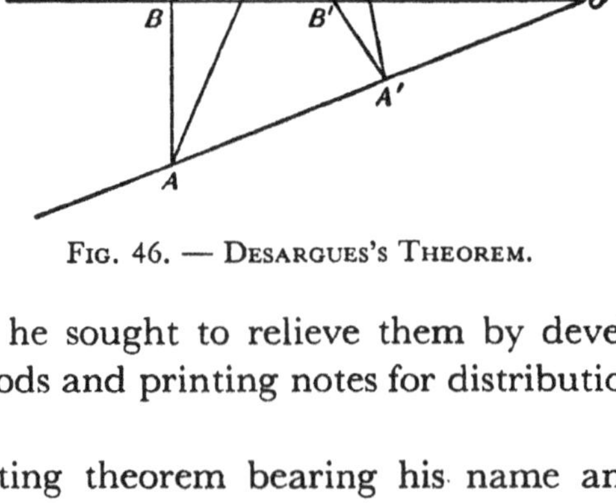

FIG. 46. — DESARGUES'S THEOREM.

in geometry, he sought to relieve them by developing more general methods and printing notes for distribution among his friends.

An interesting theorem bearing his name and typical of projective geometry is as follows: — If two triangles ABC and

A′B′C′ are so related that lines joining corresponding vertices meet in a point O, then the intersections of corresponding sides will lie in a straight line A″B″C″. It remained for Monge, the inventor of descriptive geometry (p. 382) and others more than a century later to carry this development forward.

THEORY OF NUMBERS AND PROBABILITY: FERMAT, PASCAL

But little younger than Descartes and Cavalieri was Pierre de Fermat (1601–65) a man of quite exceptional position in mathematical history. Devoting to mathematics such leisure as his public duties afforded (he was a councilor at Toulouse), he nevertheless published almost nothing, many of his results being known to us only in the form of brief marginal notes without proof. In editing Diophantus he enunciated numerous theorems on integers (for which he often left no proof), for example,

> An odd prime can be expressed as the difference of two square integers in one, and only one, way.
>
> No integral values of x, y, z can be found to satisfy the equation $x^n + y^n = z^n$, if n be an integer greater than 2.

This latter seemingly simple theorem has been verified for so wide a range of values of n, that its truth can hardly be doubted, but no general proof has yet been given in spite of a prize of 100,000 marks once awaiting him who either proved or disproved it. Fermat should be credited with a substantial share in the invention of the new analytic geometry, in which he had certainly done independent work for some years before Descartes's publication (in his posthumous *Isagoge*, 1679).

In other papers he discusses problems of maxima and minima, and passing to concrete phenomena, enunciates the interesting principle: that Nature, the great workman which has no need of our instruments and machines, lets everything happen with a minimum of outlay — an idea not indeed strange to some of the Greeks. The law of refraction of a ray of light he deals with correctly as a particular case of the principle of economy, a principle which exerted a potent influence the scientific philosophy of the following century. Thus example Euler says in 1744:

>ince the organization of the world is the most excellent, nothing 'ound in it, out of which some sort of a maximum or minimum ›perty does not shine forth. Therefore no doubt can exist, that all ion in the world can be derived by the method of maxima and nima as well as from the actual operating causes.

Fermat's work in the theory of probability is fundamental. : discusses the case of two players, A and B, where A wants o points to win and B three points. Then the game will ·tainly be decided in the course of four trials. Take the ters *a* and *b*, and write down all the combinations that can formed of four letters. These combinations are 16 in number, namely *aaaa*, *aaab*, *aaba*, *aabb*, *abaa*, *abab*, *abba*, *abbb*, *baaa*, *ab*, *baba*, *babb*, *bbaa*, *bbab*, *bbba*, *bbbb*. Now every combination which *a* occurs twice or oftener represents a case favorable A, and every combination in which *b* occurs three times or :ener represents a case favorable to B. Thus, on counting ɜm, it will be found that there are 11 cases favorable to A, d 5 cases favorable to B; and, since these cases are all ually likely, A's chance of winning the game is to B's ance as 11 is to 5.

Blaise Pascal (1623–62), like Descartes, devoted but a fracɪn of his great talent to mathematical science.

> I have spent much time in the study of the abstract sciences,—t the paucity of persons with whom you can communicate on such ɔjects gave me a distaste for them. When I began to study man, aw that these abstract studies were not suited to him, and that in ring into them, I wandered farther from my real track than those ıo were ignorant of them, and I forgave men for not having atıded to these things. But I thought at least I should find many mpanions in the *study of mankind*, which is the true and proper ıdy of man. Again I was mistaken. There are yet *fewer students of an than of Geometry.*

Learning geometry surreptitiously at 12 years, he had at 16 ·itten an essay on conic sections and at 19 constructed the st computing machine. While most of his later life was devoted to religion, theology, and literature, he undertook a wide range of physical experimentation, and made important contributions to the then new theories of numbers and probability, besides a discussion of the cycloid. The juvenile essay on conic sections (1640) contains the beautiful theorem since named for him that the three intersections of the opposite sides of a hexagon inscribed in a conic section lie in a straight line. Of geometry and logic Pascal says:

> Logic has borrowed the rules of geometry without understanding its power. . . . I am far from placing logicians by the side of geometers who teach the true way to guide the reason. . . . The method of avoiding error is sought by every one. The logicians profess to lead the way, the geometers alone reach it, and aside from their science there is no true demonstration. — (Moritz, p. 202.)

His work on probability connected itself with the problem of two players of equal skill wishing to close their play, of which Fermat's solution has been given above.

> The following is my method for determining the share of each player when, for example, two players play a game of three points and each player has staked 32 pistoles.
>
> Suppose that the first player has gained two points and the second player one point; they have now to play for a point on this condition, that if the first player gain, he takes all the money which is at stake, namely 64 pistoles; while if the second player gain, each player has two points, so that they are on terms of equality, and if they leave off playing, each ought to take 32 pistoles. Thus if the first player gain, then 64 pistoles belong to him, and if he lose, then 32 pistoles belong to him. If therefore the players do not wish to play this game, but separate without playing it, the first player would say to the second, "I am certain of 32 pistoles, even if I lose this point, and as for the other 32 pistoles, perhaps I shall have them and perhaps you will have them; the chances are equal. Let us then divide these 32 pistoles equally, and give me also the 32 pistoles of which I am certain." Thus the first player will have 48 pistoles and the second 16 pistoles.

By similar reasoning he shows that if the first player has gained two points and the second none, the division should be 56 to 8; while if the first has gained one point, the second none, it should be 44 and 20.

The calculus of probabilities, when confined within just limits, ought to interest, in an equal degree, the mathematician, the experimentalist, and the statesman. From the time when Pascal and Fermat established its first principles, it has rendered, and continues daily to render, services of the most eminent kind. It is the calculus of probabilities, which, after having suggested the best arrangements of the tables of population and mortality, teaches us to deduce from those numbers, in general so erroneously interpreted, conclusions of a precise and useful character; it is the calculus of probabilities which alone can regulate justly the premiums to be paid for assurances; the reserve funds for the disbursements of pensions, annuities, discounts, etc. It is under its influence that lotteries and other shameful snares cunningly laid for avarice and ignorance have definitely disappeared. — ARAGO. (Moritz, p. 257.)

With this work connected itself his arithmetic triangle (*Traité du triangle arithmétique*, 1654) in which successive diagonals contain the coefficients which occur in expansions by the binomial theorem, which Newton was soon to generalize.

1	1	1	1	1	1
1	2	3	4	5	
1	3	6	10		
1	4	10			
1	5				
1					

This triangle, which still carries Pascal's name, was, however, nothing new. It occurs already in Stifel (1543) and the idea goes back to the Pythagoreans.

Pascal published in 1645 his idea of an arithmetical machine, writing the Chancellor in regard to it:

Sir: If the public receives any advantage from the invention which I have made to perform all sorts of rules of arithmetic in a manner as novel as it is convenient, it will be under greater obligation to your Highness than to my small efforts, since I should only have been able to boast of having conceived it, while it owes its birth absolutely to the honor of your commands. The length and difficulty of the ordinary means in use have made me think on some help more prompt and easy to relieve me in the great calculations with which I have been occupied for several years in certain affairs which depend on the occupations with which it has pleased you to honor my father for the service of his Majesty in Normandy. I employed for this investigation all the knowledge which my inclination and the labor of my first studies in mathematics have gained for me, and after profound reflection, I recognized that this aid was not impossible to find.

MECHANICS AND OPTICS: HUYGENS

Most notable among the successors of Galileo in mechanics before we reach Newton was Christiaan Huygens of Holland (1629–95) who combined mathematical power with exceptional practical ingenuity. He first (in 1655) explained as a ring the excrescences of Saturn which had been misunderstood by Galileo and others, publishing his discovery in the occult form $a^7c^5d^1e^5g^1h^1i^7l^4m^2n^9o^4p^2q^1r^2s^1t^5u^5$. (*Annulo cingitur tenui, plano, nusquam cohærente ad eclipticam inclinato.*) He also discovered Saturn's largest moon. About the same time he made his great invention of the pendulum clock. Accepting a call to Paris by Colbert at the founding of the French Academy, he remained there from 1666 to 1681.

In optics he developed and maintained even in opposition to the authority of Newton the undulatory or wave theory which only found general acceptance a century later. The velocity of light Galileo had failed to measure by means of signal lanterns, and Descartes had likewise been unable to ascertain it by comparing the observed and computed instants of a lunar eclipse. Huygens points out that even this latter test does not prove instantaneous transmission. Roemer had observed in 1676 that the eclipse of a satellite of Jupiter was delayed about seven minutes from the computed time when seen from a remote point of the Earth's orbit and advanced by a similar amount when the distance was least. On this basis Huygens estimated the velocity of light at 600,000 times that of sound — a result about one-third too small.

The medium in which light waves travel Huygens named the ether, attributing to its particles three properties in comparison with air: extreme minuteness, extreme hardness, extreme elasticity. On this basis he worked out a consistent

FIG. 47. — CHRISTIAAN HUYGENS (1629–1695).
From *Oeuvres complètes*, 1899.

theory for reflection and refraction. His discussion of the newly discovered phenomenon of double refraction in Iceland spar has been characterized as an "unsurpassed example of

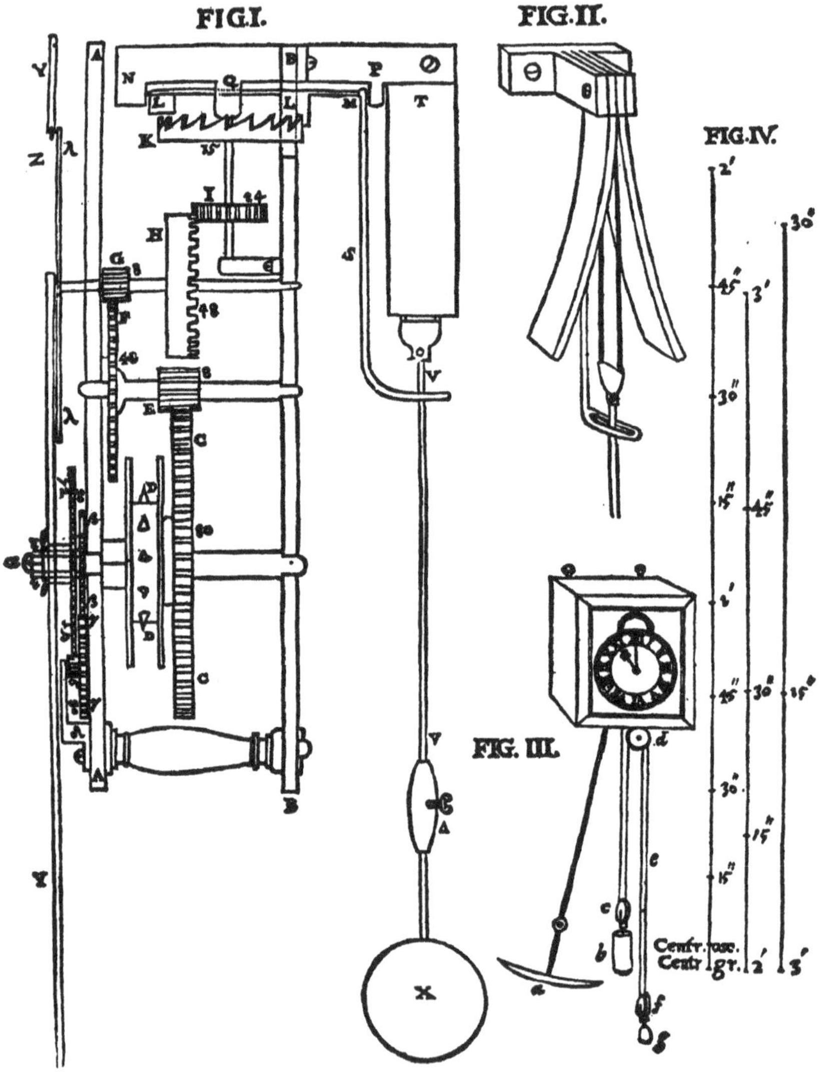

FIG. 48. — HUYGENS'S CLOCK. From Huygens, *Horologium oscillatorium*, 1673.

the combination of experimental investigation and acute analysis." The attendant phenomenon of polarization did not escape him, but his theory of wave motion was not sufficiently developed to enable him to explain the matter adequately.

In 1673 Huygens published his great work on the pendulum (*Horologium oscillatorium sive de motu pendulorum*), displaying wonderful skill in his geometrical treatment of the mechanical problems involved. The use of wheel mechanisms with weights for measuring time had been more or less familiar for several centuries, but no effective means for regulating this motion had been devised. Galileo, for example, observing the regularity of pendulum vibrations, had depended on repeated impulses by hand to maintain the motion. Huygens first made the fortunate combination of the two elements, without however inventing the modern escapement. He invented the cycloidal pendulum for which the time of an oscillation would be independent of the amplitude and made precise determinations of the length of the seconds-pendulum at Paris and the corresponding value of the important constant g. The most remarkable achievement in his treatise on the pendulum is the correct analysis of the compound pendulum based on the definition:

> The centre of oscillation of any figure whatever is that point in the line of gravity, whose distance from the point of suspension is the same as the length of the simple pendulum having the same time of vibration as the figure.

In the course of the discussion he formulates the important law, afterwards somewhat generalized by others:

> Whenever any heavy bodies are set in motion under the action of their own weight, their common centre of gravity cannot rise higher than it was at the beginning of the motion.

In computing the position of the center of oscillation he arrives at a fraction of the form $\frac{\Sigma mr^2}{\Sigma mr}$, where m denotes the mass of a particle, r its distance from the point of suspension. The numerator is the so-called "moment of inertia," the denominator the "statical moment" of later mechanics. He shows that the point of suspension and the center of oscillation are interchangeable.

Finally he discusses the theory of centrifugal force, proving that it varies as the square of the velocity and inversely as the radius. This subject he also treated more fully in a special monograph, published after his death when Newton had already given a more general theory. His theorems are:

1. When equal bodies move with the same velocity in unequal circles, the centrifugal forces are to each other inversely as the diameters, so that in the smaller circle the said force is greater.
2. When equal movable bodies travel in the same or equal circles with unequal velocities, the centrifugal forces are to each other as the squares of the velocities.
3. In his posthumous work he shows that centripetal force depends also upon what he calls *quantitates solidas*, which we now call mass.

By experiments on a revolving sphere of clay which as he anticipated assumed a spheroidal form, he explains the observed polar flattening of Jupiter. He infers that the earth must also be flattened, and makes a numerical estimate in anticipation of future verification. He explains the effect on a clock pendulum of transporting it from Paris to an equatorial locality, where its weight is opposed by an increased centrifugal force.

Like Wallis (p. 326) and Sir Christopher Wren he accepted the invitation of the Royal Society to attack the general problem of impact. This led ultimately to the publication eight years after his death of his paper "On the Motion of Bodies under Percussion." The theorems enunciated deal with various cases of central impact, one of the most notable being:

> By mutual impact of two bodies the sum of the products of the masses into the squares of their velocities is the same before and after impact.

— the first formulation (1669) of the most comprehensive law of mechanics, the Conservation of *vis viva*.

Huygens visited England in 1689, but made no use of Newton's new calculus in his published work. His theory of plane involutes and evolutes [1] in the *Horologium* bore an important relation to the calculus, and he made notable contributions to the theory of probability, on which he wrote the first formal treatise (1657), later taken over in James Bernoulli's *Ars Conjectandi* (1713). In his *History of the Mathematical Theories of Attraction and the Figure of the Earth*, Todhunter says of Huygens:

> To him we owe the important condition of fluid equilibrium, that the resultant force at any point of the free surface must be normal to the surface at that point; and this has indirectly promoted the knowledge of our subject. But Huygens never accepted the great principle of the mutual attraction of particles of matter; and thus he contributed explicitly only the solution of a theoretical problem, namely the investigation of the form of the surface of rotating fluid under the action of a force always directed to a fixed point. — TODHUNTER, I, p. vi.

WALLIS AND BARROW

Before attempting to discuss the extraordinary work of Sir Isaac Newton in the whole field of mathematical science a few words should be added concerning two slightly older English mathematicians, John Wallis (1616–1703), Savilian professor at Oxford, and Isaac Barrow (1630–77), Lucasian professor at Cambridge.

Wallis in his *Arithmetic of The Infinites* (1655) developed Cavalieri's summation ideas effectively, employing the new Cartesian geometry and a process equivalent to integration for simple algebraic cases. In particular, he explains negative and fractional exponents in the modern sense, and then proceeds to find the area bounded by OX, the curve $y = ax^m$, and any ordinate $x = h$ — or as we should say, he integrates the function ax^m. He develops ingenious methods of interpolation. He was able to express π in the form of an infinite product:

$$\frac{\pi}{2} = \frac{2}{1}\cdot\frac{2}{3}\cdot\frac{4}{3}\cdot\frac{4}{5}\cdot\frac{6}{5}\cdot\frac{6}{7}\cdot\frac{8}{7}\cdots\cdot$$

[1] An *evolute* is (in the plane) the locus of the centers of curvature of a given curve, called the *involute* of this locus.

This marks a definite turn in the history of the old problem of squaring the circle. Before such work as Wallis's the only way to study π had been with the aid of perimeters of polygons.

In Wallis's *Treatise of Algebra* he says:

> It is to me a theory unquestionable, That the Ancients had somewhat of like nature with our Algebra; from whence many of their prolix and intricate Demonstrations were derived. . . . But this their Art of Invention, they seem very studiously to have concealed: contenting themselves to demonstrate by Apagogical Demonstrations, (or reducing to Absurdity, if denied,) without showing us the method, by which they first found out those Propositions, which they thus demonstrate by other ways. . . .

He formulates the ideas of absolute space and time later adopted by Newton. Space and time are absolute and eternal because God is omnipresent and everlasting.

His analytical *Conic Sections* (1655) made Descartes's geometrical ideas much more intelligible, and his *Algebra* (1685) marks an important step forward in its systematic use of formulas. He also wrote *A Summary Account . . . of the General Laws of Motion*, enunciating the formulas for velocity after impact of masses m_1 and m_2 with velocities v_1 and v_2:

$$v = \frac{m_1v_1 + m_2v_2}{m_1 + m_2}.$$

Barrow, after varied adventures, became first Lucasian professor at the University of Cambridge, but resigned his chair six years later to his pupil Newton. His work on optics and geometry contains a notable discussion of the tangent problem and of what he calls the differential triangle, so important in the modern approach to the differential calculus. His point of view as to numbers is illustrated by the following passage:

> Now as to what pertains to these Surd numbers (which, as it were by way of reproach and calumny, having no merit of their own are also styled Irrational, Irregular, and Inexplicable) they are by many denied to be numbers properly speaking, and are wont to be banished from arithmetic to another Science (which yet is no science), viz., algebra. — MORITZ, p. 281.

ISAAC NEWTON

Isaac Newton was born within a year after Galileo's death, a century after that of Copernicus — December 25, 1642 (O.S.)[1] in a Lincolnshire village. Destined at first to become a farmer, he was fortunately sent at 17 to Cambridge university, where he quickly and eagerly mastered the mathematical work of Euclid, Descartes and Wallis, and Kepler's *Dioptrics*. His discovery of the general binomial theorem dates from this time, and he even ventured to attack the great problem of gravitation by carefully comparing the motion of the moon with that of a falling body near the earth — for which however his data were not yet sufficiently accurate.

Newton took his B.A. degree in the Lent Term, 1665. In that spring the plague appeared, and for a couple of years he lived mostly at home, though with occasional residence at Cambridge. Probably at this time his creative powers were at their highest. His use of fluxions may be traced back to May, 1665; his theory of gravitation originated in 1666; and the foundation of his optical discoveries would seem to be only a little later. In an unpublished memorandum made some years later (cancelled, but believed to be correct in the part here quoted), he thus described his work of this time: "In the beginning of the year 1665 I found the method of approximating Series and the Rule for reducing any dignity of any Binomial into such a series. The same year, in May, I found the method of tangents of Gregory and Slusius, and in November had the direct method of Fluxions, and the next year in January had the Theory of Colours, and in May following I had entrance into the inverse method of Fluxions. And the same year I began to think of gravity extending to the orb of the Moon, and . . . from Kepler's Rule of the periodical times of the Planets being in a sesquialterate proportion of their distances from the centers of their orbs I deduced that the forces which keep the Planets in their orbs must (be) reciprocally as the squares of their distances from the centers about which they revolve: and thereby compared the force requisite to keep the Moon in her orb with the force of gravity at the surface of the earth, and found them answer pretty nearly. All this was in the two plague years of 1665 and 1666, for in those days I was in the prime of my age for invention, and minded Mathema-

[1] January 5, 1643, new style.

ticks and Philosophy more than at any time since." — W. W. R. BALL, *Mathematical Gazette*, July, 1914.

In 1669 Newton succeeded his teacher Barrow as professor in Cambridge, in 1671 he became a Fellow of the Royal Society, of which he was made president in 1703. He was interested also in the social, political, and theological questions; in 1699 he became master of the mint. He was highly honored during his lifetime both in England and abroad. He died in 1727 and was buried in Westminster Abbey.

OPTICS

Interesting himself in the telescope, Newton succeeded in eliminating the disturbing chromatic aberration due to un-

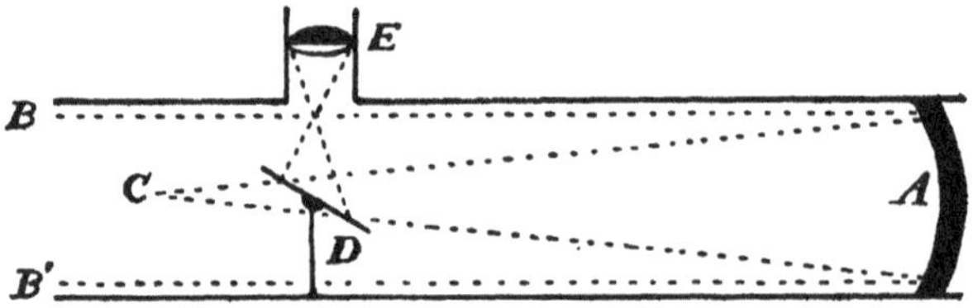

FIG. 49. — OPTICAL DIAGRAM OF NEWTON'S TELESCOPE. *A*, concave mirror; *B*, *B'*, light rays; *C*, focus; *D*, plane mirror; *E*, eye-piece. After Harvey-Gibson, *Two Thousand Years of Science*, p. 53. Permission of The Macmillan Company.

equal refraction of the different colors by constructing a reflecting telescope with a concave mirror in place of a convex lens. On the other hand, turning his attention to the colors of the solar spectrum, he wrote his *Opticks or a Treatise of the Reflections, Refractions, Inflections and Colours of Light*, published in 1704. Disclaiming any intention of setting up speculative hypotheses, he discusses the observed phenomena of refracted light, speaking of his discovery of the different refrangibility of the rays of light as "in my judgment the oddest if not the most considerable detection which hath hitherto been made in the operations of nature." While he does not insist upon it, he seems always to have the underlying idea that light itself consists of minute particles — the degree of fineness corresponding with the color — a theory which held the field — thanks to his potent authority with his too subservient followers — against the undulatory theory of Huygens until the nineteenth century. In the experiments on which this work is based Newton not only decomposed light by a refracting

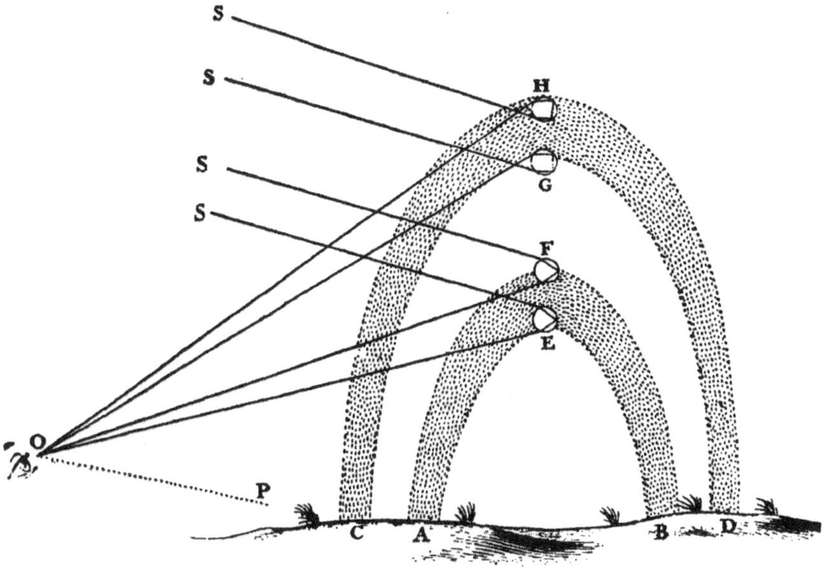

FIG. 50. — NEWTON'S THEORY OF THE RAINBOW. From Newton, *Opticks*, 1704.

prism or series of prisms, but also succeeded in recombining the component colors to reproduce the original white.

> The colors of objects, he says, are nothing more than their power to reflect one or another kind of ray. And in the rays again is nothing other than the power to transmit this motion into our organ of sense, in which last finally arises the sensation of these motions in the form of colors.

He solves at last the problem of the rainbow.

THE THEORY OF GRAVITATION: PRINCIPIA

In 1682 Newton returned to his attempt of 16 years earlier to explain the moon's motion by means of the assumed influence of gravitation. During this long interval French geographers, testing the supposedly spherical shape of the earth, had made a new and more precise triangulation — with the first use of telescopic instruments. He had also in the meantime overcome an important mathematical difficulty by proving that the attraction of a spherical body on a mass outside it was the same as if its mass were concentrated at its center. Stirred to the inmost depths of his usually calm nature by his realization that he was approaching a solution of the great problem, he had to beg a friend to complete his calculations. The new astronomy founded by Copernicus, built up by Tycho Brahe, Kepler, and Galileo, was now to be completely formulated and mathematically interpreted by Newton's crowning discovery of a single mechanical principle governing the whole.

It was a question of verifying the correctness of this principle by applying it to all measured or measurable astronomical phenomena. The investigation was gradually extended to the Moon, the planets, the moons of Jupiter, the tides, and even the comets. Everywhere the law was verified that attraction varies as the product of the masses and inversely as the square of the distance.

The whole theory was elaborated in Newton's monumental *Philosophiae Naturalis Principia Mathematica* published in 1687. He begins this treatise with a series of definitions and laws:

Definitions

1. The quantity of matter is the measure of the same, arising from its density and bulk conjointly.

2. The quantity of motion is the measure of the same, arising from the velocity and quantity of matter conjointly.

3. The . . . innate force of matter is a power of resisting, by which every body, as much as in it lies, continues in its present state, whether it be of rest, or of moving uniformly forwards in a right line.

4. An impressed force is an action exerted upon a body, in order to change its state, either of rest, or of uniform motion in a right line.

5. A centripetal force is that by which bodies are drawn or impelled, or any way tend, towards a point as to a centre.

These and succeeding definitions are followed by the famous LAWS OF MOTION:

I. Every body continues in its state of rest, or uniform motion in a right line, unless it is compelled to change that state by forces impressed upon it.

II. The change of motion is proportional to the motive force impressed; and is made in the direction of the right line in which that force is impressed.

III. To every action there is always opposed an equal reaction: or, the mutual actions of two bodies upon each other are always equal, and directed to contrary parts.

COROLLARY I continues: A body acted on by two forces simultaneously will describe the diagonal of a parallelogram in the same time as it would describe the sides by those forces separately. — SIR ISAAC NEWTON, *Mathematical Principles*. (Cajori, 1934, pp. 1–14.)

Of these laws, Karl Pearson in his *Grammar of Science* remarks:

The Newtonian laws of motion form the starting-point of most modern treatises on dynamics, and it seems to me that physical science, thus started, resembles the mighty genius of an Arabian tale emerging amid metaphysical exhalations from the bottle in which for long centuries it has been corked down. — p. 325.

Passing to variable forces he discusses in particular the motion of a body acted on by a central attractive force — i.e., a force attracting it towards a fixed point. He derives the law by which equal areas are described in equal times, and shows that conversely, if equal areas are so described in a plane, the determining force must be a central one. Turning to the consideration of orbits, he deals with the hypothesis of an elliptical orbit with the attracting force at one of the foci, and shows that the attractive force must vary inversely as the square of the distance from that focus. The same result is obtained for the other conic sections. These theorems applying to particles, he next shows that the action of a homogeneous sphere on an external particle is the same as if its mass were concentrated at its center, so that the action of two such spheres on each other is subject to the laws already derived.

Comparing planetary motions with those of projectiles which had been treated by Galileo, Newton says:

> That the planets can be held in their paths is evident from the motions of projectiles. A stone thrown is deflected from the straight line by its weight and falls describing a curved line to the earth. If thrown with greater velocity it goes farther, and it could happen that it described a curve of 10, 100, 1000 miles, and at last went outside the boundaries of the earth, and never fell back. . . .

The force of gravity for small distances being sensibly constant in direction, causes motion in a path approximately parabolic. For greater and greater ranges the change of direction of the force must be taken account of and the path recognized as elliptical or hyperbolic. The laws as stated deal with the relations of only two mutually attracting bodies. Newton of course appreciates that such a case is purely ideal and that, since every body attracts every other, the result of dealing with only two is merely a first approximation to the reality.

All planets he says are mutually heavy, therefore, for example; Jupiter and Saturn will attract each other in the vicinity of their conjunction and perceptibly disturb each other's motion. Similarly the Sun will disturb the motion of the Moon, and Sun and Moon will disturb our ocean.

Newton prefaced these applications of the theory with four rules which should guide scientific men in making hypotheses. These in their final shape, are to the following effect: (1) We should not assume more causes than are sufficient and necessary for the explanation of observed facts. (2) Hence, as far as possible, similar effects must be assigned to the same cause; for instance, the fall of stones in Europe and America. (3) Properties common to all bodies within reach of our experiments are to be assumed as pertaining to all bodies; for instance, extension. (4) Propositions in science obtained by wide induction are to be regarded as exactly or approximately true, until phenomena or experiments show that they may be corrected or are liable to exceptions. The substance of these rules is now accepted as the basis of scientific investigation. Their formal enunciation here serves as a landmark in the history of thought. — W. R. R. BALL, *Mathematical Gazette*, July, 1914.

Shrinking always from publicity [1] and controversy, Newton

[1] In one instance he authorized publication of one of his works "so it be without my name to it: for I see not what there is desirable in public esteem, were I

like Copernicus had gradually perfected his great work, but, like Copernicus, Newton might never have published it but for the fortunate urgency of a faithful disciple, Edmund Halley.

NEWTON'S MATHEMATICS: FLUXIONS

Newton's services to mathematics itself were not less original and momentous than to celestial mechanics.

His extraordinary abilities . . . enabled him within a few years to perfect the more elementary . . . processes, and to distinctly advance every branch of mathematical science then studied, as well as to create several new subjects. There is hardly a branch of modern mathematics which cannot be traced back to him and of which he did not revolutionize the treatment.

In pure geometry Newton did not establish any new methods, but no modern writer has ever shown the same power in using those of classical geometry, and he solved many problems in it which had previously baffled all attempts. — BALL.

Newton's greatest mathematical achievement was, of course, the invention of the fluxional, or infinitesimal, calculus. During his lifetime, however, he published very few of his results in this field. For the exposition of mathematical problems requiring the calculus he used, in the *Principia*, the old Greek method, probably not to repel readers by too much new mathematics. In his treatise on *The Method of Fluxions and Infinite Series* (said to have been written in 1671, but not published until "translated from the Author's Latin Original not yet made publick" by John Colson in 1736), he says:

1. Having observed that most of our modern Geometricians neglecting the synthetical Method of the Ancients, have applied themselves chiefly to the analytical Art, and by the Help of it have overcome so many and so great Difficulties, that all the Speculations of Geometry seem to be exhausted, except the Quadrature of

able to acquire and maintain it: it would perhaps increase my acquaintance, the thing which I study chiefly to decline." Again in 1675 he writes, "I was so persecuted with discussions arising out of my theory of light, that I blamed my own imprudence for parting with so substantial a blessing as my quiet, to run after a shadow."

Curves, and some other things of a like Nature which are not yet brought to Perfection: To this End I thought it not amiss, for the sake of young Students in this Science, to draw up the following Treatise; wherein I have endeavored to enlarge the Boundaries of Analyticks, and to make some Improvements in the Doctrine of Curved Lines.

— a sufficiently modest introduction of perhaps the most important step in the progress of mathematical science.

Something further as to the evolution of his theory of Fluxions may be indicated, without too much technical detail, by the following passages from Brewster:

Having met with an example of the method of Fermat, in Schooten's Commentary on the Second Book of Descartes, Newton succeeded in applying it to affected equations, and determining the proportion of the increments of indeterminate quantities. These increments he called *moments*, and to the velocities with which the quantities increase he gave the names of *motions*, *velocities* of *increase*, and *fluxions*. He considered quantities not as composed of indivisibles, but as generated by motion; and as the ancients considered rectangles as generated by drawing one side into the other, that is, by moving one side upon the other to describe the area of the rectangle, so Newton regarded the areas of curves as generated by drawing the ordinate into the abscissa, and all indeterminate quantities as generated by continual increase. Hence, from the flowing of time and the moments thereof, he gave the name of *flowing quantities* to all quantities which increase in time, that of *fluxions* to the velocities of their increase, and that of *moments* to their parts generated in moments of time.

Newton then applies the propositions to the solution of twelve problems, many of which were at that time entirely new, for example:

To draw tangents to curve lines.

To find the points distinguishing between the concave and convex portions of curved lines.

To find the point at which lines are most or least curved.

To find the nature of the curve line whose area is expressed by any given equation.

The nature of any curve line being given, to find its area

when it may be done; or two curved lines being given, to find the relation of their areas when it may be.

An account of Newton's important work in analytic geometry and the theory of algebraic equations lies outside the range of the present work.

Much of Newton's reluctance to publish his more revolutionary theories may be attributed to his distaste for controversy, and he was unfortunately involved not only in such issues as to priority as his own reticence invited, but also in defending himself against attacks on philosophic grounds. The character of some of these may be illustrated by the following passages from an eminent critic, Bishop George Berkeley (1734):

> He who can digest a second or third fluxion, a second or third difference, need not, methinks, be squeamish about any point in Divinity.
>
> And what are these fluxions? The velocities of evanescent increments. And what are these same evanescent increments? They are neither finite quantities, nor quantities infinitely small, nor yet nothing. May we not call them the ghosts of departed quantities? — *The Analyst*, Sects. 7 and 35.

In regard to the controversy between the friends of Newton and those of Leibniz as to priority in the invention of the calculus, Newton himself says in a celebrated scholium:

> The correspondence which took place about ten years ago, between that very skilful geometer G. G. Leibnitz and myself, when I had announced to him that I possessed a method of determining maxima and minima, of drawing tangents, and of performing similar operations, which was equally applicable to surds and to rational quantities, and concealed the same in transposed letters, involving this sentence, (*Data Aequatione quotcunque Fluentes quantitates involvente, Fluxiones invenire, et vice versa*), this illustrious man replied that he also had fallen on a method of the same kind, and he communicated his method, which scarcely differed from my own, except in the forms of words and notation (and in the idea of the generation of quantities). — *Principia*, Book II, Ed. 2. (Brewster, *Memoirs*, II, p. 29.)

Among Newton's contributions to mathematics should also be mentioned the binomial series (1669) and the first classification of plane curves of the third degree (1704). In his *Arithmetica Universalis* (written about 1670, published 1707) he enriched the theory of algebraic equations.

The exalted estimation in which Newton's genius has been held in later times may be illustrated by the following passages.

The great Newtonian Induction of Universal Gravitation is indisputably and incomparably the greatest scientific discovery ever made, whether we look at the advance which it involved, the extent of the truth disclosed, or the fundamental and satisfactory nature of this truth. — WHEWELL, *Hist. Ind. Sci.*, Ed. 3, 1857, II, 136.

The efforts of the great philosopher . . . were always superhuman; the questions which he did not solve were incapable of solution in his time. — ARAGO. (Moritz, p. 167.)

Newton was the greatest genius that ever existed, and the most fortunate, for we cannot find more than once a system of the world to establish. — LAGRANGE. (Moritz, p. 167.)

His own attitude is sufficiently indicated in his statements:

I do not know what I may appear to the world, but, to myself, I seem to have been only like a boy playing on the seashore, and diverting myself in now and then finding a smoother pebble or a prettier shell than ordinary, whilst the great ocean of truth lay all undiscovered before me.

If I have seen farther than Descartes, it is by standing on the shoulders of giants. — MORITZ, p. 170.

LEIBNIZ

Newton's great contemporary and scientific rival Gottfried Wilhelm Leibniz (1646–1716) has been called the Aristotle of the seventeenth century. Born in 1646 at Leipsic, he took his doctor's degree at 20 and was immediately offered a university professorship. Alchemy, diplomacy, philosophy, mathematics, all shared his energetic attention. In the last he invented a calculating machine and the differential calculus. In regard to the machine Felix Klein says:

merely the formal rules of computation are essential, for only these can be followed by the machine. It cannot possibly have an intuitive conception of the meaning of the numbers. It is thus no accident that a man so great as Leibnitz was both father of purely formal mathematics and inventor of the first calculating machine.

After travels in Germany, France, Holland, and England, where he made the acquaintance of many great men of science, he became, in 1676, librarian to the dukes of Hanover, for whom he wrote many historical volumes. Later he accepted an invitation of the prince of Brandenburg to go to Berlin, where he founded the Academy of Sciences (1700). He was also instrumental in the organization of similar bodies in St. Petersburg, Dresden, and Vienna. In 1714 he returned to Hanover, fell in disfavor with the court, and died in obscurity. Through an enormous correspondence he stayed in steady contact with the most illustrious persons in Europe on almost all political and scientific questions. His advanced ideas on education may be inferred from his remark:

We force our youths first to undertake the Herculean labor of mastering different languages, whereby the keenness of the intellect is often dulled, and condemn to ignorance all who lack knowledge of Latin.

Leibniz's first publication of the calculus was in 1684, "A new Method to find Maxima and Minima," in the periodical, *Acta Eruditorum*, which he founded in 1682. This paper contains the symbols dx, dy, dy/dx, and the elementary rules of differentiation, as $d(uv) = vdu + udv$. It was followed by a series of other papers, one of 1686 with the principles of the integral calculus, in which the notation $\int f(x)dx$ appears.

But for the overshadowing genius of Newton, Leibniz's service to the progress of science would have been even greater than it actually was. Comparing their work in mathematics where their competition was keenest, it should be appreciated that while Newton's work in mathematical science was incomparably greater in range, it was Leibniz who gave to the differential calculus the better form and notation out of which our own has grown.

It appears that Fermat, the true inventor of the differential calculus, considered that calculus as derived from the calculus of finite differences by neglecting infinitesimals of higher orders as compared with those of a lower order . . . Newton, through his method of fluxions, has since rendered the calculus more analytical, he also simplified and generalized the method by the invention of his binomial theorem. Leibnitz has enriched the differential calculus by a very happy notation. — LAPLACE.

It is one of the curiosities in the history of mathematics, remarks Cajori, that both Newton and Leibniz used "infinitely small quantities" rather than the more precise notion of limit already worked out by others. Leibniz's sense of mathematical form was also well exemplified by his work of 1666 on permutations and combinations, *Dissertatio de arte combinatoria*, many ideas of which subsequently led to the invention of determinants.

He established important contributions to the application of the calculus to geometry (point of inflection, osculating plane), and inspired his pupils, especially the brothers Bernoulli, to further remarkable discoveries. He worked also on formal logic and was one of the founders of the mathematical treatment of this field.

Appreciating the great importance of the best symbolism he conducts an extensive correspondence with other mathematicians and makes many alternative experiments. Passing on to his broader design of mathematical logic and emulating Descartes's overconfidence, Leibniz says exultantly of his broad scheme of mathematical logic, "I dare say that this is the last effort of the human mind, and, when this project shall have been carried out, all that men will have to do will be to be happy, since they will have an instrument that will serve to exalt the intellect, not less than the telescope serves to perfect their vision."

Among Leibniz's symbols which are still in current use are dx, dy, his sign of integration $\int$, his colon for division, his dot for multiplication, his signs for similar and congruent.

Leibnitz believed he saw the image of creation in his binary arithmetic, in which he employed only two characters, unity and zero. Since God may be represented by unity, and nothing by zero, he imagined that the Supreme Being might have drawn all things from nothing, just as in the binary arithmetic all numbers are expressed by unity with zero. — LAPLACE.

HALLEY: PREDICTION OF COMETS

In applying Newton's theories to known comets his friend and disciple Edmund Halley (1656–1742) made the astonishing discovery that some of them instead of visiting the solar system once for all, actually described elliptical orbits of vast extent and great eccentricity about the sun. Among these he found one which having appeared in 1531, 1607, and 1682, should, if his identifications were correct, return in 1759. This bold prediction was fulfilled, and Halley's comet has not only reappeared in 1835 and 1910, but has even been traced back almost to the beginning of our era. A second similar prediction of Halley awaits verification in the year 2255.

In physics, Halley enunciated for spherical lenses and mirrors the correct formula $\frac{1}{f} = \frac{1}{a_1} + \frac{1}{a_2}$ and that for the barometric determination of altitudes. His mathematical work included graphical discussion of the cubic and biquadratic equations, a method of computing logarithms, and an edition of Apollonius from both Greek and Arabic sources. In his paper on "An Estimate of the Degrees of the Mortality of Mankind, drawn from curious Tables of the Births and Funerals at the City of Breslau; with an Attempt to ascertain the Price of Annuities upon Lives" (*Phil. Trans.* **17**, 596–610, 654–56, 1693), he laid the foundations of a new and important branch of applied mathematics. Having in boyhood occupied himself with magnetic experiments, in middle life he travelled in the tropics and made the first magnetic map, published in 1701 under the title "A general chart, showing at one view the variation of the compass." Drawing curves on this chart through points of declination, he invented a graphical method of wide future usefulness. From naval captain he became professor of geometry at Oxford, then astronomer royal till his death in 1742. One of his most notable achievements in astronomy was the discovery of actual changes in the apparent relative positions of the fixed stars, Aldebaran, Arcturus, and Sirius — answering a question centuries old.

REFERENCES FOR READING

BREWSTER, DAVID, *Memoirs of Sir Isaac Newton*, 1855.

CLARK, G. N., *Science and Social Welfare in the Age of Newton*, 1937.

DESCARTES, RENÉ, *Discourse on Method*, Trans. by John Veitch, 1850 (Everyman's Library, 1912).

NEWTON, SIR ISAAC, *Mathematical Principles of Natural Philosophy*, Ed. by F. Cajori, 1934.

SULLIVAN, J. W. N., *Isaac Newton*, 1938.

From previous lists: Ball, *Hist. Math.*, Ch. XV, XVI; Berry, *Hist. Astron.*, Ch. VIII, IX, X; Lodge, *Pioneers*, Ch. VII, VIII, IX; Mach, *Sci. Mech.*

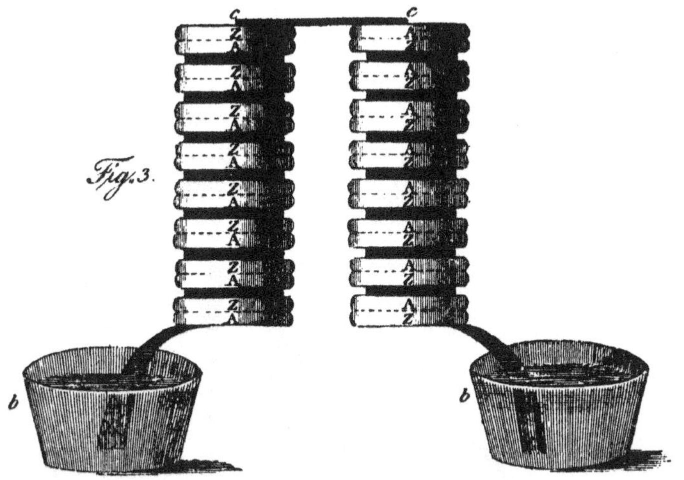

Chapter XIV

Natural and Physical Science in the Eighteenth Century

The seventeenth and eighteenth centuries mark the period in which, owing to the use of the several vernacular languages of Europe in the place of the medieval Latin, thought became nationalized. Thus it was that . . . people could make journeys of exploration in the region of thought from one country to another, bringing home with them new and fresh ideas. Such journeys . . . were those of Voltaire to England in 1726 . . . of Adam Smith in 1765 to France. — MERZ, I, p. 16.

THE INCREASING INTEREST IN SCIENCE

In the eighteenth century, especially in its latter half, chemistry, geology, botany, zoology, and physics, began to make deep impression on the learned world, while astronomy and mathematics ventured upon bolder and more far-reaching generalizations than they had ever before made. Science as a special discipline, or as a branch of learning worthy of the highest consideration, had as yet scarcely begun to make itself felt, but the names of Newton and Descartes were frequently heard in the salons of Paris and keen observers like Voltaire perceived the rising of a new tide in the affairs of men. A growth of popular interest might naturally have been expected after the great discoveries of the sixteenth and seventeenth centuries. What was not looked for was the concurrence of those political and social upheavals ever since rightly known as revolutions; viz., the French Revolution, the American Revolution and, probably most important of all, the Industrial Revolution.

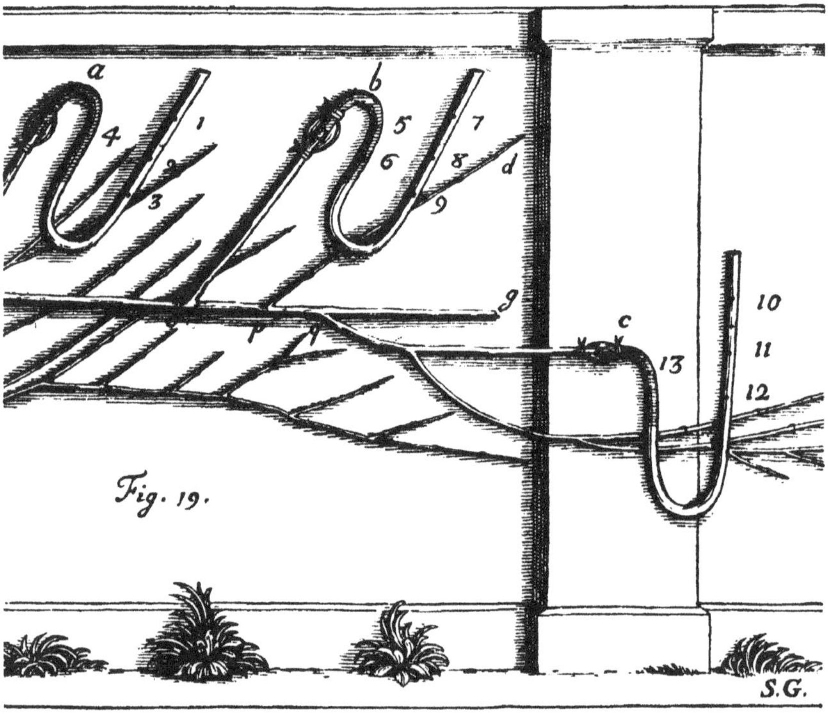

FIG. 51. — THE EARLIEST MANOMETER. *a, b, c,* "mercury gages" attached to branches of a vine; 1, 2, 3, . . . etc., graduation marks. After Hales, *Vegetable Staticks*, Fig. 19.

A NATURAL PHILOSOPHER: HALES

To the seventeenth century knowledge of natural phenomena, Stephen Hales (1677–1761) added fresh ideas and

NOTE: The earliest artificial source of a continuous electric current was Volta's electric pile. The two-column form is shown on the opposite page. A is a silver disk, Z one of zinc. A moist pad separates each pair of disks from the next pair. The metallic strip *cc* joins the columns in series, and the opposite end of each one is connected to a basin of water *b*. After Volta, *Phil. Trans.*, 1800, pl. 17.

new quantitative methods that became of great importance to the younger investigators of the eighteenth century in several departments of science. A country clergyman, he devoted the leisure intervals of forty years to scientific research, and the results may be found in his *Statical Essays*, 1727 and

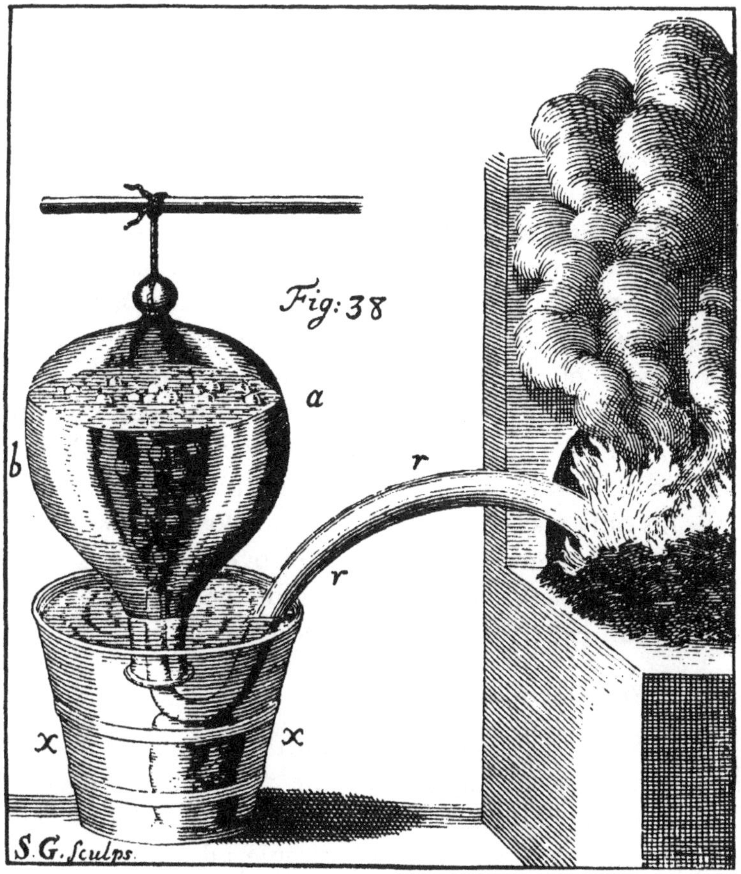

FIG. 52. — THE PNEUMATIC TROUGH. *ab*, "chymical receiver" filled with water and inverted in a vessel of water, *xx*, over the open end of a "leadon syphon fixt to the nose" of the iron retort *rr*. After Hales, *Vegetable Staticks*, Fig. 38.

1733. The first volume, *Vegetable Staticks* is a consecutive record of experiments that mark him as the founder of experimental physiology of plants. The second volume, *Hæmastatics*, contains the greatest single contribution since Harvey to the knowledge of the circulation of the blood. Hales determined arterial pressure in several animals of different size, by the height of the blood in a fine, straight, glass tube connected to a large artery. He also measured venous pressure and discussed the factors affecting the rate and volume of flow.

Throughout his work Hales applied the new physics to the solution of physiological problems. With Grew's anatomy of plants as a basis, he attacked the problems of the flow of sap and the transpiration of water from the leaves. In order to measure sap-pressure, he devised the mercury-gauge, or manometer, Fig. 51.

To the problem of plant-nutrition Hales applied a suggestion of Newton, that a gas may be fixed in a solid combination. He distilled plants to see if he could recover air probably absorbed through the leaves, and he measured the gas given off by means of the pneumatic trough, a bell-jar inverted over water and connected by a bent tube with the retort, Fig. 52. Hales investigated a great variety of other substances in this way, but the gases obtained were all alike to him, his only test being obedience to Boyle's law (p. 303). He advanced chemistry by his improved methods of handling gases and his clear enunciation of the principle that the same gas can exist free or combined.

ON DIFFERENT KINDS OF AIR

At the beginning of the eighteenth century, air and water were regarded as elements, and the phlogiston theory dominated chemical thought until nearly the end of the century. The first to prove that air is not a simple substance was Joseph Black (1728–99), then a medical student at Glasgow in search of a new alkali while preparing his remarkable thesis of 1755. Here the influence of Hales is to be seen in the first detailed examination of a definite chemical action, and its reversal, with all changes in weight accounted for. Black began with magnesia, a "mild alkali" ($MgCO$), treated by heat or with acid, and compared it to limestone and chalk, long known, when heated, to yield quicklime, a "caustic alkali" (CaO), then supposed to be a compound of limestone and phlogiston. After many quantitative experiments on the relations between mild and caustic alkalis, Black concluded that when the calcareous earths are heated they give off a large amount of air, which when bubbled through limewater made from quicklime forms a chalky precipitate. To this kind of air (now known as *carbon-dioxide*) he gave the name "fixed air"; and by this *test* he showed that (1) it occurs normally as a small constituent of atmospheric air, from which by this process it can be completely removed, that it is (2) expired from the lungs and is a product (3) of fermentation of beer and (4) of the combustion of charcoal. He identified it with the *gas sylvester* of van Helmont (p. 305). Black, by his clear recognition of "fixed air" as something with distinct qualities, and by his quantitative methods, opened a new field in chemistry.

T. O. Bergman (1735–84) of Upsala, supplemented the work of Black on "fixed air" by weighing the new gas, finding it heavier than air, and he discovered that it is very soluble in water, and is acid to litmus (a test recommended by Boyle). He accordingly named it "aerial acid."

Another gas, now called *hydrogen*, was probably first produced and found to be inflammable by Boyle, who obtained it from iron filings treated with mineral acid. The first accurate experiments on the properties of this gas were described in a memoir, "On Factitious Airs" (*Phil. Trans.*, 1766) by Hon. Henry Cavendish (1731–1810), a shy, eccentric member of a noble family, having a passion for statistics and accurate observation, combined with inventive genius and the possession of great wealth. He compared "fixed air" with this new kind, which, for the reason that it took fire whenever flame was applied to it, and also because he believed it to be a cause of explosions in mines, he called "inflammable air." He identified it with phlogiston and inferred that it was derived from the metal, according to Stahl's theory. His experiments established the existence of two gases different from common air, and showed a great advance in method.

In the following year (1767) a liberal Non-Conformist

named Joseph Priestley (1733–1804), who had been teaching languages at Warrington Academy, came as a pastor to Leeds and took a house next to a brewery. He had written books on oratory and law and an excellent *History of Electricity* that secured him the F.R.S. The brewery afforded an abundance of "fixed air," and Priestley amused himself with this, obtaining by his experiments results so fascinating that he forsook electricity and became the father of pneumatic chemistry.

Starting with little knowledge of chemistry, he devised his own apparatus and greatly improved the pneumatic trough by using quicksilver in place of water. In all, he isolated and described nine different gases, all but three of them new to science. He made three discoveries of fundamental importance: (1) The first was "nitrous air" (*nitric oxide*, NO) from metals treated with "spirit of nitre" (nitric acid); and the use of this gas as a *test* for "the fitness of the air for respiration" — really a valuable quantitative test for oxygen, the production of red fumes and the diminution in volume showing the proportion of oxygen present. (2) Secondly, he showed that *plants tend to keep air fit for respiration.* Cooling having failed, he tried a sprig of mint (Aug. 17, 1771) in air vitiated by combustion, and ten days later found that a candle burnt perfectly in this air. By later experiments he showed this property of plants to depend on their green coloring matter and on exposure to sun-light. (3) His greatest discovery (*oxygen*) followed the acquisition of a large burning glass, with which he tried the effects of heating a great variety of substances in a bell-jar over mercury. From *mercurius calcinatus per se* (red mercuric oxide) he obtained much gas (Aug. 1, 1774) and was surprised to find that in it a candle burned with a remarkably vigorous flame. He obtained the same gas from red lead. This was all he knew about it when, a few months later in Paris with Lord Shelburne, he told Lavoisier of his discovery. Soon after, he thoroughly examined the new gas, called it "dephlogisticated air," and communicated his results to the Royal Society, March 15, 1775. This was the first public announcement of the discovery of oxygen — so named by Lavoisier, who made it the key-stone of the new chemistry.

After eight years of research while serving as librarian for Lord Shelburne, Priestley resumed the ministry in 1780 at Birmingham. Here he completed his six volumes *On Different Kinds of Air*, 1775–86. But his numerous political and theological writings and speeches aroused such hostility that in 1791 his meeting house was burned and his residence sacked by a mob: and a few years later he found it expedient to move with his family to Pennsylvania. His last scientific paper was a defense of the phlogiston theory.

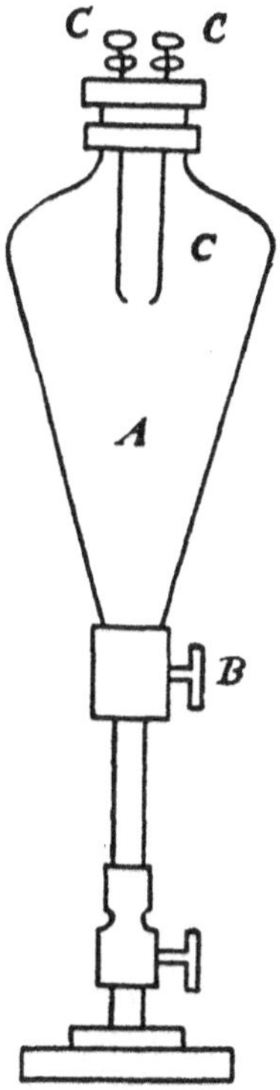

Fig. 53. — The "Eudiometer" Described by Cavendish. *A*, chamber for exploding mixed gases; *B*, tap; *C*, *C*, electrodes. After Harvey-Gibson, *loc. cit.*, p. 175. Permission of The Macmillan Company.

Priestley just missed a fourth great discovery when he happened to explode "inflammable air" with common air by an electric spark, and noticed dew covering the inside of the closed vessel. Cavendish repeated this experiment, collected the "dew," and found "that almost all the inflammable air, and about one-fifth of the common air, are turned into pure water." The next step was to explode inflammable air and dephlogisticated air together and obtain the combining values: 2.02 to 1, as described in the *Experiments on Air*, 1784. One puzzle remained — the water produced was acid. This effect was explained, 1785, by the accidental presence of "phlogisticated air" (nitrogen) in the explosion chamber. Cavendish, not only had *synthesized water*, but unwittingly had accomplished the *fixation of atmospheric nitrogen.* His results, however, were expressed in the language of the phlogiston theory, to which he adhered all his life.

During the time of these investigations in England, a great amount of chemical work was accomplished in Sweden by Carl Wilhelm Scheele (1742–86), who "remained a poor apothecary all his life, yet was really one of the first chemists of Europe." He is to be credited with the discovery of chlorine, glycerine, and numerous organic acids; and, from notes published after his death, it appears that he had prepared oxygen before Priestley. He called it "empyreal air."

A NEW CHEMISTRY: LAVOISIER

The study of the laws governing chemical reactions was advanced by Etienne-François Geoffroy (1672–1731) with his tables of affinities of each base for various acids. These tables were improved upon by Bergman, whose work on "fixed air" has been mentioned and whose methods have earned for him the title of father of analytical chemistry.

He prepared a table of affinities governing both "wet reactions" (in a solution) and "dry reactions" (in a fusion), under the title, "Attractiones electivae simplices" (*Nova Acta Reg. Soc. Sci. Upsaliensis*, **2,** 161–250, pls. 7–8, 1775).

This work, together with the researches on gases of Black, Cavendish, Priestley, and Scheele, laid the foundations upon which Antoine-Laurent Lavoisier (1743–94) began the erection of the present lofty structure of modern chemistry. Lavoisier, endowed with critical insight, creative imagination, and capacity for sound judgment, was educated at the Collège Mazarin in Paris, attended scientific lectures in the Jardin du Roi, and for three years assisted Guettard in his pioneer work in geology (p. 359). His first communication to the Académie Royale des Sciences, 1765, showed his unremitting use of the balance and his method of analysis and synthesis. It stated that when gypsum is heated to make plaster of Paris, pure water is given off exactly equal in weight to that required to set the same amount of plaster. It is said that modern chemistry was born with this experiment. His experiments to determine the degree of purity of water obtainable by distillation led to other experiments, 1770, that forever silenced the ancient dogma of the transmutation of water into earth. The "earth" was found to be derived from the solvent action of hot water on glass.

The law of *conservation of matter* (or *of mass*) was soon formulated in Lavoisier's mind. He tested it when, in collaboration with several other investigators, he experimented with diamonds, supposed by some to have magical power and to vanish when heated. Lavoisier and his associates, in 1772, found a diamond to be uninjured by heat if air was excluded, and in the presence of air to be converted, like charcoal, into a gas with all the characteristics of Black's "fixed air." With the diamond in a bell-jar over water or mercury they were able to subject it to a high temperature by placing it in the focus of the sun's rays from a large lens that these investigators used also to study the effect of heat on various metals.

Lavoisier amplified the experiments of the English chemists, in his *Opuscules physiques et chimiques*, 1774. He found that when metal is burned in a closed space the volume of air decreases in proportion to the increase in weight of the calx, and that the process stops when part of the air has become exhausted. How to recover the active part of the air from the calx was suggested in conversation by Priestley, and in April, 1775, Lavoisier read to the Academy his most famous memoir, "On the Nature of the Principle that combines with Metals in Calcination and that increases their Weight," in which he showed that what united with a metal to form a calx was air, purer and more respirable, if possible, than atmospheric air; and that, when a calx was reduced in the presence of charcoal, "fixed air" resulted from a combination of air from the metal and from the charcoal. After this first reading, Lavoisier repeated Priestley's experiments, which resulted in his discovery that charcoal when heated with the calx of mercury completely disappears. Therefore, when this memoir was revised for publication and read again in August, 1778, he was able to say definitely that "'fixed air' is the result of combination of the eminently respirable portion of the air with charcoal," and he proposed names for the new gases and their compounds. This *new theory of combustion* marked the end of the phlogiston theory, which received its death blow in Lavoisier's memoir of 1783, "Réflexions sur le Phlogistique"

(*Mém. Acad. R. Sci.*, for 1783, p. 505, 1786), "one of the most notable documents in the history of chemistry."

Detailed examination of the changes in air during breathing and the effects produced in air by calcining or burning, led Lavoisier to a new conception of respiration, as set forth in his report to the Academy — "Experiments on Respiration of Animals . . ." 1777.[1] He concludes that respiration is a process of combustion that may take place (1) in the lungs or (2) in the blood with an exchange of oxygen and carbonic acid (names adopted later) in nearly equal proportion through the lungs.

This led to the study of animal heat, which might be a fluid ("*calorique*") or might be a mode of motion and, in that case, "the consequence of the conservation of active forces. Whatever its cause, it is measurable." In collaboration, Lavoisier and Laplace (p. 376) made an ice-calorimeter. They measured (1) the amount of "fixed air" produced in a given time by a guinea-pig breathing in "pure air" (oxygen) in a receiver over mercury; (2) the quantity of ice melted by a guinea-pig in the calorimeter for the same length of time; (3) the amount of ice melted when the same amount of "fixed air" was produced by combustion of charcoal — concluding, 1780, that "respiration is . . . perfectly analogous to the combustion of charcoal," and is the source of heat in the body.

When he heard that Cavendish had composed water "by the combustion of two airs," Lavoisier immediately synthesized water and then *reversed* the process by the original method of passing steam over red-hot iron, which took up the oxygen and allowed collection of pure "inflammable air," which he named *hydrogen gas*. He concluded, 1783, that water is a product of combustion; and also of respiration, 1785. In 1789 and 1790, further studies of metabolism, with Seguin taking the place of the guinea-pig, demonstrated the relation between heat-production and muscular work, digestion, and exposure to cold; and the regulatory activity of the skin.

[1] A classic of scientific method translated in full by C. M. Taylor, 1934, pp. 66–74.

Lavoisier outlined the main facts of respiration as known to-day and took the first steps toward a universal theory of *conservation of energy*.

Lavoisier's work, "the first great synthesis of chemical principles," culminated in the *Traité élémentaire de chimie*, 1789, in which the new theory was expounded systematically for beginners, in the new language previously adopted in collaboration with other chemists. The elements were named [1] oxygen, hydrogen, azote (nitrogen), carbon, etc., with reference to their properties; and the compounds, according to their composition. The *chemical equation* was suggested for the expression of experimental results, and the doctrine of *conservation of matter* was stated clearly: "Nothing creates itself . . . in every operation or reaction, there is an equal quantity of matter before and after . . . there is only exchange or modification." The book had an enthusiastic reception. Modern chemistry had come of age.

Lavoisier's last scientific work was with the commission to establish the units of the *decimal system* of weights and measures decreed by the National Assembly. Unfortunately for him and for science, early in his career he had bought a share in the hated Ferme Générale, a private corporation commissioned to collect taxes, and was active in its management. With other officers of the Ferme, he was guillotined during the Terror of 1794. "The Republic has no need of scholars," said the judge.

BEGINNINGS OF MODERN IDEAS OF SOUND

That air is really the intermediary for sound was proved in 1705 by Hauksbee's experiment of placing a clock in a vacuum. After Galileo, the studies, mathematical and experimental, of Newton, Euler, and Joseph Sauveur (1653–1716), who first drew attention to overtones or harmonics, brought acoustics to the point where it was taken up and given much of its present form by E. F. F. Chladni (1756–1827) of Wit-

[1] Oxygen, acid-former with sulphur or phosphorus; hydrogen, water-former; azote, not supporting life; carbon, charbon, French for charcoal.

tenberg — "the Father of Modern Acoustics." The work of Chladni at the beginning of the nineteenth century laid broad and deep the foundations of acoustics as we know that science today, and upon that foundation Helmholtz and Tyndall in the middle of that century reared a large part of the modern superstructure. Chladni carried much further the experiments of Galileo on vibrating plates, substituting a violin bow for the chisel, and sand for dust on the plates, obtaining thereby a wonderful variety of figures. He also devised a simple method of counting the number of vibrations corresponding with each note.

THE BEGINNINGS OF MODERN IDEAS OF HEAT. LATENT AND SPECIFIC HEAT. CALORIMETRY

Heat was generally regarded during the eighteenth century as one of the *imponderable elements* — a subtle fluid, "the matter of heat." It was included with the other three (light, magnetism, and electricity) under a new name, "caloric," by the French chemists in the reformed nomenclature of 1787. High temperature meant the presence, and low temperature, the absence of caloric.

The mercury thermometer was made a convenient tool by a German instrument maker, D. G. Fahrenheit. In 1714, he introduced two fixed points in the scale, zero being the temperature of a mixture of ice and salt. A scale of eighty degrees between the freezing and boiling points of water devised in 1730 by R.-A. F. de Réaumur (1683–1757), a distinguished zoologist, is still used in many countries; and Anders Celsius of Upsala in 1742 introduced the centigrade scale now universally used in scientific work.

The first to distinguish between temperature and quantity of heat was Black, famous for his researches on "fixed air" (p. 345), then professor of chemistry in Glasgow and later at Edinburgh. His work on heat, begun in 1761 but not published until 1803, four years after his death, introduced *calorimetry*, the quantitative study of heat; but there is no evidence that Black ever used the ice-calorimeter usually attributed to him. His method of measuring heat-exchanges in terms of a temperature change for an equal mass of water, in degrees Fahrenheit, was thoroughly satisfactory at the time.

Black saw that different kinds of matter, e.g., water and mercury, require different quantities of heat to raise their temperatures an equal number of degrees. His editor tells us that Black estimated "capacity for heat" (called *specific heat* by J. C. Wilcke) "by mixing two bodies in equal masses, but of different temperatures; and then stated their capacities as inversely proportional to the changes of temperature of each by the mixture."

Black discovered that heat may be applied to water containing ice without raising the temperature, and later he found the same to be true of boiling water. But once the ice was all melted, or the escape of steam stopped in a closed vessel, the temperature began to rise. The heat that somehow became imperceptible or concealed during the melting of ice or the evaporation of water, Black named *latent heat.* James Watt (1736–1817) assisted Black in the experiments on the latent heat of steam and used the results in his work on the steam-engine (see below). They sought a quantitative ratio between latent heat and quantity of steam, and concluded that the heat lost in expansion of steam equals the amount used to produce it.

The much discussed question of the "weight of heat" was taken up by Sir Benjamin Thompson, Count Rumford (1753–1814), F.R.S. (born in Woburn, Mass.). While Minister of War in Bavaria, he observed a great amount of heat produced during the boring of brass cannon. This led to the classic paper (*Phil. Trans.* 1798, **88,** 80), in which he argued that heat is *a form of motion.* This was followed by a magnificent attack, with use of the balance, on the material theory of heat (*Phil. Trans.* 1799, **89,** 179), concluding, "that all attempts to discover any effect of heat upon the apparent weight of bodies will be fruitless." But the material theory survived until the middle of the next century.

EIGHTEENTH CENTURY RESEARCHES ON LIGHT

These derive from the great work of Newton, and especially of Huygens, in the previous century, and continue with the fruitful inventions of the achromatic telescope by Hall in 1733, and the work of Dollond, an English optician, upon achromatic lenses, leading up to the construction in 1758 of achromatic telescope objectives. The achromatic telescope now became a serviceable instrument, but the compound microscope had to wait more than half a century longer for correspondingly serviceable achromatic objectives. It was not until the opening of the nineteenth century that much further progress was made in our knowledge of light. It is rather for progress in sound, in heat, and in electricity, that eighteenth century physical science is chiefly notable.

BEGINNINGS OF MODERN IDEAS OF ELECTRICITY AND MAGNETISM

At the opening of the century, the term "electricity" meant the power of certain substances called "electricks," when rubbed, to attract light objects, and this power was attributed to "effluvia" emitted from the electrified body. Boyle had shown in 1675 that rubbed amber can attract in a vacuum, and in the same year Jean Picard had observed flashes of light in a vacuum chamber above mercury. These experiments were continued by Francis Hauksbee, "one of the most active experimental philosophers of his age," and one of the first to study capillary action. In his experiments "on mercurial phosphorus" he showed that light is emitted by mercury when agitated in a vacuum. He attributed this to electricity and compared it to lightning.

A discovery by Stephen Gray (1729) created a difficulty for the effluvium theory. He found that the electric virtue of a glass tube could be conveyed to other bodies, and through wire or wet string to bodies many feet away. It was found at once that bodies differ in *conductivity*, and the conductors were identified by Desaguliers with Gilbert's "non-electricks."

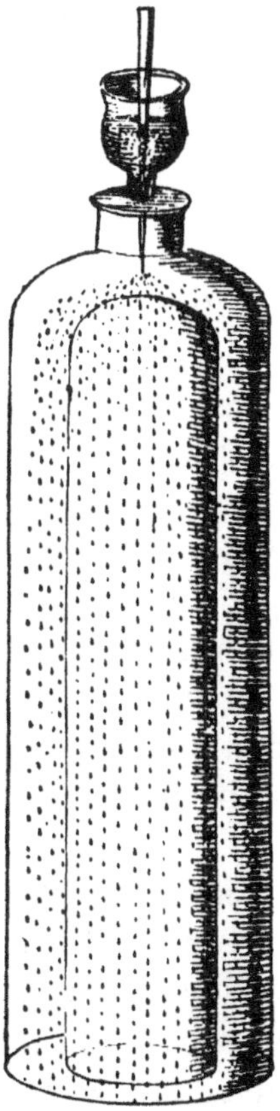

FIG. 54. — AN EARLY VACUUM TUBE. "Experiment II, Shewing that Mercury will appear as a Shower of Fire, whilst descending in Vacuo from the top to the bottom of a tall Receiver." The mercury under atmospheric pressure in the funnel is driven violently against the inner cylinder. After Hauksbee, *Phil. Trans.*, No. 303, 1705.

Another surprising discovery, made by C. F. Dufay (1735), was that gold-leaf electrified by rubbed glass was repelled by the same, but attracted by rubbed resin. He named the *two kinds* of electricity, *vitreous* and *resinous*, from the substances which produced them.

The acquisition of a recently improved frictional electric machine started Benjamin Franklin (1706–90) on investigations that brought him international reputation and many university degrees. Gray had observed *induction.* It was discovered independently in Holland, and the Leyden jar was invented there about 1745. Franklin made the first complete study of the Leyden jar, in 1747–48, and explained the action of "Mr. Muschenbroek's wonderful bottle." Previously, in a letter dated July 11, 1747, he had introduced the terms *positive* and *negative*, being the first to regard an electric charge as a disturbance of equilibrium. He recognized but one kind of electricity, which he believed to be an element present in moderate amount in all bodies. When glass was rubbed it would draw electricity from the rubber and become "electrised positively, or *plus*," while the rubber became "electrised negatively, or *minus*." A body "having only the middle quantity of electrical fire may receive a spark" from a positively charged body and give one to a body negatively charged (*Phil. Trans.* **45**, 98–100).

Franklin communicated his results usually in personal letters not intended for publication. These and other papers, gathered by an anonymous editor, were printed in London, the fifth edition in 1774, under the title, *Experiments and Observations on Electricity made at Philadelphia in America.* From this volume we learn that he found the inside surface of the Leyden jar to be positive and the outside surface negative. He observed that "bodies having less than the common quantity of electricity repel each other, as well as those that have more," and that "glass cannot be electrified minus." But a brimstone globe, used in the place of glass in the electrical machine, charges negatively. For a row of Leyden jars connected in series, he invented the term electric battery. Not knowing that French scientists had carried out his suggestion that electricity could be drawn from a cloud by an elevated rod, Franklin and his son, during a thunder storm, in June, 1752, charged a Leyden jar with a key tied to a kite-string — one of the greatest triumphs of science, for by correlating the thunderbolts of Zeus with the shocks of a Leyden jar, he, to a great extent explained one of the oldest known and most awe-inspiring phenomena of nature.

A pioneer in the study of the magnetic field was C. A. Coulomb (1736–1806), who invented the very sensitive torsion balance, with which he fixed the units of electricity or magnetism and demonstrated in 1784 that magnetic poles attract or repel each other with forces inversely as the square of the distance. Abraham Bennett applied the principle of induction in the invention of the gold-leaf electroscope, 1787.

At the time when a lucky accident brought the irritability of nerves to the attention of a physiologist, Luigi Galvani (1737–98) of Bologna, the only sources of electricity then known produced merely a *spark* or a shock. In 1786 he reported that when a muscle and nerve were connected by two metals in contact with each other, the muscle would contract as if under the discharge of an electric machine. He attributed the effect to "animal electricity." This notion was violently attacked by Alessandro Volta (1745–1827), professor of physics at Pavia and long a student of electricity. He claimed that the electricity came from the contact of two metals, unaware of the electrolysis in the animal juices, as well as in the moist pads that separated the successive layers of copper and zinc in his own "electric pile." His great discovery that the latter could produce a *continuous electric current*, announced to the Royal Society in 1800, aroused immense interest, and soon was followed by the construction of powerful batteries of the modern type. Thus began the application of electricity to human needs, and hence came our words: galvanic, and voltaic, galvanometer, voltmeter, etc.

THE BEGINNINGS OF MODERN IDEAS OF THE EARTH

Throughout the eighteenth century the Biblical account of the Flood and the Creation in six days dominated men's thought of the Earth and its history. The prolonged controversy as to the nature and origin of "figured stones" stimulated much collecting of fossils and the production of excellent illustrations. The "diluvian doctrine" that fossils are the remains of animals buried by the Flood was warmly supported by the Church; it also explained the existence of extinct elephants in Siberia, known before the scientific expedition under P. S. Pallas (1741–1811), sent out by Catherine II of Russia to observe the transit of Venus of 1769 and to obtain information in all departments of natural history. Pallas made the surprising discovery of vast numbers, not only of fossil elephants, but also of rhinoceros and buffalo in northern Siberia. More important was a memoir by Pallas (in the reports of the expedition, 1772–76) on the formation of mountains showing his clear recognition of a *geological sequence*.

A central figure in the brilliant group that made Paris the metropolis of the scientific world, was G.-L. Leclerc de Buffon (1707–88), distinguished for his spirit of bold generalization and his clear and eloquent style. The first volume of his *Histoire Naturelle* (1749) contained his Theory of the Earth, in which the existence of fossil shells was cited to show conditions of the earth's surface differing from the present; and following Descartes and Leibniz, with keen insight he linked the earth's history with that of the solar system. In 1778 appeared his masterpiece, *Époques de la Nature*, in which he divided the history of the globe into six epochs.

> In breadth and grandeur of conception Buffon far surpassed the earlier writers . . . It was his great merit to have pointed out that the history of our earth is a long chronological record, the memorials of which are to be read in the frame-work of the globe itself, and to have himself applied the historical method to its interpretation. Nor were his services less conspicuous in breaking down the theological barrier which, after so many centuries, still blocked the way towards a free and unfettered study of the crust of the earth. — GEIKIE, *Founders of Geology*, p. 96.

Contemporary with Buffon in Paris was another pioneer in geology of a very different type — J.-E. Guettard (1715–86), who employed the young Lavoisier for several years as an assistant and had much influence on his career. He was an accurate observer, with much originality, who opened paths in a number of fields, while shunning theory. In a memoir of 1746 he produced the first maps ever constructed to show the superficial distribution of minerals and rocks. While in central France seeking material for a more detailed map, he discovered (1752) that the highlands of Auvergne dominated by the Puy-de-Dôme, "a hill made famous by Pascal" (p. 302), is a group of old volcanoes. He thus became the originator of the Vulcanist party. He was one of the founders of palæontological geology, collecting and describing hundreds of fossils. In his work on the *Degradation of Mountains*, he was one of the first to demonstrate the effects of water in carving the topography of the land.

Guettard's work in Auvergne was followed up in 1763 by Nicholas Desmarest (1725–1815). Among the old lavas, he found basalt frequently showing the columnar structure of the well-known Giant's Causeway on the coast of Ireland. Desmarest's memoir of 1774 is a landmark in the history of geology. By showing widespread volcanic action in countries where there was no human record of such events, he accomplished one of the triumphs of geology. Desmarest became the founder of volcanic geology, and he originated a fruitful theory of the origin of valleys by the action of streams of water, making the first attempt to trace the history of a landscape and to explain it by causes now at work.

The idea of *geological succession*, a chronological sequence of strata, was of slow growth. During the early and middle eighteenth century, many observers published descriptions of the successive strata of particular localities. They noted the general character of the rocks and the kind of fossils in them, whether marine or land forms. H.-B. de Saussure (1740–99) was the first to study the Alps and to adopt the terms *geology* and geologist (1779). G. Arduino (1713–95) applied to the rocks of Italy the *threefold classification* still in use — Primary, Secondary, and Tertiary. L. G. Lehmann (1756) produced the first *sections* to show the order of strata, and G. C. Füchsel (1762) originated the idea of a *formation* as a continuous series of strata representing a certain epoch in the history of the globe. Finally Bergman, the celebrated chemist (p. 346) explained the structure of the earth (1769) as formed by uniform layers of different thickness precipitated or deposited from a *universal ocean.*

These ideas were adopted by A. G. Werner (1750–1817), whose small treatise of 1774, introducing order and a definite nomenclature into the description of minerals, made him at once the most renowned mineralogist in Europe. The next year, he was appointed to teach in the School of Mines at Freiberg in Saxony, founded in 1765, the first of its kind, where he had been a student before going to Leipzig. There he wielded immense influence, chiefly through the writings of his enthusiastic pupils. He greatly advanced petrographic knowledge and did great service to geology by insisting on the doctrine of geological succession; but the progress of geology was checked when his passion for order and precision led him to adopt the idea of a world of rock formations precipitated chemically by a universal ocean that covered the highest mountains, with the definite sequence observed in Germany. His defense of the idea of this aqueous origin of granite and basalt led directly to the acrimonious controversy between his followers, the Nepunists, and the Vulcanists, who adopted the views of Desmarest and of Hutton.

James Hutton (1726–97) of Edinburgh was perhaps the first to grasp the fundamental idea of modern geology that the past can be explained by the present. Living in a region full of inspiration for the geologist, he saw evidence of gigantic convulsions of the earth's surface, and his search for their causes continued for thirty years before he published his memoir on *A Theory of the Earth*, 1785, which is a turning point in the history of geology. By observation and reading he had collected facts and he tabulated them in such a way that they would reveal their story. Hutton's system included erosion and deposition, distortion of strata, volcanic action and molten intrusion due to internal heat and pressure, and alteration of sediments by heat under pressure suggested to him by Joseph Black (p. 345). Geikie says, "The whole of the modern doctrine of earth sculpture is to be found in the Huttonian theory." Hutton's writings were defective in style and arrangement, but the summary of his system, with original additions, by his friend John Playfair (1748–1819), *Illustrations of the Huttonian Theory of the Earth*, 1802, is one of the classics of scientific literature.

EIGHTEENTH CENTURY PROGRESS IN BOTANY AND ZOOLOGY

In the biology of the eighteenth century the foremost figure is Carolus Linnaeus (Carl von Linné, 1707–78) of Upsala, the founder of modern systematic botany and zoology. J. P. de Tournefort had shown (1700) how to distinguish genera of plants. Linnaeus, who believed every species to have been specially created, at the beginning of the world, established the "*binomial nomenclature*," by which each species receives a double Latin name — a noun (generic name) and a qualifying word, as *Homo sapiens*, his name for Man. Linnaeus also developed a brief, clear, and accurate mode of description for each group. In his *Systema Naturae* (Ed. 1, 1735) every known plant and animal, including Man, eventually had a place in a Class, Order, Genus, and Species. The tenth edition (1758) is accepted as the starting point of modern nomenclature. His teaching aroused great interest in the discovery and naming of new species. In his classification of plants, Linnaeus used an artificial system based on the characters of the sexual organs in the flower. A more natural system of botany was worked out at the Jardin du Roi in Paris by B. de Jussieu (1699–1777) and published by A.-L. de Jussieu (1748–1836) in his *Genera plantarum*, 1774.

Large, beautifully illustrated books, are characteristic of eighteenth century zoology. The great work of Réaumur (p. 353) on insects (1737–48) is still a source of information. The most popular author was Buffon (p. 358), director of the Jardin du Roi, who, in his celebrated *Histoire Naturelle* (1749–1804) dealing with Man, mammals, and birds, sought to portray living things as part of the unity of Nature. He supposed organisms to have a common origin from "organic molecules," and, in opposition to Linnaeus, he regarded species as capable of modification in structure. The superb anatomical drawings of mammals and descriptive text by Louis Daubenton (1716–1800) are notable features of the work.

The question of the origin of the embryo was debated throughout the century. The chief advocate of Ovism and of Preformation was Charles Bonnet (1720–93) of Geneva, who found evidence for his doctrine in his important investigation, instigated by Réaumur, of the parthenogenesis of Aphids (cf. p. 297), *Traité d'insectologie*, 1745. Also in the same work he gives a linear "scale of natural beings," placing Man at the top and the polyps in the middle to connect animals with plants — thus illustrating his belief, with Buffon, in gradations between all parts of the universe. In 1759, C. F. Wolff (1733–94) published his doctor's thesis, *Theoria generationis*, which is a landmark in the history of embryology. He was the first to compare the development of plants and animals. He found in both cases a process of Epigenesis (cf. p. 296), leaves and flowers forming gradually from undifferentiated material in the bud, and the parts of the chick arising in the same way in the egg, minute ampullæ, or sacs, appearing as the first step in organization. In a later memoir (1768–69), Wolff described the development of the digestive system from its origin as a straight tube. But these fundamental discoveries were overshadowed by the prestige of Bonnet until the next century.

Other eighteenth century naturalists applied the experimental method with important results. Réaumur's experiments on birds proved digestion to be a chemical process. The idea of Spontaneous Generation was revived by J. T. Needham (1748) with the approval of Buffon. In infusions cooled after boiling, Needham found an abundance of micro-organisms — proof of origin from organic molecules. Chief among the experimenters was Lazzaro Spallanzani (1729–99) of Modena and Pavia. He repeated Needham's experiments, but with air excluded, and found no growth so long as the flasks were kept sealed. He studied spermatozoa, fertilized frog's eggs artificially, and by his experiments showed that the fertilization of the egg is not due to an immaterial "*aura spermatica*," as Harvey had thought, but is effected by contact with the seminal fluid, without which the egg perishes. Influenced by Bonnet, he would not credit this effect to spermatozoa, although he found the residue from filtration to be the active component. In experiments on digestion (1782), Spallanzani proved that the gastric juice is secreted by the stomach and that its action differs essentially from fermentation of wine.

J. G. Koelreuter (1733–1806) of Carlsruhe made the first truly scientific experiments on hybridization of plants (1761–66). He studied the normal sexual processes and distinguished flowers that are self-fertilized, pollinated by the wind, and by insects, and he described their relations. Jan Ingen-Housz, a Dutch physician living at times in London, repeated (1779) Priestley's experiments, and later (1796) with the new chemistry of Lavoisier at his command, laid the foundations of our knowledge of plant nutrition and respiration. He established the fact that plants derive their carbon from the carbonic-acid-gas of the air and give off oxygen from their green parts when exposed to light, also that they require oxygen and give off carbon dioxide in the dark. Here the subject was taken up by Théodore de Saussure (1767–1845) in Geneva. He introduced the new quantitative method of Lavoisier, confirmed Ingen-Housz, and proved that plants derive their nitrogen and salts from the soil through their roots. In his treatise of 1804 the theory of plant nutrition takes its modern form.

The idealistic philosophy of Immanuel Kant (1724–1804) produced a group of German "Nature-philosophers." One of the most important of them was the poet, J. W. von Goethe (1749–1832), who invented the word *morphology* to denote the science that concerns the plan of structure, as distinguished from descriptive anatomy, of living things. His most important scientific work was on the metamorphosis of plants. In this he shows the floral parts to be modified leaves, and he describes an ideal plant.

PROGRESS IN THE MEDICAL SCIENCES

The century opens with Herman Boerhaave (1668–1738) beginning to teach medicine at Leyden (1701), where he carried on the tradition of Sydenham. Boerhaave adopted no system of medicine, but endeavored to bring all the sciences to the service of the patient. In his *Institutiones medicae* (1708) he introduced the term "physiology" in its modern sense, and established physiology as an academic discipline in the medical curriculum. One of his most famous pupils was Albrecht von Haller (1708–77) of Bern, professor at Göttingen 1736–53, whose *Elementa Physiologiae*, 1757, marks the beginning of modern physiology. Haller's greatest achievement was the establishment (1739 and 1743) of the doctrine of muscular *irritability*, a capacity of the muscle to respond to stimulation independently of the nerve, although normally excited by "nervous force." Irritability is recognized now as a fundamental property of living matter.

In the meantime, G. B. Morgagni (1682–1771), founder of modern anatomical pathology, was collecting material at Padua for his great book, *Seats and Causes of Disease*, published in 1761. This work, correlating with extraordinary completeness clinical detail and post-mortem findings, set a new standard in the science of pathology by its emphasis on detail and thoroughness.

In spite of these scientific teachers, many systems of medicine arose in the eighteenth century through hasty guesses at the causes of disease. The last and most enduring of these was Homœopathy, announced in 1796 and developed during the next decade by Samuel Hahnemann (1755–1843), an eminent German physician, as a protest against the futility of the systems then prevailing. It introduced the small dose and had important influence upon medical thought of the next century.

In London the energetic teaching of John Hunter (1728–93) was giving surgery its place as a branch of scientific medicine based firmly on physiology and pathology, as well as on anatomy. One may still see in London Hunter's extraordinary collection of illustrative preparations. A diagnostic method of first importance was introduced in 1761 by Leopold Auenbrugger (1722–1809) of Vienna. This was *percussion*, tapping, to outline the position of organs and lesions. The century culminated in the epoch-making work of Hunter's pupil, Edward Jenner (1749–1823). Immunization by inoculation from a mild case of small-pox had come from Constantinople in 1721. Jenner followed up a rural belief that immunity to small-pox was conferred by cow-pox. He began in 1778 to collect cases, in 1796 inoculated a boy with cow-pox, and later found the boy immune to small-pox. In his *Inquiry into Causes and Effects of the Variolae Vaccinae*, 1798, Jenner demonstrated, through accurate observation, careful generalization, and thorough verification, the value of *vaccination* and created a new science of Preventive Medicine.

THE INDUSTRIAL REVOLUTION. INVENTIONS. POWER

Far-reaching in their consequences as were the French Revolution of 1793 and the American Revolution of 1776, it is the Industrial Revolution, especially after 1770, with which the student of the history of science has chiefly to deal. Before the Industrial Revolution, almost everywhere before 1760 or even 1770, whatever machinery existed was run mostly by hand or foot, and hence was easily operated in the separate homes of the workers. But within the next thirty years the factory system had come, with coöperative labor in or about some central power-plant, and with machinery driven by water power or steam. With this change, which increased the output of the individual, and took work and workers out of the home, a revolution began which is still affecting every country and has modified the very structure of human society.

The chief factor in this change was the improvement of the steam-engine by Watt, whose work with Black on heat has been mentioned. To Newcomen's engine, Watt added in 1765 or 1769 a separate condenser with an exhaust pump. In 1782 he made the engine *double-acting* by admitting steam alternately at opposite ends of the piston, and by an automatic valve allowed *expansion* of steam to complete the stroke. Thus in the hands of Watt, heat became an efficient source of power. Henceforth machinery was to become the handmaid of toil, and to bring with it not only factory industry in place of home industry but, before long, improved means of transportation, effecting a virtual shrinkage of the world and a far closer contact of mankind.

Almost coinciding with the introduction of water power and steam power, came a great burst of invention. The spinning "jenny" and the "water frame" came almost hand in hand with the "mule" and the "power loom" (1785); while, as if to meet these on the cotton field, the cotton "gin" (engine), was invented in 1793 by Eli Whitney of Connecticut to replace the slow and tedious process of separating the cotton fiber, or staple, from its seed — hitherto laboriously done by hand. Applied chemistry also began to appear, e.g., in the manufacture of illuminating gas by Murdoch at Salford, England, in 1792, while the discoveries of Galvani and Volta at the very end of the century opened up that new era of electricity in the midst of which we dwell today.

THE INFLUENCE OF SCIENCE UPON THE SPIRIT OF THE EIGHTEENTH CENTURY

Writers on the literature of the eighteenth century, after condemning it because of its comparative barrenness in great works of art or literature, are apt to find the reason in one or another aspect of the growth of science. Professor Dowden, for example, in his essay on Goethe, remarks that

> Rousseau's emancipation of the heart, was felt in the eighteenth century to be a blessed deliverance from the eager, yet too arid, speculation of the age,

although he admits that:

> Humanity, as Voltaire said, had lost its title-deeds, and the task of the eighteenth century was to recover them.

Dowden's unusually charitable judgment of the century is more or less typical of literary opinion generally.

For the scientist, on the other hand, few centuries in all history are more important, for the eighteenth was not only rich in scientific performance but still more pregnant with promise. And even in art — if in that term music be included — and literature, a century which produced a Haydn, a Mozart and a Beethoven, with a Burns, a Voltaire, a Wordsworth and a Goethe, need not fear comparison.

We have mentioned above the first School of Mines: viz., that at Freiberg, in Saxony (1765). The first School of Civil Engineering was established in Paris (1747). In this century also were established new universities, e.g., Yale (1701), Göttingen (1737), Princeton (1746), Bonn (1777) and Brussels (1781).

The "physiocrats" and the "encyclopaedists" of the French school of practical philosophers also deserve notice, for they were professedly inspired by science and seeking to apply it to human society. Nor should we forget the service to social science of Count Rumford, who for the first time grappled boldly with problems so far apart as the control of mendicity, of smoky chimneys, and of poverty. Perhaps his greatest service was as leading founder of the Royal Institution in London, chartered in 1800 by George III.

The nebular hypothesis of Laplace, through its central idea of natural development rather than sudden and special (artificial) creation of the solar system was an important preparation of men's minds for theories of transformation or evolution. Hutton's Theory of the Earth enforced the same idea for the familiar earth, while the metamorphosis of the parts of the flower, pointed out by Goethe, helped to pave the way for acceptance of the idea of gradual modification of organs and even of organisms into others. To these matters we shall return in our discussion of Evolution.

REFERENCES FOR READING

Aykroyd, W. R., *Three Philosophers* (*Lavoisier, Priestley, Cavendish*), 1935.
Cole, F. J., *Early Theories of Sexual Generation*, 1930.
Dalton, J. C., *Experimental Method in Medical Science*, 1882.
Foster, M. L., *Life of Lavoisier*, 1926.
Geikie, Sir A., *Founders of Geology*, Ed. 2, 1905.
McKie, Douglas, *Antoine Lavoisier the Father of Modern Chemistry*, 1935.
McKie, D. and N. H. de V. Heathcote, *Discovery of Specific and Latent Heat*, 1935.
Miall, L. C., *Early Naturalists*, 1912.
Partington, J. R., *The Composition of Water*, 1928.

From previous lists: Singer, *Hist. Med.;* Taylor, *Nature of Air.*

Note: The diagram on the opposite page was used to explain a method for finding the area of a spherical triangle, *ABC*, on a unit sphere. *BCDB'C'E*, *BAB'A'*, *CAC'A'* represent great circles.

$$s = A + B + C - \pi.$$

"This beautiful theorem, discovered by Le Gendre, is much used in geodetical operations," Nathaniel Bowditch, *Mécanique Céleste by the Marquis de La Place*, Boston, 1839, Vol. I, p. 743.

Chapter XV

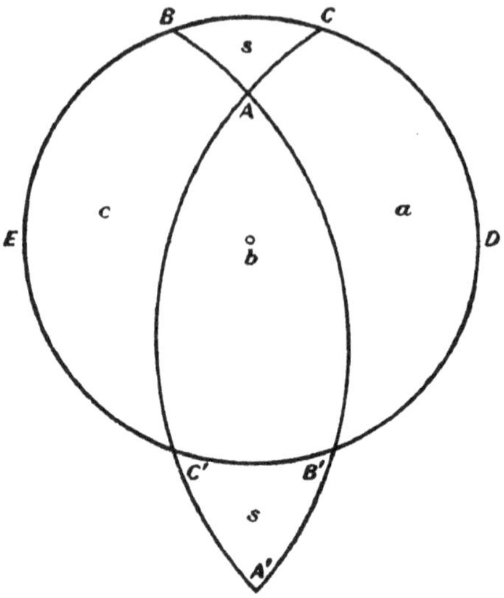

Modern Tendencies in Mathematical Science

Thought-economy is most highly developed in mathematics, that science which has reached the highest formal development, and on which natural science so frequently calls for assistance. Strange as it may seem, the strength of mathematics lies in the avoidance of all unnecessary thoughts, in the utmost economy of thought-operations. The symbols of order, which we call numbers, form already a system of wonderful simplicity and economy. When in the multiplication of a number with several digits we employ the multiplication table and thus make use of previously accomplished results rather than to repeat them each time, when by the use of tables of logarithms we avoid new numerical calculations by replacing them by others long since performed, . . . — we see in all this but a faint reflection of the intellectual activity of a Lagrange or Cauchy, who with the keen discernment of a military commander marshals a whole troop of completed operations in the execution of a new one. — E. MACH. (Moritz, p. 11.)

The iron labor of conscious logical reasoning demands great perseverance and great caution; it moves on but slowly, and is rarely illuminated by brilliant flashes of genius. It knows little of that facility with which the most varied instances come thronging into the memory of the philologist or historian. — HELMHOLTZ. (Moritz, p. 22.)

Nature herself exhibits to us measurable and observable quantities in definite mathematical dependence; the conception of a function is suggested by all the processes of nature where we observe natural phenomena varying according to distance or to time. — MERZ, II, 676.

I often say that when you can measure what you are speaking about and express it in numbers, you know something about it; but when you cannot measure it, when you cannot express it in numbers, your knowledge is of a meager and unsatisfactory kind; it may be the beginning of knowledge, but you have scarcely in your thoughts advanced to the stage of science. — KELVIN.

We have now reached a period of maturity in the evolution of mathematical science beyond which any attempt to follow its details would involve technical discussions outside the range of this work. The present chapter will be devoted to a brief survey of certain modern tendencies in mathematics, mechanics, mathematical physics, and astronomy. The most notable single fact in the centuries under discussion is the increasing specialization resulting from the great expansion of scientific knowledge. It is no longer possible for the individual scholar to command the range at once of philosophy, mathematics, physics, chemistry, and the natural sciences. It has even become more and more difficult to have a general knowledge of any one of these broad fields.

MATHEMATICS AND MECHANICS IN THE EIGHTEENTH CENTURY

The invention of the infinitesimal calculus by Newton and Leibniz was comparable in its relations and consequences with the discovery of a new world by Columbus two centuries earlier. As in that case the discovery was not an absolutely sudden one; other explorers had anticipated or imagined, but only genius of that highest order which we call inspired, gained the complete revelation. The years next following the great discovery were naturally a period of eager and wide-ranging exploration, of optimistic self-confident pioneering. Such was the power of the new method, that one might rashly hope no secret of nature could long resist its attack. As circumnavigation of the globe was not long in following the discovery of America, so the cycle of mathematical knowledge might be completed. The parallel has failed. The calculus grew out of the insistent grappling by mathematicians with problems which had defied the feebler tools of the earlier mathematics. One obstacle after another has been gradually surmounted by the invention of new and more powerful methods of ever increasing generality, just as increasingly powerful telescopes have revealed unnumbered new suns; and no boundary or limit to this evolutionary progress can be foreseen or imagined. On the other hand, as the new world has been gradually settled, civilized, and cultivated, so the fields of mathematics which were opened up in the eighteenth century have been critically examined in the nineteenth, with much revision of fundamentals.

The main features of eighteenth century mathematics were: — the working out of the differential and integral calculus into substantially the form they have ever since retained; the beginnings of differential equations as a natural outgrowth of integral calculus, and the beginnings of the calculus of variations; the systematic application of the new ideas to mechanics, and in particular to celestial mechanics. The century was also notable for important discoveries in astronomy and physics, including for example that of the aberration of light; a vigorous attack on "the problem of three bodies"; and the earlier telescopic work of the Herschels, culminating in the discovery of a new planet, Uranus.

Among the leading mathematicians of the period were Maclaurin of Scotland, various members of the Swiss Bernoulli family, Euler also a native of Switzerland, Lagrange of Italy, and in France, Clairaut, d'Alembert, and Laplace. In spite of the unique supremacy of Newton, the absence of Britons from this list is notable. The bitter personal controversy between Newton's adherents and those of Leibniz produced or aggravated an unfortunate division between the English and the continental mathematicians. For the former, persistence in Newton's inferior notation became a matter of national pride, and progress was correspondingly retarded. Of the mathematicians named above, the Bernoullis and Euler on the continent and Maclaurin in Scotland bore a leading part in the systematization of the calculus, while Lagrange and Laplace were preëminent in the development of analytical and celestial mechanics respectively.

Colin Maclaurin (1698–1746) was professor of mathematics at Edinburgh. His *Treatise of Fluxions* (1742) was "the first logical and systematic exposition of the method of fluxions," and the applications to problems contained in it were characterized by Lagrange as the "masterpiece of geometry, comparable with the finest and most ingenious work of Archimedes." Maclaurin's point of view may be illustrated by the following passage:

> Magnitudes were supposed to be generated by motion; and, by comparing the increments that were generated in any equal successive parts of the time, it was first determined whether the motion was uniform, accelerated, or retarded. . . . When the motion was accelerated, this increment was resolved into two parts; that which alone would have been generated if the motion had not been accelerated, but had continued uniform from the beginning of the time, and that which was generated in consequence of the continual acceleration of the motion during that time. . . . In the method of infinitesimals, the element, by which any quantity increases or decreases, is supposed to be infinitely small, and is generally expressed by two or more terms, some of which are infinitely less than the rest, which being neglected as of no importance, the remaining terms form what is called the *difference* of the proposed quantity. The terms that are neglected in this manner, as infinitely less than the other terms of the element, are the very same which arise in consequence of the acceleration, or retardation, of the generating motion, during the infinitely small time in which the element is generated. . . . The conclusions are accurately true, without even an infinitely small error. . . .

In this book we find the so-called series expansion of Maclaurin, which forms a standard topic of modern elementary calculus. Maclaurin himself, however, explicitly renounces the authorship, which, according to him, goes to Brook Taylor

(1685–1731). Taylor, indeed, published in 1715 the so-called series of Taylor, in our notation:

$$f(x + h) = f(x) + \frac{h}{1}f'(x) + \frac{h^2}{1.2}f''(x) + \cdots.$$

The Bernoulli family is one of the most outstanding examples of a family of which members of several generations became famous as men of science. The brothers John (1667–1748), professor at Groningen, and James (1645–1708), professor at Basle, were the most famous pupils of Leibniz. In bitter personal rivalry they discovered many theorems on the calculus, concerning catenaries, geodetic lines, brachistochrones, etc. James's posthumous *Ars conjectandi* (1713) is a landmark in the theory of probability. Their nephew Daniel Bernoulli (1700–82), also professor at Basle, made such good use of the new mathematical methods in attacking previously unsolved problems of mechanics, that he has been called the founder of mathematical physics. He recognized the importance of the principle of the conservation of force anticipated in part by Huygens.

Leonhard Euler (1707–83), while Swiss by birth, spent most of his life at the courts of St. Petersburg and Berlin. In spite of partial and ultimately complete blindness, his scientific productivity was enormous, and one of the most ambitious scientific undertakings of our own time is the publication of his complete works in 45 volumes, by international coöperation. Our college mathematics — algebra, analytic geometry, and the calculus — owes its present shape largely to his work.

Euler's *Introductio in analysin infinitorum* (1748) contains algebra and some calculus. There are expansions in series of e^x, $\sin x$, $\cos x$, and the fundamental formula

$$e^{ix} = \cos x + i \sin x.$$

The trigonometry is given here, for the first time, in modern notation. Euler's fundamental works on calculus are the *Institutiones calculi differentialis* (1755) and the *Institutiones cal-*

culi integralis (1768–74). His *Complete Introduction to Algebra* dates from 1770.

Euler's *Complete Introduction to Algebra* was one of the most influential books on algebra in the eighteenth century, and not the least because it is written with extraordinary clearness and in easily intelligible form. Euler was at that time already totally blind. He picked out a young man whom he had brought with him from Berlin as an attendant and who could reckon tolerably, but who otherwise had no understanding of mathematics. He was a tailor by trade and of moderate intellectual capacity. To him Euler dictated this book, and the amanuensis not only understood everything well but in a short time acquired the power to carry out difficult algebraic processes by himself with much facility. It was this book which completing the development begun by Vieta made algebra an international mathematical shorthand. — *Zeits. math. nat. Unterricht*, **43**, 154.

Euler formulated the idea of function which has proved so fundamental in modern mathematics, both pure and applied. His work also contains the first systematic treatment of the calculus of variations (1744), which is defined as "the method of finding the change caused in an expression containing any number of variables when one lets all or any of the variables change" or more geometrically "a method of finding curves having a particular property in the highest or the lowest degree."

In other fields Euler "was the first to treat the vibrations of light analytically and to deduce the equation of the curve of vibration as dependent upon elasticity and density. . . . He deduced the law of refraction analytically and explained that the rays of greater wavelength must suffer the least deviation. . . . He studied dispersion in the search for a corrective for chromatic aberration, which Newton had declared unattainable. . . . It was this investigation that induced Dolland to construct his achromatic lenses. . . . Euler was thus the only physicist of the eighteenth century who advanced the undulatory theory." — HOPPE (quoted by C. A. Noble), *Bull. Amer. Math. Soc.*, **14**, 136.

Euler gained a share of the prize of £20,000 offered by the British parliament for a method of determining longitude at sea, half of the same prize falling to John Harrison the maker of a ship's chronometer sufficiently accurate for the same purpose (1762).

He was probably the most versatile as well as the most prolific of mathematicians of all time. There is scarcely any branch of modern analysis to which he was not a large contributor, and his extraordinary powers of devising and applying methods of calculation were employed by him with great success in each of the existing branches of applied mathematics; problems of abstract dynamics, of optics, of the motion of fluids, and of astronomy were all in turn subjected to his analysis and solved. — BERRY, p. 292.

PROGRESS IN THEORETICAL MECHANICS

Euler, in his *Mechanica* (1736), gave a consistent exposition of Newton's mechanics. In 1744 he developed Newton's ideas in theoretical astronomy in a book on the motions of planets and comets. The rapid development of mechanics in the eighteenth century culminated in the great classical treatises of J. d'Alembert (1717–83) — *Traité de dynamique* — and J. L. Lagrange (1736–1813) — *Mécanique analytique*, systematizing and coördinating the theories and results thus far obtained. D'Alembert, working out ideas based on Huygens's theory of the center of oscillation, formulated a very general dynamical principle since known under his name.

To him is attributed the celebrated epigram concerning Benjamin Franklin, "He snatched the thunderbolt from heaven, the sceptre from tyrants" (*Eripuit coelo fulmen sceptrumque tyrannis*).

Lagrange, a native of Turin, also spent many years in Berlin and his later life in Paris, where he became professor at the École polytechnique as established in 1795. His chief work, the *Mécanique analytique*, is a masterly discussion of the whole subject, showing by the aid of the new mathematical methods its dependence on a few fundamental principles. The significance and importance of this work are within its field comparable with those of Newton's *Principia*. On the death of his royal patron, Frederick the Great, in 1787, he was invited from Berlin not only to Paris, but to Spain and to Naples, accepting the first-named opportunity. Lagrange's works include also very important contributions to differential equations and the calculus of variations. In his notable *Théorie des fonctions analytiques* (1797) he attempts to base the calculus on the notion of the finite alone.

His great powers of analysis were applied with success to problems in astronomy and cartography. In contrast with the predominantly geometrical and synthetic methods of Newton, Lagrange's methods are mainly analytical.

When we have grasped the spirit of the infinitesimal method, and have verified the exactness of its results either by the geometrical method of prime and ultimate ratios, or by the analytical method of derived functions, we may employ infinitely small quantities as a sure and valuable means of shortening and simplifying our proofs. — LAGRANGE. (Moritz, p. 324.)

CELESTIAL MECHANICS

Pierre Simon marquis de Laplace (1749–1827) was of Norman antecedents, and played a great part in the scientific activity under Napoleon. In the five volumes of his *Mécanique céleste* (1799–1825), he produced a permanent monument to his own genius. It was his lofty ambition to

offer a complete solution of the great mechanical problem presented by the solar system, and bring theory to coincide so closely with observation that empirical equations should no longer find a place in astronomical tables.
He regarded analysis merely as a means of attacking physical problems, though the ability with which he invented the necessary analysis is almost phenomenal. As long as his results were true he took very little trouble to explain the steps by which he arrived at them; he never studied elegance or symmetry in his processes, and it was sufficient for him if he could by any means solve the particular question he was discussing. — W. W. R. BALL, pp. 415, 420.

Nathaniel Bowditch, the American translator of his great work, remarks significantly:

I never come across one of Laplace's "*Thus it plainly appears*" without feeling sure that I have hours of hard work before me to fill up the

FIG. 55. — Pierre Simon de Laplace (1749–1827). After Quetelet, *Théorie des Probabilités*, 1855.

chasm and find out and show *how* it plainly appears. — (Moritz, p. 161.)

In the words of the historian Todhunter,

a complete evolution of the history will restore the reputation of Laplace to its just eminence. The advance of mathematical science is on the whole remarkably gradual, for with the single exception of Newton there is very little exhibition of great and sudden developments; but the possessions of one generation are received, augmented and transmitted by the next. It may be confidently maintained that no single person has contributed more to the general stock than Laplace. — *Hist. math. Theories of Attraction*, I, xvii.

THE PERTURBATION PROBLEM

Newton had worked out the theory of a single planet or satellite revolving about its primary. The consequent discrepancies between the observed and the computed positions were held by some to indicate inexactness in his hypothetical laws. Laplace occupied himself with a thorough study of the great problem of three bodies,[1] and without fully solving it, accounted to a great extent for the discrepancies in question. In particular he maintained the stability of the solar system. His *Mécanique céleste* has been characterized as an infinitely extended and enriched edition of Newton's *Principia.*

In his confidence in the extending range of mathematical methods Laplace says:

Given for one instant an intelligence which could comprehend all the forces by which nature is animated and the respective positions of the beings which compose it, if moreover this intelligence were vast enough to submit these data to analysis, it would embrace in the same formula both the movements of the largest bodies in the universe and those of the lightest atom: to it nothing would be uncertain, and the future as the past would be present to its eyes. The human mind offers a feeble outline of that intelligence, in the per-

[1] Given at any time the positions and motions of three mutually gravitating bodies, to determine their positions and motions at any other time — a particular case of the actual more general problem: Given the 18 known bodies of the solar system, and their positions and motions at any time, to deduce from their mutual gravitation by a process of mathematical calculation their positions and motions at any other time; and to show that these agree with those actually observed.

fection which it has given to astronomy. Its discoveries in mechanics and in geometry, joined to that of universal gravity, have enabled it to comprehend in the same analytical expressions the past and future states of the world system. — *Théorie anal. des prob.* (Moritz, p. 328.)

THE NEBULAR HYPOTHESIS

In his *Exposition du système du monde* (1796), "one of the most perfect and charmingly written popular treatises on astronomy ever published, in which Laplace never uses either an algebraical formula or a geometrical diagram," he presents the arguments for his nebular hypothesis along the following general lines:

In spite of the separation of the planets they bear certain remarkable relations to each other;

All the planets travel about the sun in the same direction and almost in the same plane;

The satellites also travel about their planets in this same direction and almost in the same plane;

Finally, sun, planets and satellites revolve in the same sense about their own axes and this rotation is approximately in the orbital plane.

These agreements cannot be accidental. Laplace seeks the cause in the existence of an original vast nebulous mass forming a sort of atmosphere about the sun and extending beyond the outermost planet. Initial or acquired rotation of the nebula attended by gradual cooling and contraction has caused the centrifugal separation of masses analogous to Saturn's rings, out of which planets have gradually condensed, throwing off their own satellites in the process. This hypothesis had already been proposed in substance by Kant in 1755, and even earlier by Emanuel Swedenborg (1688–1772), whose extensive scientific writings range from mathematics to anatomy and physiology. In his *Principia rerum naturalium*, 1734, he was seeking the fundamental principles upon which every thing in the world is constructed, and he suggested that the planets had been formed from the sun by centrifugal force depending on a continually increasing vortical motion of the solar mass, which he regarded as having been less extended than Laplace afterward assumed. The later history of the nebular hypothesis will be touched on below.

MODERN ASTRONOMY. TELESCOPIC DISCOVERIES

The immense impetus given to astronomy by the revolutionary discoveries of Copernicus, Tycho Brahe, Galileo, Kepler, and Newton, followed in the eighteenth century by the complete working out of the mathematical consequences of the gravitation theory by Laplace and others, placed the science in advance of all its rivals and seemed to make it a model for their imitation.

A different and most far-reaching tendency appears with the work of the Herschels. Friedrich Wilhelm Herschel (1738–1822), a poor German musician emigrating to England and devoting his spare time unremittingly to astronomy, with the help of his capable sister laid the foundations of modern physical astronomy. In 1781 he amazed himself as well as the scientific world by discovering beyond Saturn a new planet, Uranus — taking it at first for a comet. Constructing more and more powerful telescopes he discovered several satellites of Uranus and two of Saturn. He also determined a motion of the solar system as a whole, towards a point in the constellation Hercules. He catalogued more than 800 double stars and more than 2,000 nebulas, recognizing among the latter, as he believed, different stages of the evolution of other planetary systems. He observes:

> This method of viewing the heavens seems to throw them into a new kind of light. They are now seen to resemble a luxuriant garden, which contains the greatest variety of productions, in different flourishing beds; and one advantage we may at least reap from it is, that we can, as it were, extend the range of our experience to an immense duration. For, to continue the simile I have borrowed from the vegetable kingdom, is it not almost the same thing, whether we live successively to witness the germination, blooming, foliage, fecundity, fading, withering, and corruption of a plant, or whether

a vast number of specimens selected from every stage through which the plant passes in the course of its existence, be brought at once to our view? — BERRY, p. 340.

With a reminiscence of Descartes, he says:

I determined to accept nothing on faith, but to see with my own eyes what others had seen before me. . . . When I had carefully and thoroughly perfected the great instrument in all its parts I made systematic use of it in my observations of the heavens, first forming a determination never to pass by any, the smallest, portion of them without due investigation. — *Ibid.*, p. 325.

To the eighteenth century also belong elaborate and costly expeditions to observe transits of Venus, as a means for determining the distance from the sun to the earth.

REACTION OF MATHEMATICS ON CHEMICAL AND PHYSICAL SCIENCE

Besides the extension of mathematical ideas and methods to mechanics, astronomy, optics and other branches of physics, chemistry was now also becoming a quantitative science. So Scheele begins a work published in 1777:

To resolve bodies skilfully into their components, to discover their properties and to combine them in different ways, is the chief purpose of chemistry.

Richter, in his *Stoichiometry* (1792–1802), even speaks of chemistry as a branch of applied mathematics. Already the pioneer Robert Boyle had written (1661):

I confess, that after I began . . . to discern how useful mathematicks may be made to physicks, I have often wished that I had employed about the speculative part of geometry, and the cultivation of the specious Algebra I had been taught very young, a good part of that time and industry that I had spent about surveying and fortification (of which I remember I once wrote an entire treatise) and other parts of practick mathematicks.

Mathematicks may help the naturalists, both to frame hypotheses, and to judge of those that are proposed to them, especially such as relate to mathematical subjects in conjunction with others.

Even in natural science Stephen Hales says in 1727:

> And since we are assured that the all-wise Creator has observed the most exact proportions, *of number, weight and measure,* in the make of all things; the most likely way therefore to get any insight into the nature of these parts of the creation, must in all reason be to number, weigh and measure. And we have much encouragement to pursue this method, of searching into the nature of things, from the great success that has attended any attempts of this kind.

Summing up these tendencies, a recent writer remarks:

> In the eighteenth century mathematics was regarded by many scholars as the ideal, the completeness and exactness of whose methods should be arrived at by other less highly developed branches. So Laplace's popularized version of his celestial mechanics met an eager need, and even Voltaire undertook the championship of the Newtonian philosophy. Logic and even ethics were drawn into the mathematical retinue. For Maupertuis the good is a positive quantity, the bad a negative. Joys and griefs make up human life according to the laws of algebraic addition, and it is the business of statesmen to see that the positive balance is as large as possible. The great Buffon adds to his natural history a supplement on moral arithmetic. Mathematics aims at the leadership both in natural science and in human affairs.

In spite of this predilection in learned and polite society, educational curricula remained weak and conservative. Powerful progressive tendencies growing out of the French Revolution found expression, however, in the founding of the École polytechnique — under the leadership of Monge — which has ever since been an important center of mathematical activity. Its curriculum included, in the first year, analytic geometry of space and descriptive geometry; in the second, mechanics of solids and liquids; in the third, theory of mechanics.

The leader of the École polytechnique, Gaspard Monge (1746–1818), was not only a great organizer, but also an outstanding geometer. His *Géométrie descriptive* (1795), the result of his lectures at the École polytechnique, has remained a standard text on descriptive geometry, notably on orthographic projection. His *Applications de l'analyse à la géométrie*

(1795) is an important contribution to differential geometry. Many nineteenth century textbooks have been the result of lectures originally given at the École polytechnique.

NINETEENTH CENTURY MATHEMATICS

As in the century following Newton, France became the great center of mathematical activity, so in the nineteenth century the leadership passed to Germany, under the inspiration of Gauss and Riemann of Göttingen, Jacobi of Königsberg, Weirstrass of Berlin — to mention but a few of the most prominent men. Outside of Germany, conspicuous names are Cauchy, Galois, Hermite, Legendre, and Poincaré in France, Cayley and Sylvester in England, Abel in Norway, and Lobatchewski in Russia.

Characteristic of this period are: the development of a general theory of functions based on unifying coördinating principles, compensating the powerful specializing tendencies, and a profound critical revision of the previously accepted axioms, leading for example to the development of a Non-Euclidean geometry. In the science generally there is systematic development of instruction and research, notably in the German universities; of publication, by the establishment of mathematical journals, and the preparation of encyclopedias; numerous national societies are formed, and international congresses held — offsetting more or less completely the extension of scientific research into new parts of the world and the loss of Latin as an international language of the learned. These tendencies are naturally not confined to the mathematical sciences. In some respects mathematics has merely enjoyed its share in the prosperity of a more scientific age, in some it has perhaps suffered, at any rate relatively, from the powerful stimulus given the natural sciences by the working out of evolutionary theories. From a position of acknowledged primacy among a small number of recognized sciences, it has come to be regarded as but one of many.

While it is impossible here even to enumerate the different

branches of mathematical science developed during this period, certain features of historical interest may be touched upon.

GAUSS

Carl Friedrich Gauss (1777–1855), the "prince of mathematicians" (*princeps mathematicorum*), lived a retired life as director of the astronomical observatory at Göttingen. He enriched mathematics in a large variety of ways, and in many respects set the pace for the expansion of mathematical research in the nineteenth century. His most profound discoveries were achieved in his youth. In 1795 he discovered the method of least squares, in 1796 the constructibility of the regular polygon of 17 sides by compass and ruler, and in 1797 fundamental results in elliptic functions and the first proof of the fundamental theorem of algebra (the hypothesis of Girard, see below). In 1801 he published his classic on number theory, *Disquisitiones arithmeticae.* Gauss was equally prolific in astronomy and geodesy, and contributed to electrodynamics. In 1801 he computed the orbit of planetoids, published in *Theoria motus*, 1809. He worked also in differential geometry, complex functions and potential theory, and in Non-Euclidean geometry. With Weber he is one of the inventors of the electric telegraph, 1833.

PROBABILITY: CURVE OF ERROR

Eighteenth century mathematicians had shown considerable interest in the theory of probability. But we are indebted to Buffon, a celebrated naturalist, for one of the most remarkable contributions. It is known as "the needle problem":

Take a stick of length $2L$, and from a short height drop it upon a table on which are drawn parallel lines a distance D, greater than $2L$, apart. Then the stick will either cross one of these lines or it will not. Suppose that this experiment be tried N times and the observation made that the stick crosses one of the lines C times. Then π, the ratio of a circumference to its diameter, can be computed within probable limits of error from these numbers by the formula

$$\pi = 4\frac{NL}{CD}.$$

Laplace's mathematical treatment of probability not only gave precision to astronomical results, but found similar application in many fields. His *Théorie analytique des probabilités* (1812) is a landmark in the history of the subject. He says in his introduction:

> The most important questions of life are, for the most part, really only problems of probability. Strictly speaking one may even say that nearly all our knowledge is problematical; and in the small number of things which we are able to know with certainty, even in the mathematical sciences themselves, induction and analogy, the principal means for discovering truth, are based on probabilities, so that the entire system of human knowledge is connected with this theory.
>
> It is remarkable that a science (probabilities) which began with the consideration of games of chance, should have become the most important object of human knowledge.
>
> The theory of probabilities is at bottom nothing but common sense reduced to calculus; it enables us to appreciate with exactness that which accurate minds feel with a sort of instinct for which ofttimes they are unable to account. — (Moritz, pp. 340, 342.)

From the theory of probability, C. F. Gauss (1777–1855) deduced the so-called law of error, represented by a curve of normal distribution and since applied to the graphical representation of a very wide range of phenomena not only in physical and biological science, but in sociology and public health. While the action of individual molecules or men may seem quite arbitrary, this fundamental statistical method enables us to predict average behavior or mass action. Pioneer work in this field was done by the Belgian astronomer, Adolphe Quetelet (1796–1874). Deeply influenced by Laplace, he published in 1828 his *Instructions populaires sur le calcul des probabilitées*, one of the earliest popularizations of the subject. His principal work, *Sur l'homme et le développement de ses facultés, ou Essai de physique sociale*, 1835, one of the greatest books of the nineteenth century, was the first attempt to apply mathematical analysis to the study of man — not alone of his body but of his behavior and morality. He showed that

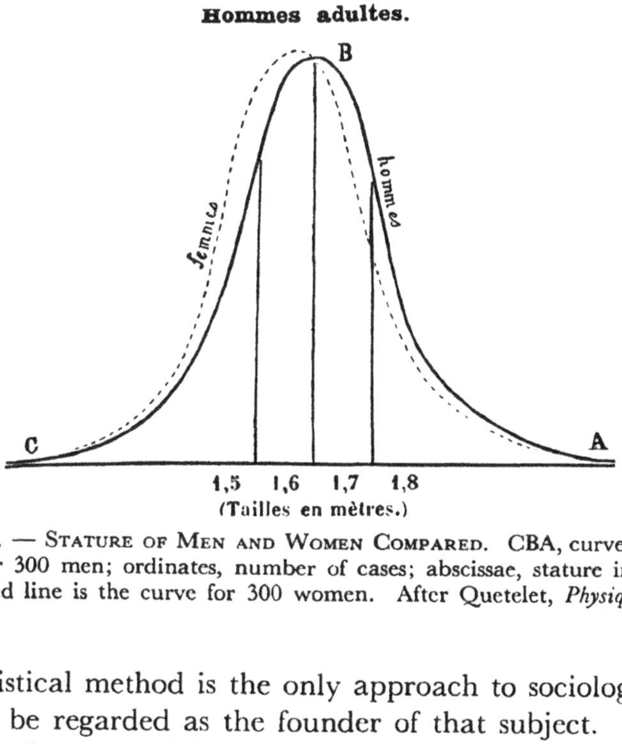

FIG. 56. — STATURE OF MEN AND WOMEN COMPARED. CBA, curve of distribution for 300 men; ordinates, number of cases; abscissae, stature in meters. The dotted line is the curve for 300 women. After Quetelet, *Physique sociale*, 1869.

the statistical method is the only approach to sociology, and he may be regarded as the founder of that subject. Besides other works on probability and statistics, he wrote also on the history of science in Belgium.

I like to think of the constant presence in any sound Republic of two guardian angels: the Statistician and the Historian of Science. The former keeps his finger on the pulse of Humanity, and gives necessary warning when things are not as they should be. The Historian of Science . . . will not allow Humanity to forget its noblest traditions or to be ungrateful to its greatest benefactors. If the Statistician is like a physician, the Historian is like a priest . . .

Quetelet realized these two needs ahead of the great mass of the people, even the best educated, and such prophetic vision of the intellectual necessities of the future, if that is not genius, what else is it and what else is genius? — GEORGE SARTON, *Isis*, **23**, 22–23, 1935.

Since the time of Quetelet the statistical method, no longer confined to astronomy and sociology, has been found to be the best approach to a vast number of problems in biology, chemistry, and physics.

NON-EUCLIDEAN GEOMETRY

This new development in geometry dates back to attempts made to prove one of Euclid's postulates, which states that through a point outside a line one and only one line can be drawn parallel to it. Such attempts already date back to antiquity and were resumed in the seventeenth and eighteenth centuries. After many failures Gauss and later Boylai and Lobachevsky finally decided to accept this postulate as impossible of a demonstration and to proceed in building a geometry with another postulate. They selected the postulate that through a point outside of a line more than one line can be drawn in the plane of point and line which does not intersect this line. This geometry was called Non-Euclidean. Gauss never published his result. Johann Bolyai, a young Hungarian officer, published his ideas in 1832. The first publications of Lobachevsky, a professor in Kazan, date from 1829–1830. Later other forms of Non-Euclidean geometries were published and also the proof that these geometries are as logically consistent as Euclidean geometry.

In an address on "The History of Mathematics in the Nineteenth Century," James Pierpont remarks:

> Each century takes over as a heritage from its predecessors a number of problems whose solution previous generations of mathematicians have arduously but vainly sought. It is a signal achievement of the nineteenth century to have triumphed over some of the most celebrated of these problems.
>
> The most ancient of them is the quadrature of the circle, which already appears in our oldest mathematical document, the Papyrus Rhind, B.C. 2000. Its impossibility was finally shown by Lindemann, 1882.
>
> But of all problems which have come down from the past, by far the most celebrated and important relates to Euclid's parallel axiom. Its solution has profoundly affected our views of space and given

rise to questions even deeper and more far-reaching, which embrace the entire foundation of geometry and our space conception. — *Bull. Amer. Math. Soc.*, **11,** 155.

There is no doubt that it can be rigorously established that the sum of the angles of a rectilinear triangle cannot exceed 180°. But it is otherwise with the statement that the sum of the angles cannot be less than 180°; . . . I have been occupied with the problem over thirty years and I doubt if anyone has given it more serious attention, though I have never published anything concerning it. — Gauss (1824).

I will add that I have recently received from Hungary a little paper on Non-Euclidean geometry, in which I rediscover all *my own ideas* and *results* worked out with great elegance. . . . I consider the young geometer von Bolyai a genius of the first rank. — Gauss (1832). (Moritz, pp. 355, 357.)

The gradual adoption of new and revolutionary ideas on this subject may be further illustrated by the following passages:

The characteristic features of our space are not necessities of thought, and the truth of Euclid's axioms, in so far as they specially differentiate our space from other conceivable spaces, must be established by experience and by experience only. — R. S. Ball.

Everything in physical science, from the law of gravitation to the building of bridges, from the spectroscope to the art of navigation, would be profoundly modified by any considerable inaccuracy in the hypothesis that our actual space is Euclidean. The observed truth of physical science, therefore, constitutes overwhelming empirical evidence that this hypothesis is very approximately correct, even if not rigidly true. — Bertrand Russell.

What Vesalius was to Galen, what Copernicus was to Ptolemy, that was Lobatchewski to Euclid. There is, indeed, a somewhat instructive parallel between the last two cases. Copernicus and Lobatchewski were both of Slavic origin. Each of them has brought about a revolution in scientific ideas so great that it can only be compared with that wrought by the other. — W. K. Clifford.

Geometrical axioms are neither synthetic *a priori* conclusions nor experimental facts. They are conventions: our choice, amongst all possible conventions, is guided by experimental facts; but it remains free, and is only limited by the necessity of avoiding all contradiction. . . . In other words, axioms of geometry are only definitions in disguise. That being so what ought one to think of this question: Is the Euclidean Geometry true? The question is nonsense. One might as well ask whether the metric system is true and the old measures false; whether Cartesian co-ordinates are true and polar co-ordinates false. — HENRI POINCARÉ, *Nature*, **45**, 407.

The recent utilization of Non-Euclidean geometry by Einstein and others in the theory of relativity may be mentioned in this connection.

IMAGINARY NUMBERS

The solution of algebraic equations had always been hampered by the seeming impossibility of performing the inverse processes involved. The equation $x + 5 = 0$ could not be solved before negative numbers were known; and the equations $2x = 5$ and $x^2 = 2$ would be equally insoluble without fractions and irrational numbers. Cardan (1545) first introduced imaginary numbers, and Bombelli, a professor at Bologna, used them extensively in 1572. Albert Girard, a Huguenot mathematician living in Holland, expressed as early as 1629 the theorem that every algebraic equation of degree n must have n roots. Much of the mystery surrounding imaginaries disappeared with Euler and Gauss. Gauss introduced the geometrical interpretation of imaginaries in the plane, and proved, for the first time exactly, the hypothesis of Girard (1799). This work was of far-reaching importance, not merely for higher mathematics, but even for a subject so concrete as electrical engineering.

That this subject [of imaginary magnitudes] has hitherto been considered from the wrong point of view and surrounded by a mysterious obscurity, is to be attributed largely to an ill-adapted notation. If for instance, $+1$, -1, $\sqrt{-1}$ had been called direct, inverse, and lateral units, instead of positive, negative, and imaginary (or even impossible) such an obscurity would have been out of the question. — GAUSS. (Moritz, p. 282.)

GROUPS

The great notion of Group, . . . though it had barely merged into consciousness a hundred years ago, has meanwhile become a concept of fundamental importance and prodigious fertility, not only affording the basis of an imposing doctrine — the Theory of Groups — but therewith serving also as a bond of union, a kind of connective tissue, or rather as an immense cerebro-spinal system, uniting together a large number of widely dissimilar doctrines as organs of a single body. — C. J. KEYSER. (Moritz, p. 290.)

Concluding the address quoted above, Pierpont says:

What strikes us at once in our survey of mathematics in the last century is its colossal proportions and rapid growth in nearly all directions, the great variety of its branches, the generality and complexity of its methods, an inexhaustible creative imagination, the fearless introduction and employment of ideal elements, and an appreciation for a refined and logical development of all its parts. — *Bull. Amer. Math. Soc.*, **11**, 158.

THE DISCOVERY OF NEPTUNE

In a century filled with remarkable scientific achievement, no single triumph has been more conspicuous, or in some respects more dramatic, than the discovery of the planet Neptune by Adams and Leverrier. From the time of Newton the perturbations of the planets had been the subject of continual observation and study. Improved telescopes demanded — and at the same time facilitated — more extended and refined computations. Descrepancies between computed and observed positions indicated disturbing forces of known or in some cases unknown origin. In particular, irregularities — never exceeding two minutes of arc — in the motion of the most recently discovered planet Uranus, led the young Cambridge graduate John Couch Adams (1819–92) and the eminent French astronomer Leverrier (1811–77) to independent attacks on the formidable problem of determining the mass and position of a hypothetical new planet which could cause the observed effects on Uranus. Unfortunately for Adams the necessary coöperation on the part of the observatories was not promptly available, so that the actual discovery connected itself with the somewhat later work of Leverrier. The discovery was naturally accepted as an extraordinary illustration of the power of mathematical astronomy and a convincing proof of the Newtonian theory of gravitation.

While the telescope serves as a means of penetrating space, and of bringing its remotest regions nearer us, mathematics, by inductive reasoning, has led us onwards to the remotest regions of heaven, and brought a portion of them within the range of our possibilities; nay, in our own times — so propitious to the extension of knowledge — the application of all the elements yielded by the present conditions of astronomy has even revealed to the intellectual eyes a heavenly body, and assigned to it its place, orbit, mass, before a single telescope has been directed towards it. — HUMBOLDT. (Moritz, p. 245.)

COSMIC EVOLUTION

Reference has been made to the nebular hypothesis included by Laplace in his extended discussion of the solar system. During the nineteenth century this theory has been subjected to searching scrutiny from many points of view and much doubt has been cast on its validity.

The following summary of present opinion is given by Hale in his *Stellar Evolution:*

The nebular hypothesis of Laplace still remains as the most serious attempt to exhibit the development of the solar system. Attacked on many grounds, and showing signs of weakness that seem to demand radical modification of Laplace's original ideas, it nevertheless presents a picture of the solar system which has served to connect in a general way a mass of individual phenomena, and to give significance to apparently isolated facts that offer little of interest without the illumination of this governing principle.

We are now in a position to regard the study of evolution as that of a single great problem, beginning with the origin of the stars in the nebulas and culminating in those difficult and complex sciences that endeavor to account, not merely for the phenomena of life, but for the laws which control a society composed of human beings.

As a complement to the preceding may be added the following from another specialist in planetary evolution:

It is to the glory of astronomy that in it were initiated the two most fundamental intellectual movements in the history of mankind, viz. the establishment of the possibility of science and of the doctrine of evolution. Our intellectual ancestors in the valleys of the Euphrates and the Nile and on the hills of Greece looked up into the sky at night and saw order there and not chaos. By painstaking observations and calculations they discovered the relatively simple laws of the motions of the heavenly bodies, whose invariable and exact fulfilment led to the belief that the whole universe in all its parts is orderly and that science is possible. In the modern world this conclusion is so commonplace that its immense value is apt to be overlooked, but a study of the superstitions and the hopeless stagnation of those portions of mankind which have not yet made the discovery gives us some measure of its worth. The modern supplement to the conception that the universe is not a chaos is that not only is it an orderly universe at any instant, but that it changes from one state to another in a continuous and orderly fashion. This doctrine that science is extensive in time, as well as in space, is the fundamental element in the theory of evolution and the completion of the conception of science itself. . . . For half a century now, the doctrine of evolution has been a fundamental factor in the elaboration of all scientific theories, and its influence has spread to every field of intellectual effort. It has been the good fortune of mankind that his skies have sometimes been free of clouds and that he has been able to observe those relatively simple yet majestic and impersonal celestial phenomena which have not only led to so important results as the founding of science and the doctrine of evolution, but have strongly colored his poetry, philosophy and religion, and have stimulated him to the elaboration of some of his most profound mathematical theories. — F. R. MOULTON.

The astronomer just quoted is one of the authors of the planetesimal hypothesis, under which the primary factor in cosmic evolution is the tidal attraction between two stars passing within a short distance of each other — presumably a rare phenomenon.

DISTANCE OF THE STARS

Among other astronomical discoveries bearing a notable relation to the history of mathematical science is that of measurable stellar parallax by F. W. Bessel (1784–1846). One of the traditional objections to the Copernican theory had been the fact that no change could be detected in the relative position of the stars, such as would apparently result from revolution of the earth in a vast orbit. Now with more and more powerful instruments it turned out that there were stars near enough to show precisely the displacement discovered.

In 1838 Bessel attacked this ancient problem successfully by making accurate observations of the relative positions of the double star (61 Cygni) and its celestial neighbors. He obtained for the distance of the star 657,000 times the mean distance from the earth to the sun. Such inconceivably vast distances have been expressed conveniently since then in a unit called the light-year, i.e., the distance a ray of light travels in an entire year at 186,000 miles per second. The distance of 61 Cygni is 10.7 such units.

MATHEMATICAL PHYSICS

The further progress of applied mathematics in the nineteenth century has been interestingly summarized by R. S. Woodward in a presidential address to the American Mathematical Society, from which the following extracts are quoted.

Next came the splendid contributions of George Green under the modest title of "An essay on the application of mathematical analysis to the theories of electricity and magnetism." It is in this essay that the term "potential function" first occurs. Herein also his remarkable theorem in pure mathematics, since universally known as Green's theorem, and probably the most important instrument of investigation in the whole range of mathematical physics, made its appearance.

We are all now able to understand, in a general way at least, the importance of Green's work, and the progress made since the publication of his essay in 1828. But to fully appreciate his work and subsequent progress one needs to know the outlook for the mathematico-physical sciences as it appeared to Green at this time and to realize his refined sensitiveness in promulgating his discoveries. "It must certainly be regarded as a pleasing prospect to analysts," he says in his preface, "that at a time when astronomy, from the state of perfection to which it has attained, leaves little room for further applications of their art, the rest of the physical sciences should show themselves daily more and more willing to submit to it." . . . "Should the present essay tend in any way to facilitate the application of analysis to one of the most interesting of the physical sciences, the author will deem himself amply repaid for any labor he may have bestowed upon it; and it is hoped the difficulty of the subject will incline mathematicians to read this work with indulgence, more particularly when they are informed that it was written by a young man who has been obliged to obtain the little knowledge he possesses, at such intervals and by such means as other indispensable avocations which offer but few opportunities of mental improvement, afforded." Where in the history of science have we a finer instance of that sort of modesty which springs from a knowledge of things?

Just as the theories of astronomy and geodesy originated in the needs of the surveyor and navigator, so has the theory of elasticity grown out of the needs of the architect and engineer. From such prosaic questions, in fact, as those relating to the stiffness and the strength of beams, has been developed one of the most comprehensive and most delightfully intricate of the mathematico-physical sciences. Although founded by Galileo, Hooke, and Mariotte in the seventeenth century, and cultivated by the Bernoullis and Euler in the last century, it is, in its generality, a peculiar product of the present century. It may be said to be the engineers' contribution of the century to the domain of mathematical physics, . . .

As [Karl] Pearson has remarked, "There is scarcely a branch of physical investigation, from the planning of a gigantic bridge to the most delicate fringes of color exhibited by a crystal, wherein it does not play its part." It is, indeed, fundamental in its relations to the theory of structures, to the theory of hydromechanics, to the elastic solid theory of light, and to the theory of crystalline media. — *Bull. Amer. Math. Soc.*, **6**, 133–163.

REFERENCES FOR READING

BELL, E. T., *Men of Mathematics*, 1937.

CLERKE, A. M., *History of Astronomy during the Nineteenth Century*, Ed. 4, 1902.

——, *The Herschels and Modern Astronomy*, 1895.

DAMPIER, SIR WILLIAM, *History of Science*, 1930, Ch. V.

HALE, G. E., *Study of Stellar Evolution*, 1908.

MERZ, J. T., *History of European Thought in the Nineteenth Century*, 1896, Chs. IV, V.

POINCARÉ, HENRI, *Science and Hypothesis*, 1905.

WALKER, HELEN M., *Studies in the History of Statistical Method*, 1929.

WHITEHEAD, A. N., *Introduction to Mathematics*, 1911.

From previous lists: Lodge, *Pioneers;* Mach, *Sci. Mech.*

NOTE: The diagram on the opposite page illustrates one of Faraday's experiments in electrolysis. It is "a convenient arrangement for chemical decomposition by common electricity." *a, b,* pieces of tinfoil on a glass plate; *c,* wire to positive conductor of electric machine; *g,* to negative conductor; *p, n,* decomposing poles of platina wire dipping into copper sulphate solution. "Twenty turns of the machine precipitated so much copper on *p* that it looked like copper wire." There was no change at *n*. Faraday, "Identity of Electricities Derived from Different Sources," *Phil. Trans.*, **123,** pl. II, 1833.

Chapter XVI

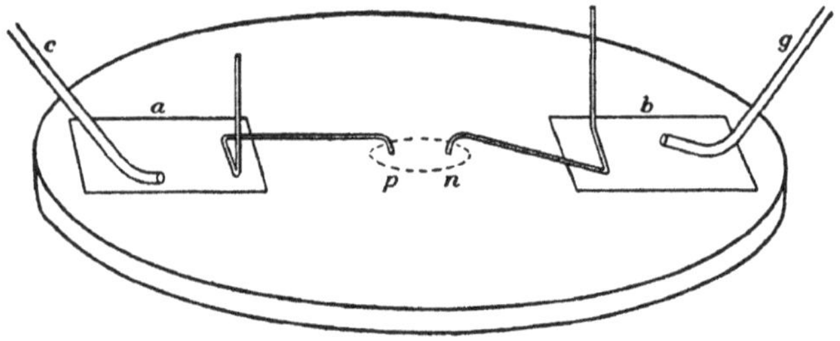

Some Advances in the Physical and Chemical Sciences during the Nineteenth Century

The law of gravitation embraced cosmical and some molar phenomena, but led to vagueness when applied to molecular actions. The atomic theory led to a complete systematization of chemical compounds, but afforded no clue to the mysteries of chemical affinity. And the kinetic or mechanical theories of light, of electricity and magnetism, led rather to a new dualism, the division of science into sciences of matter and of the ether. . . . A more general term had to be found under which the different terms could be comprised, which would give a still higher generalization, a more complete unification of knowledge. One of the principal performances of the second half of the nineteenth century has been to find this more general term, and to trace its all-pervading existence on a cosmical, a molar, and a molecular scale . . . this greatest of all exact generalizations — the conception of energy. — MERZ, II, p. 96.

The founders of the exact sciences of electricity and magnetism . . . turned their undivided attention to the determination of the law of force, according to which electrified and magnetized bodies attract or repel each other. In this way the true laws of these actions were discovered, and this was done by men who never doubted that action took place at a distance, without the intervention of any medium, and who would have regarded the discovery of such a medium as complicating rather than as explaining the undoubted phenomena of attraction. — J. CLERK MAXWELL, "On Action at a Distance," *Proc. Royal Institution*, **7**, 49, 1873.

ACCELERATION OF THE PHYSICAL SCIENCES

The later eighteenth and the whole of the nineteenth centuries are characterized by increasingly rapid development of the physical sciences, which become more and more completely differentiated, and more and more important in their influence upon industry and civilization. We shall attempt to enumerate some phases of this varied development which are most general in their character and most far-reaching in their consequences.

At the beginning of this century, as shown in the previous chapter, mathematics was in a stage of triumphant expansion, in which the related sciences of astronomy and mechanics participated. General physics and chemistry were still in the preliminary stage of collecting and coördinating data, with attempts at quantitative interpretation, while in their train the natural sciences were following somewhat haltingly.

The most notable advance in physical science during the century is the gradual working out of the great fundamental principle of the *conservation of energy*, affecting profoundly the whole range of phenomena. Of equal — or even greater — importance is the gradual realization of progressive development — *evolution* — not only in plant and animal life but even in the inorganic world. Physics is gradually enriched by experimental researches and by the working out of mathematical theories of heat, light, magnetism, and electricity. Chemistry, largely hitherto a collection of unrelated facts, becomes more and more coördinated with physics and mathematics by means of the spectroscope, the principle of the conservation of energy, the atomic theory, the kinetic theory of gases, and the study of molecular structure. On the other hand, its relations with the organic world are made more clear through the investigation of the compounds of carbon.

All other sciences, pure and applied, as well as the industries, profit unexpectedly and almost inconceivably by these nineteenth century advances in physics and chemistry. The discovery of a new planet Neptune, as related in Chapter XV, is soon rivalled by the startling achievements of the new physical and chemical astronomy.

Reserving for the following chapter a sketch of the development of the natural sciences under the influence of the theory of evolution, we proceed to outline briefly some of the more notable advances in the physical sciences.

MODERN PHYSICS

Leading features in the development of physics in the nineteenth century have been: — the working out of consistent theories of light and radiant heat as wave phenomena of a peculiar hypothetical medium called the "ether"; the extensive investigation of electrical and magnetic phenomena and the development of an electro-magnetic theory even so far as to include optics; the working out of a kinetic theory of gases with important relations to chemical as well as physical theory; the elaboration of general theories of matter, force, and energy, all culminating in the crowning discovery of the great unifying principle of the Conservation of Energy.

HEAT: CARNOT, JOULE

The invention of the thermometer has been traced in Chapter XII. To the nineteenth century belongs the determination of an absolute scale [1] as distinguished from the arbitrary one previously employed.

The idea that heat is not a substance but a mode of molecular motion arose in the seventeenth and eighteenth centuries, but was first given a substantial experimental basis by Count Rumford in the closing years of the eighteenth century, as we have seen, Chapter XIV, p. 354.

Rumford made a cylinder of gun-metal rotate in a box containing water, and by the friction of a revolving borer driven by horsepower the water was heated to boiling in two and a half hours.

Deeply impressed he exclaims:

What is heat? Is there any such thing as an *igneous fluid?* . . . Anything which any *insulated* body, or system of bodies, can continue to furnish *without limitation,* cannot possibly be a material substance:

[1] The absolute scale is based on the indirect determination of a temperature (− 273° Centigrade = − 459° Fahrenheit) at which the internal activity which constitutes heat is supposed to cease.

and it appears to me to be extremely difficult, if not quite impossible, to form any distinct idea of anything, capable of being excited, and communicated, in the manner the heat was excited and communicated in these experiments, except it be MOTION. — *Phil. Trans.*, **88,** 98–99, 1798.

The "mechanical equivalent of heat" — i.e., the work required to heat one pound of water one degree — was roughly calculated.

Epoch-making in the theory of heat were the researches of Sadi Carnot (1796–1832), whose words follow:

Wherever there is a difference of temperature followed by return to equilibrium the generation of power may take place. Water vapor is one means, but not the only one. . . . A solid body, for example a metal bar, gains and loses in length when it is alternately heated and cooled, and thus is able to move bodies fastened to its ends. . . .

The whole process he pictures as a *cycle* in which a certain portion of the heat applied is converted into work, a certain other portion being lost. Thus the new science of thermodynamics was born. The thorough and complete investigation of the "mechanical equivalent of heat" belongs to J. P. Joule (1818–89) of Manchester, England, a pupil of Dalton the chemist.

I shall lose no time in repeating and extending these experiments, being satisfied that the grand agents of nature are, by the Creator's fiat, *indestructible;* and that whatever mechanical force is expended, an exact equivalent of heat is *always* obtained. — JOULE.

LIGHT; WAVE THEORY, VELOCITY: YOUNG, FRESNEL

As stated in Chapter XIII Huygens had supported a wave theory of light, while Newton accepted an emission theory. That sound was propagated by atmospheric waves was well known. There was a troublesome contrast however in the phenomenon of shadows. How could wave-propagation be reconciled with sharply defined shadows? Why should not light "go round a corner" as well as sound? These difficulties were met by Thomas Young (1773–1829) who revived Huygens's wave theory, which was definitively established by A. J. Fresnel's researches on diffraction, beginning in 1815.

The determination of the velocity of light by observations of the moons of Jupiter has been mentioned already (p. 322). About 1850 this problem was solved by a new method devised by A. H. L. Fizeau (1819–96). A ray of light passes between the teeth of a wheel to a mirror and back again. During the time required by the ray to pass thus out and back, the gap through which it has passed may have been just replaced by a tooth, in which case the light will be intercepted. By measuring the speed of the wheel when it is just such as to replace a gap by the next tooth, the speed of the light ray may be determined. The result agreed with that obtained by the astronomical method within about 0.5%. At almost the same time Léon Foucault (1819–68), by an ingenious laboratory device, proved that light travels more slowly in water than in air — a result incompatible with the emission theory.

The sporadic beginnings of a genuine kinetic view of natural phenomena, after having been cultivated . . . by Huygens and Euler, and early in the nineteenth century by Rumford and Young, were united into a consistent physical theory by Fresnel, who has been termed the Newton of optics, and who consistently, and all but completely, worked out one great example of this kind of reasoning. He has the glory of having not only established the undulatory theory of light on a firm foundation, but still more of having impressed natural philosophers with the importance of studying the laws of regular vibratory motion and the phenomena of periodicity in the most general manner. . . .

In astronomy and optics the suggestion of common sense, which regards the earth as stationary and light as an emission travelling in straight lines, had indeed allowed a certain amount of definite knowledge . . . to be accumulated. A real physical theory, however, was impossible until the notions suggested by common sense were completely reversed, and an ideal construction put in the place of a seemingly obvious theory. This was done in astronomy at one stroke by Copernicus; in optics only gradually, tentatively, and hesitatingly. . . .

Newton himself had pronounced the pure emission theory to be insufficient . . . and only a preliminary formulation. . . .

Young boldly generalized the undulatory theory by maintaining that "a luminiferous ether pervades the universe, rare and elastic in a high degree," that the sensation of different colors depends on the different frequency of vibration excited by light in the retina. . . .

In January, 1817, long before Fresnel had made up his mind to adopt a similar conclusion . . . Young announced in a letter . . . that in the assumption of transverse vibrations, after the manner of the vibrations of a stretched string, lay the possibility of explaining polarization. . . . — MERZ, II, v. p., 7–28.

THE SPECTROSCOPE AND SPECTRUM ANALYSIS

Nothing would have seemed more remote to the chemists of the eighteenth century than the possibility of applying their analytical methods to the stars. In the spectrum of Newton "dark lines" had been seen in 1815 by Joseph Fraunhofer (1787–1826), also in 1802 by Wollaston; and Foucault, noticing in the spectrum of the electric arc between carbon electrodes certain bright lines resembling dark ones in the solar spectrum, found when the two spectra were superimposed that these bright and dark lines coincided (1849).[1]

However, it was the employment of flame tests in qualitative chemical analysis by R. W. Bunsen (1811–99) that led to the development of the "spectroscope" in 1859–60. This is essentially no more than a pair of telescopes so attached to the prism producing the spectrum as to give parallel rays from a slit and to facilitate minute scrutiny of the images of the latter. With this instrument, Gustav Kirchhoff (1824–87) demonstrated (1859) the absorption of radiations by vapor of the same substances which emit them; and, in this way, not only accounted for the dark Fraunhofer lines in the solar spectrum, but also recognized two great applications of this principle — the detection of the chemical constituents of celestial bodies, and the discovery of unknown elements by the study of the flames of terrestrial substances. In this way the element Caesium was discovered by Bunsen in 1860 and Rubidium by Bunsen and Kirchhoff later in the same year

[1] G. G. Stokes, *Phil. Mag.* (4) **19**, 193–197, 1860.

In 1862, only three years after Kirchhoff and Bunsen's application of the spectroscope to the study of the sun, Huggins measured the position of the lines in the spectra of about forty stars, with a small slit spectroscope attached to an 8-inch telescope. In 1876 he successfully applied photography to a study of the ultra-violet region of stellar spectra, and in 1879 published his paper "On the Photographic Spectra of Stars." The results were arranged and discussed with reference to their bearing on stellar evolution. — G. E. HALE, *Stellar Evolution.*

The first application of the spectroscope to the chromosphere of the sun was made in 1868 by Pierre Janssen and Sir Norman Lockyer, independently, revealing the chemical composition of the solar prominences as chiefly hydrogen, calcium, and helium, the last previously unknown. In 1889 E. C. Pickering (1846–1919) discovered a double star by means of the spectroscope.

ELECTRICITY AND MAGNETISM: FARADAY, AMPÈRE, MAXWELL

Electricity and magnetism had been studied independently during the two centuries since William Gilbert's *De Magnete.* H. C. Oersted (1777–1851) first discovered the connection between them, his great paper is dated July, 1820.

A. M. Ampère (1775–1836) stimulated by Oersted's discovery of the effect of the electric current on magnets, published in the same year an interesting discussion of electrodynamics and soon after enunciated his celebrated law:

> Two parallel and like directed currents attract each other, while two parallel currents of opposite directions repel each other.

Six years later appeared his *Theory of Electrodynamics, Deduced from Experiment only,* which is a foundation stone of the modern science of electricity.

G. S. Ohm (1787–1854) established the quantitative relation between electric current and electro-motive force in the same conductor (1827), introduced the notion of electrical resistance, and described its variation in different conductors.

Michael Faraday (1791–1867) rescued electricity from the

mysterious notion of currents acting on each other through empty space, by the fruitful conception of a magnetic field, of which a new and comprehensive mathematical theory was gradually worked out by James Clerk Maxwell (1831–79). Faraday's discoveries were so far-reaching that they have even been coupled with the law of the conservation of energy and Darwin's theory of descent as the greatest scientific ideas of the latter half of the century.

> Atoms and lines of force have become a practical — shall I say a popular? — reality, whereas they were once only the convenient method of a single original mind for gathering together and unifying in thought a bewildering mass of observed phenomena, or at most capable of being utilized for a mathematical description and calculation of actual effects. — MERZ, II, 77.

Yet Helmholtz says of Faraday:

> It is indeed remarkable in the highest degree to observe how, by a kind of intuition, without using a single formula, he found out a number of comprehensive theorems, which can only be strictly proved by the highest powers of mathematical analysis. . . . I know how often I found myself despairingly staring at his descriptions of lines of force, their number and tension, or looking for the meaning of sentences in which the galvanic current is defined as an axis of force. . . . — *Vorträge und Reden*, II, 277.

Faraday apprehended the principle of the conservation of energy even before it had come to clear expression as common property, saying, for example, in refuting the theory that electricity could be generated by metallic contact alone:

> But in no case, not even in those of [electric fishes], is there a pure creation or a production of power without a corresponding exhaustion of something to supply it.

Like Young, Dalton, and Joule, Faraday did not belong to the orthodox Cambridge school then dominant in English mathematical and physical science, and recognition of the significance of his ideas was consequently retarded.

> What the atomic theory has done for chemistry, Faraday's lines of force are now doing for electrical and magnetic phenomena. . . .

Yet the circumstances under which Faraday's work was done were those of penury. — MERZ, I, 246.

ELECTRO-MAGNETIC THEORY OF LIGHT

In 1845 Faraday writes:

I . . . have at last succeeded *in magnetising and electrifying a ray of light, and in illuminating a magnetic line of force.* . . . Employing a ray of light, we can tell, *by the eye*, the direction of the magnetic lines through a body; and by the alteration of the ray and its optical effect on the eye, can see the course of the lines just as we can see the course of a thread of glass, . . . — *Phil. Trans.*, **Year 1846,** p. 2 and footnote p. 1.

I have deduced the relation between the statical and dynamical measures of electricity, and have shown by a comparison of the electro-magnetic experiments of Kohlrausch and Weber with the velocity of light as found by Fizeau, that the elasticity of the magnetic medium in air is the same as that of the luminiferous medium, if these two coexistent, coextensive and equally elastic media are not rather one medium. . . . We can scarcely avoid the inference that light consists in the transverse undulations of the same medium which is the cause of electric and magnetic phenomena. — J. C. MAXWELL, *Phil. Mag.*, (4) **23,** 15 and 22, 1862.

We must not listen to any suggestion that we may look upon the luminiferous ether as an ideal way of putting the thing. A real matter between us and the remoter stars I believe there is, and that light consists of real motions of that matter, motions just such as are described by Fresnel and Young, motions in the way of transverse vibrations. — KELVIN, *Baltimore Lectures*, 1884, p. 8.

H. R. Hertz (1857–94), a pupil of Helmholtz, first proved in 1888 the existence of those undulations which now bear his name, showing also that these travel with the rapidity of light, and that they are, like light and heat waves, susceptible of reflection, refraction, and polarization. Until he measured their length and velocity, no great progress was made in verifying those relations experimentally. Such more recent applications of Hertz's ideas as radio-telegraphy and radio-telephony testify to their immense practical as well as theoretical importance.

With the establishment of the electro-magnetic theory of

light, what we may call the undulatory series became complete. Sound had long been known to be due to waves or "undulations" and the wave theory of heat and of light was accepted, so that it had only remained to prove the existence of electro-magnetic undulations, and to show that such waves moved with the velocity and other characteristics of light.

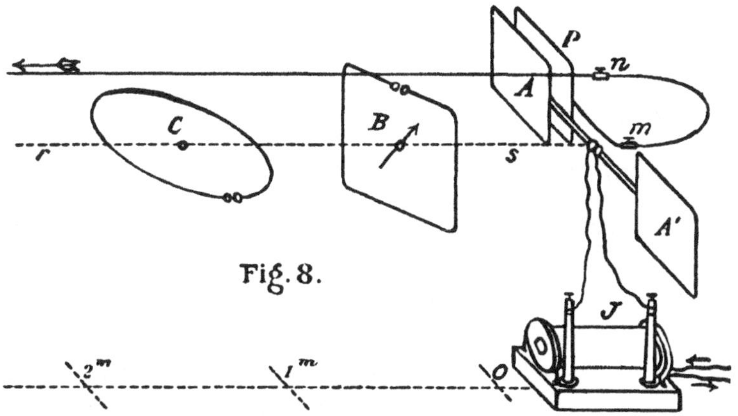

FIG. 57. — APPARATUS USED BY HERTZ TO STUDY ELECTRIC WAVES. *AA'*, primary conductor of two brass plates and copper rod with spark-gap in the middle; *J*, induction coil; *B*, *C*, two forms of tuned secondary circuit, copper wire with adjustable spark-gap; *rs*, base line of experiments; *mn*, wire connected with plate *P* to compare velocity of waves in wire with those in air. After Hertz, *Wiedemann's Annalen*, **34**, pl. 4, 1888.

This is a result which Maxwell deduced mathematically from the experiments of Oersted (1820) and Faraday (1831) and which Hertz confirmed experimentally in 1888.

The velocity of propagation of an electro-magnetic disturbance in air . . . does not differ more from the velocity of light in air . . . than the several calculated values of these quantities differ among each other. — MAXWELL.

ETHER DRIFT; RELATIVITY

The ether postulated by the undulatory theory must, it seemed, possess extraordinary qualities, the study of which connects itself with theories which have revolutionized modern physics. In 1887 A. A. Michelson and E. W. Morley made their famous experiment to ascertain if the earth moved relatively to the hypothetical ether. If the earth in its orbital motion traverses the ether it should take longer for a beam of light to traverse a path to and fro in the direction of this translatory motion than the same to and fro distance in a path at right angles to this line of motion. The negative results of the experiment carefully made and many times repeated appeared to prove that there was no measurable ether drift. The far-reaching consequences of this discovery for the theory of relativity and of physics in general lie beyond the range of the present chapter.

X-RAYS, RADIOACTIVITY

The discovery in 1895 of X-rays by W. C. Röntgen (1845–1923) opened a long vista of modern physics for the twentieth century. It was soon followed by the sensational discoveries of Henri Becquerel (1852–1908) and by Pierre and Marie Curie of the so-called radioactive substances and of the electron by J. J. Thomson. The atom was about to lose its supposed fundamental indestructibility and the transmutation of elements, the immemorial dream of the alchemists was approaching realization.

THE QUANTUM THEORY

While the classical mechanics of Newton gave a perfectly satisfactory method of treatment for matter in bulk, it was found toward the close of the century to be inapplicable to the minute study of radiant energy. Radiations from a hot body are not all of one wave length, and W. Wien in 1896 made an important advance by applying the theory of probability to such radiation. Meanwhile evidence was accumulating that the emission and absorption of radiation was not, as formerly believed, a continuous process. This result was due mainly to data obtained in the outstanding researches on black-body radiation made at Charlottenburg by Otto Mummer and Ernst Pringsheim during the years 1897 to 1899, taking very exact measurements with the bolometer invented by S. P. Langley (1834–1906). Their results when plotted —

intensity of radiation against wave length — were in accord with Wien's law only in the region of the shorter wave lengths.

The third Lord Rayleigh (1842–1919) suggested in 1900 a law that agreed with experimental results on long wave lengths, but failed when applied to the other end of the spectrum. Discarding the theory of continuity, Max Planck, professor at Berlin, was able in 1900 and 1901 [1] to fit a mathematical probability curve to the observed curve of radiation

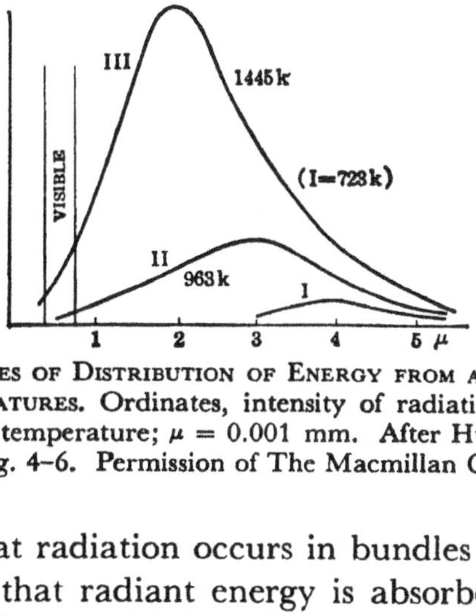

FIG. 58. — CURVES OF DISTRIBUTION OF ENERGY FROM A "BLACK BODY" AT DIFFERENT TEMPERATURES. Ordinates, intensity of radiation; abscissae, wave lengths; k, absolute temperature; μ = 0.001 mm. After Hull, *Elementary Survey of Modern Physics*, Fig. 4–6. Permission of The Macmillan Company.

by assuming that radiation occurs in bundles of energy; that is, he assumed that radiant energy is absorbed and emitted by matter in whole multiples of a small finite unit, or *quantum* ϵ, proportional to the vibration frequency of the radiation ν, multiplied by a universal constant h.

$$\epsilon = h\nu. \qquad h = 6.55 \times 10^{-27} \text{ erg-seconds.}^2$$

This general law of radiant energy, which has been compared in significance to Dalton's atomic theory of matter, has been found to have wide application and has had important influence in the extraordinary development of theoretical physics in recent years.

[1] *Vehr. deutsch.-physikal. Ges.*, 1900, 202–204; 237–245; *Drude's Annalen*, **4**, 553–566, 1901.

[2] Millikan (1916) by a very accurate method found $h = 6.56 \times 10^{-27}$ erg-seconds. $h\nu$ is pronounced "*h* new."

KINETIC THEORY OF GASES. CLAUSIUS

The modern theory of gases was born . . . when Joule in 1857 actually calculated the velocity with which a particle of hydrogen . . . must be moving, assuming that this atmospheric pressure is equilibrated by the rectilinear motion and impact of the supposed particles of the gas on each other and the walls of the containing vessel. This meant taking the atomic view of matter in real earnest, not merely symbolically, as chemists had done. — MERZ, I, 434.

The theory owes its full development, however, to the researches of Maxwell, R. J. E. Clausius, and Ludwig Boltzmann.

The great turning-point, indeed, lay in the kinetic theory of gases, which . . . had introduced quite novel considerations by showing how the dead pressure of gases and vapors could be explained on the hypothesis of a very rapid but disorderly translational movement of the smallest particles in every possible direction. — MERZ, II, 56.

THE CONCEPTION OF ENERGY

Newton's *Principia* contains by implication the modern notion of energy — but the first clear and consistent fixing of the modern terminology is found in J. V. Poncelet's *Mécanique industrielle*, 1829. The idea of *work* was thus developed from the standpoint of the engineer — notably under the influence of W. J. M. Rankine; while on the other hand, it is a not less remarkable fact that Black, Young, J. R. Mayer and Hermann Helmholtz (1821–94) all came to their scientific work through another form of applied science — medicine.

A considerable step toward the general idea of the conservation of energy was taken by Rumford in his determination of the mechanical equivalent of heat, but the final achievement is due mainly to Joule in England and Mayer and Helmholtz in Germany. In 1847 Helmholtz read before the Physical Society of Berlin one of the most remarkable papers of the century (*Die Erhaltung der Kraft*), in which he says with full justice:

I think in the foregoing I have proved that the above mentioned law does not go against any hitherto known facts of natural science, but is supported by a large number of them in a striking manner. I have tried to enumerate as completely as possible what consequences result from the combination of other known laws of nature, and how they require to be confirmed by further experiments. The aim of this investigation, and what must excuse me likewise for its hypothetical sections, was to explain to natural philosophers the theoretical and practical importance of the law, the complete verification of which may well be looked upon as one of the main problems of physical science in the near future.

Fifteen years later Helmholtz spoke of the principle as follows:

The last decades of scientific development have led us to the recognition of a new universal law of all natural phenomena, which, from its extraordinarily extended range, and from the connection which it constitutes between natural phenomena of all kinds, even of the remotest times and the most distant places, is especially fitted to give us an idea of what I have described as the character of the natural sciences, which I have chosen as the subject of this lecture.

This law is the Law of the Conservation of Force, a term the meaning of which I must first explain. It is not absolutely new; for individual domains of natural phenomena it was enunciated by Newton and Daniel Bernoulli; and Rumford and Humphry Davy have recognised distinct features of its presence in the laws of heat.

The possibility that it was of universal application was first stated by Dr. Julius Robert Mayer . . . in 1842, while almost simultaneously with, and independently of him, James Prescot Joule . . . made a series of important and difficult experiments on the relation of heat to mechanical force, which supplied the chief points in which the comparison of the new theory with experience was still wanting.

The law in question asserts, that the *quantity of force which can be brought into action in the whole of Nature is unchangeable*, and can neither be increased nor diminished. — *Popular Lectures*, Trans. by E. Atkinson, pp. 319–320.

This doctrine is now so fundamental and so familiar as to require no further comment. Lavoisier's theory of the indestructibility of matter had already become axiomatic. Henceforth (until quite recent years), energy also was to be considered constant and indestructible.

DISSIPATION OF ENERGY

It remained for William Thomson (Lord Kelvin, 1824–1907), applying the principle of the conservation of energy to the thermodynamic laws of Carnot, to deduce the other great principle of the Dissipation of Energy, which recognizes that while *total* energy is constant, *useful* energy is diminishing by the continual degeneration of other forms into non-useful or dissipated heat. All workers in this field, from Carnot to Thomson, had appreciated the impossibility of "perpetual motion." Helmholtz expresses his appreciation of Thomson's contribution to the theory in a striking passage:

We must admire the acumen of Thomson, who could read between the letters of a mathematical equation, for some time known, which spoke only of heat, volume and pressure of bodies, conclusions which threaten the universe, though indeed only in infinite time, with eternal death. — *Vorträge und Reden*, I, 43.

Thomson, more than any other thinker, put the problem into common-sense language. . . . He saw at once, when adopting Joule's doctrine of the convertibility of heat and mechanical work, that, if all processes in the world be reduced to those of a perfect mechanism, they will have this property of a perfect machine, namely, that it can work backward as well as forward. It is against all reason and common sense to carry out this idea in its integrity and completeness. ". . . If, then, the motion of every particle of matter in the universe were precisely reversed at any instant, the course of nature would be simply reversed for ever after. The bursting bubble of foam at the foot of a waterfall would reunite and descend into the water; the thermal motions would reconcentrate their energy, and throw the mass up the fall in drops re-forming into a close column of ascending water. Heat which had been generated by the friction of solids and dissipated by conduction, and radiation with absorption, would come again to the place of contact, and throw the moving body back against the force to which it had previously yielded. Boulders would recover from the mud the materials required to rebuild them into their previous jagged forms, and would become re-united to the mountain-peak from which they had formerly broken away. And if also, the materialistic hypothesis of life were true, living creatures would grow backwards, with conscious knowledge of the future, but with no memory of the past, and would become again unborn. But the real phenomena of life infinitely transcend human science, and speculation regarding consequences of their imagined reversal is utterly unprofitable. Far otherwise, however, is it in respect to the reversal of the motions of matter uninfluenced by life, a very elementary consideration of which leads to the full explanation of the theory of dissipation of energy." — MERZ, II, 131–133, quoting SIR W. THOMSON, *Proc. Royal Soc. Edin.*, **8**, 325–331, 1874.

MODERN CHEMISTRY

Main features in nineteenth century chemistry are: — the discovery of the fundamental quantitative relations of chemical reactions; the development of a consistent and definite theory of atoms, molecules, and valence; the synthesis of organic substances; the discovery of periodic relations and characteristics; the development of ideas of chemical structure; the development of electro-chemistry; the foundation of physical chemistry.

With Lavoisier, "the father of modern chemistry," the science, heretofore descriptive and empirical, had become quantitative and productive, seeking like the older sciences of astronomy and physics to make itself mathematical — an exact science. Postulating the existence of indestructible elementary substances, Lavoisier controlled and interpreted chemical reactions by careful weighing. Until he entered the field there was no generalization wide enough to entitle chemistry to be called a science.

CHEMICAL LABORATORIES: LIEBIG

In the nineteenth century chemical studies received a powerful impetus through the establishment of teaching laboratories at the universities — in which Justus Liebig (1803–73) at Giessen in 1826 was a pioneer. He writes:

At Giessen all were concentrated in the work, and this was a passionate enjoyment. . . . The necessity of an institute where the pupil could instruct himself in the chemical art, . . . was then in the air, and so it came about that on the opening of my laboratory . . . pupils came to me from all sides. . . . I saw very soon that all progress in organic chemistry depended on its simplification.

. . . The first years of my residence at Giessen were almost exclusively devoted to the improvement of organic analysis, and with the first successes there began at the small university an activity such as the world had not yet seen. . . . A kindly fate had brought together in Giessen the most talented youths from all countries of Europe. . . . Every one was obliged to find his own way for himself. . . . We worked from dawn to the fall of night. — *Deutsche Rundschau*, **66**, 30–39. (Merz, I, 190.)

To investigate the essence of a natural phenomenon, three conditions are necessary. We must first study and know the phenomenon itself, from all sides; we must then determine in what relation it stands to other natural phenomena; and lastly, when we have ascertained all these relations, we have to solve the problem of measuring these relations and the laws of mutual dependence — that is, of expressing them in numbers. In the first period of chemistry, all the powers of men's minds were devoted to acquiring a knowledge of the properties of bodies. . . . This is the alchemistical period. The second period embraces the determination of the mutual relations or connections of these properties; this is the period of phlogistic chemistry. In the third . . . we ascertain by weight and measure and express in numbers the degree in which the properties of bodies are mutually dependent. The inductive sciences begin with the substance itself, then come just ideas, and lastly, mathematics are called in, and, with the aid of numbers, complete the work. — LIEBIG, *Familiar Letters*, Ed. 4, p. 60. (Merz, I, 389.)

QUANTITATIVE RELATIONS: ATOMS; MOLECULES; VALENCE

It took . . . nearly a century . . . before the rule of definite proportions was generally established, becoming a guide for chemical analysis. . . .

The vaguer terms of chemical affinity and elective attraction, of chemical action, of adhesion and elasticity . . . gradually disappeared, when by the aid of the chemical balance each simple substance and each definite compound began to be characterized and labelled with a fixed number. — MERZ, I, 392.

J. L. Proust (1754–1826), analyzing various metallic oxides and sulphides, obtained, in opposition to Berthollet, constant percentage results, and is, therefore, to be regarded as the discoverer of the law of definite proportions. John Dalton (1766–1844) had the happy inspiration to interpret these figures in relation to weights of the combined oxygen, making the lightest element, hydrogen, the unit or measure of his system (*New System of Chemical Philosophy*, 1808). His hypothesis that elements combine in weights proportional to small whole numbers — the "law of multiple proportions," has since been verified by innumerable analyses.

It has been shown that Dalton was in the habit of regarding all physical phenomena as the result of the interaction of small particles. He was thus naturally led to the conception of definite atomic weights to be determined by experiment. In his own words:

> We can as well undertake to incorporate a new planet in the solar system or to annihilate one there as to create or destroy an atom of hydrogen. All the changes we can effect consist in the separation of atoms bound together before and in the union of those previously separated.

The atomic theory while highly serviceable has always been subjected to severe criticism. In 1840, for example, J. B. A. Dumas (1800–84) declared that it did not deserve the confidence placed in it, and that if he could he would banish the word "atom," convinced that science should confine itself to what could be known by experience. As late as 1852 Edward Frankland (1825–99) says:

> I had not proceeded far, in the investigation of the organo-metallic compounds before the facts brought to light began to impress upon me the existence of a fixity in the maximum combining value or capacity of saturation in the metallic elements which had not before been suspected. . . . It was evident that the atoms of zinc, tin, arsenic . . . had only room, . . . for the attachment of a fixed and definite number of the atoms of other elements. — *Experimental Researches*, p. 145. (Merz, I, 415.)

Independent researches have, in combination with the older chemical theories, introduced so much definiteness into this line of thought that "the Newtonian theory of gravitation is not surer to us now than is the atomic or molecular theory in chemistry and physics — so far, at all events, as its assertion of heterogeneousness in the minute structure of matter, apparently homogeneous to our senses, and to our most delicate direct instrumental tests." — MERZ, I, 424, quoting KELVIN, *Popular Lectures*, 1886, I, 4.

The three main criticisms of the atomic theory are: —

(1) that it is based on inference, not on direct observation; and is therefore only a provisional hypothesis; (2) that it takes no account of chemical forces — "affinity"; (3) that it overemphasizes analysis.

The idea of "atomicity" and "valence" . . . was not possible without the clear notion of the "molecule" as distinct from the "atom." This idea had lain dormant in the now celebrated but long forgotten law of Amadeo Avogadro (1776–1856), which was established in 1811 almost immediately after the appearance of Dalton's atomic theory.

It had been known since the time of Boyle and Mariotte that equal volumes of different gases under equal pressure change their volumes equally if the pressure is varied equally, and it was also known . . . that equal volumes of different gases under equal pressure change their volumes equally with equal rise of temperature. These facts suggested to Avogadro, and almost simultaneously to Ampère, the very simple assumption that this is owing to the fact that equal volumes of different gases contain an equal number of the smallest independent particles of matter. This is Avogadro's celebrated hypothesis. It was the first step in the direct physical verification of the atomic view of matter. — MERZ, I, 427.

SYNTHESIS OF ORGANIC SUBSTANCES

Until the middle of the nineteenth century there was an apparently fundamental separation between organic and inorganic nature. Since then they have been brought together by the general laws of energy and to some extent by the principles of evolution, as will appear in the following chapter. In 1828 Friedrich Wöhler (1800–82), of Göttingen, had indeed succeeded in preparing urea out of inorganic materials, a discovery which disproved such difference as was hitherto considered to exist between organic and inorganic bodies.

This momentous discovery by Wöhler furnished the impetus to an enormous amount of labor in preparing countless pure "organic" compounds in the laboratory, including synthetic dye stuffs and medicinals. Of special interest is the work of

Emil Fischer (1852–1919) on the constitution of the sugars and the proteins.

A PERIODIC LAW AMONG THE ELEMENTS

With gradually increasing knowledge of the fundamental constants of chemistry — the atomic weights — attempts were naturally made to connect these with the chemical and physical properties of the corresponding elements: valence, affinity, specific gravity, specific heat, etc. In 1869–71 D. I. Mendeléjeff (1834–1907), a Russian chemist, succeeded in establishing remarkable relations between these data, and on tabulating them enunciated his Periodic Law, which has resulted in the discovery of several new and hitherto unsuspected elements. As the existence of the planet Neptune (p. 390) had been predicted to fill an apparent gap in a system, so Mendeléjeff under the periodic law was able to predict the existence of other and missing elements in the series of chemical elements. And just as the prediction of Adams and Leverrier was fulfilled by the actual discovery of Neptune, so the prophecy of Mendeléjeff was justified by the discovery of gallium in 1871, scandium in 1879, and germanium in 1886. Furthermore, the periodic law enabled Mendeléjeff to question the correctness of certain accepted atomic weights, and here, also, he was justified by subsequent redeterminations.

CHEMICAL STRUCTURE

Crystallography — a science of the nineteenth century — established an important connection between chemistry and geometry. R. J. Haüy (1743–1822) made mineralogy "as precise and methodical as astronomy. . . . He was to Werner and Romé de l'Isle, his predecessors, what Newton had been to Kepler and Copernicus."

In the early years of the atomic theory W. H. Wollaston (1766–1828) had predicted that philosophers would seek a geometrical conception of the distribution of the elementary particles in space — a prophecy first practically fulfilled by the conception of the benzene ring (1865) in the mind of F. A.

Kerkulé (1829–96) and by J. H. Van't Hoff's *Chemistry in Space* (1875).

PHYSICAL CHEMISTRY: ELECTROLYTIC AND THERMODYNAMIC DEVELOPMENTS OF CHEMISTRY

In the latter part of the nineteenth century much light was thrown on a wide range of physical and chemical phenomena by the study of solutions and their electrolytic behavior. Much had already been accomplished by Sir Humphry Davy (1778–1829) in the decomposition of substances by the electric current, leading for example to the first isolation of the elements, sodium and potassium. His successor, Faraday, at the newly founded Royal Institution, showed that for a given substance the amount decomposed is dependent solely on the quantity of electricity passed through and that for different substances the amounts set free at the electrodes are proportional to their chemical equivalents. To him the name *electrolysis* is due. A closer study of the phenomena of electrolysis led Clausius to the hypothesis that the molecules of salts, acids, and bases, previously regarded as disintegrated only by the passage of the electric current, are already dissociated in ordinary solutions. To these electrically charged part-molecules Faraday gave the name *ions*. Svante Arrhenius (1859–1927) proved that salts in dilute solution are dissociated into their ions almost completely, instead of only very slightly as Clausius supposed. This theory of Arrhenius, known as the Theory of Electrolytic Dissociation, of which an account would be too technical for the present purpose, coördinates and correlates heterogeneous masses of chemical facts, which apparently bore little or no relation to one another, and refers them to a common cause.

During the latter part of the nineteenth century a study of the rate and equilibrium conditions of chemical reactions led by degrees to the formulation of the so-called law of mass action and to many important thermodynamic relations. Chemistry thus came to share with physics the possibility of utilizing the calculus, becoming thereby more fully a quantitative science.

THERMODYNAMICS: GIBBS

Heat being regarded as a form of energy subject to the general law of conservation, it was recognized that many processes would be irreversible, on account of the diffusion of the heat involved. This led to the formulation of the important second law of thermodynamics and the conception by Clausius of the *entropy* function as a function which in all processes as they occur in nature tends to increase. From this fact in small-scale systems it was argued by some that the universe continually approaches ultimate stagnation, with its total energy uniformly distributed.

A contribution of major importance was made by J. W. Gibbs (1839–1903) in his paper, "On the Equilibrium of Heterogeneous Substances," obscurely published (1877) in the *Transactions of the Connecticut Academy of Science*. This "long neglected masterpiece" was in 1892 discovered by W. Ostwald, who says in his autobiography:

> Gibbs concerns himself exclusively with energy magnitudes and their factors and holds himself completely free from all kinetic hypotheses. By so doing he wins for his results a permanence and security that place them among the highest products of intellectual attainment. It is a fact that up to the present time, not a single error either in his formulas, his results, nor yet — and this is the most remarkable — in his assumptions has been found. — *Lebenslinien*, II, p. 62.

Donnan, on the occasion of the centenary celebration of the founding of the Franklin Institute twenty years after the death of Gibbs, referred to the paper as "one of the mightiest works of genius the human mind has ever produced."

REFERENCES FOR READING

Cajori, Florian, *History of Physics*, Rev. Ed., 1929.
Hertz, Heinrich, *Electric Waves*, 1893.
Moore, F. J., *History of Chemistry*, Ed. 2, 1931.

RAMSAY, WILLIAM, *The Gases of the Atmosphere and the History of their Discovery*, Ed. 4, 1915.
ROSCOE, H. E., *John Dalton*, 1895.
SODDY, F., *Matter and Energy*, 1912.
THOMPSON, S. P., *Michael Faraday: His Life and Work*, 1898.
TILDEN, W. A., *Progress of Scientific Chemistry in Our Own Time*, 1899.
TYNDALL, JOHN, *Faraday as a Discoverer*, Ed. 4, 1884.

From previous lists: Merz, *Hist. European Thought;* Poincaré, *Sci. and Hypothesis.*

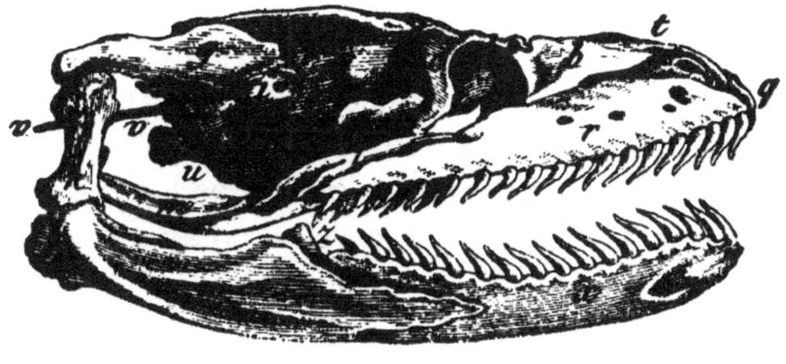

Chapter XVII

Some Advances of Natural Science in the Nineteenth Century

What the classical renaissance was to men of the fifteenth and sixteenth centuries, the scientific movement is to us. It has given a new trend to education. It has changed the outlook of the mind. It has given a new intellectual background to life. — SADLER.

The rapid increase of natural knowledge, which is the chief characteristic of our age, is effected in various ways. The main army of science moves to the conquest of new worlds slowly and surely, nor ever cedes an inch of the territory gained. But the advance is covered and facilitated by the ceaseless activity of clouds of light troops provided with a weapon — always efficient, if not always an arm of precision — the scientific imagination. It is the business of these *enfants perdus* of science to make raids into the realm of ignorance wherever they see, or think they see, a chance; and cheerfully to accept defeat, or it may be annihilation, as the reward of error. Unfortunately the public, which watches the progress of the campaign, too often mistakes a dashing incursion . . . for a forward movement of the main body; fondly imagining that the strategic movement to the rear, which occasionally follows, indicates a battle lost by science. — HUXLEY.

INFLUENCE OF EIGHTEENTH CENTURY REVOLUTIONS

If the French Revolution had done no more than to upset as it did the social equilibrium of the centuries, its effect in stimulating inquiry and generating doubt in almost every direction could not have failed to further scientific studies and promote wholesome investigation into the fundamental relations of man and nature. But even before that revolution, some of the ablest minds in France, keenly alive to the teachings of Descartes and Newton and the lessons of seventeenth century science, had rejected the current cosmogony of Moses, although they had nothing with which to replace it. In particular, the eighteenth century questioned all custom and authority, and the theory of special creation possessed no other basis.

THE UNIFICATION OF BIOLOGY

The fusion of zoology and botany into a compact science, Biology, which had been foreshadowed by Wolff and Goethe, found distinct expression at the beginning of the century. In 1802 G. R. Treviranus (1776–1837) published a book under the title *Biologie, oder Philosophie der lebenden Natur*. He says here, "The subjects of our inquiry will be the various forms and phenomena of life, the conditions and laws under which this state occurs, and the causes through which it is brought about. The science that occupies itself with these subjects we will designate by the name *biology*, or the science of life." In the same year J. B. de Lamarck (1744–1829) in his *Hydrogéologie*, inquires, "What is the influence of living beings on surface materials of the earth, and what are the general results of this influence?" Then he goes on to say that Terrestrial Physics should include three essential parts: Meteorology, Hydrogeology, and finally the fundamental ideas relative to the origin and development of the organization of living bodies — Biology. This term meant for Lamarck, evidently, what we call the theory of organic evolution.

Thus Biology was born in 1802. Then came the question of its direction of growth. Should it be idealistic and develop Stahl's idea of vital force as the guiding and crystallizing influence, or should it be materialistic? The history of nineteenth century biology is largely a history of the development

---

NOTE: On the opposite page, the picture of the skull of a large python from Java illustrates the work of a great comparative anatomist, Georges Cuvier. The lettering refers to his text. After Cuvier, *Le Règne animal*, 1817, pl. VII, Fig. 3.

of mechanistic theories. Treviranus was a vitalist. Lamarck developed a mechanical theory of life. The mechanistic view was to prevail.

BIOPHYSICS AND BIOCHEMISTRY

The first step toward a mechanistic explanation of vital phenomena was made by N. T. de Saussure, whose researches in vegetable physiology, published in 1804, were described in Chapter XIV. Until 1828 all organic substances were supposed to be produced by vital force. That was the year in which Wöhler synthesized urea from inorganic materials; and in 1832 appeared a famous paper by Wöhler and Liebig showing that a complex organic group of atoms may act as a radical and remain unchanged as a constituent of a long series of compounds — a discovery of prime importance for the understanding of chemical changes in the living body.

Henri Dutrochet (1776–1847) was the first to make exact studies of osmosis, 1826 to 1837, and to call attention to its value for the explanation of important phenomena in living organisms, formerly ascribed to vital force. Turning to photosynthesis, Dutrochet could say with certainty in 1837 that the capacity of plants to utilize carbon dioxide depends upon a definite green coloring matter, *chlorophyll*. In 1840 Liebig, vigorously opposing the "humus theory," asserted that carbon dioxide of the air is the only source of carbon in plants. But it was not until 1862 and 1864 that Julius Sachs (1832–97) was able to present experimental proof that starch is derived from the carbon dioxide absorbed and that the process depends upon sunlight acting on chlorophyll-containing particles within the cells.

GREAT SCHOOLS OF PHYSIOLOGY

In the meantime two great schools of animal physiology had arisen in France and Germany under the leadership of François Magendie (1783–1855) and Johannes Müller (1801–58). Magendie, the first modern exponent of the mechanistic philosophy, studied all systems of organs with special reference to disordered functions. He discovered the sensory and motor functions of the spinal nerve roots independently of Sir Charles Bell, and he developed an experimental technique that continues to be the basis of physiological research.

Johannes Müller was professor of anatomy in Berlin, 1832–58. An ardent vitalist, but gifted with German enthusiasm and thoroughness, he gathered about himself the most notable group of students in the history of biology and founded the most remarkable school of mechanistic physiology. He taught that only definite and methodical experiments could make the general principles of science intelligible and set them on a sure foundation. Müller's own researches were on the physiology of the nervous system, microscopic anatomy, and embryology. In 1834 he began the publication of the *Handbuch der Physiologie*, the standard textbook of physiology during the pre-Darwinian period, and he is regarded as the founder of scientific medicine in Germany. By experiments on himself he discovered (1826) the law of "specific nerve energies," that a stimulated nerve gives its own sensation and none other. With frogs, he confirmed the results of Bell and Magendie on the roots of spinal nerves, and he made important observations on the development of the sexual organs and on the histology of glands. Later he turned to comparative anatomy and marine zoology with conspicuous success.

THE ENERGY OF LIFE

Müller's most distinguished pupil, Hermann Helmholtz (1821–94) was the son of a poor but learned schoolmaster. While still an army surgeon, Helmholtz gave the vitalistic doctrine its severest shock. Beginning in 1845 with a paper on the metabolism of contracting muscle, he followed this in 1846 with a determination of the heat evolved during muscular activity. Then came the meeting of the Physical Society of Berlin, July 23, 1847, when he read his paper on the Law of Conservation of Energy, *Ueber die Erhaltung der Kraft*, in which he applied this law to living things as well as to inorganic forces. It created a great sensation, and, to the surprise of his friends, proved him to be a master of mathematical physics. This paper, published separately in 1847, marks an epoch in the history of science.

As was shown in Chapter XVI, Helmholtz was not the first to formulate the theory of conservation of energy. He was anticipated in 1842 and 1845 by a German country doctor in Schwabia, J. R. Mayer, who emphasized the biological aspect; and to a large extent by the English physicist, J. P. Joule, in his experiments on heat. But Helmholtz was the first to present the broad generalization in a convincing way. In 1850, the year after his appointment to be Director of the Physiological Institute at Königsberg, he reported the first measurement of the rate of transmission of the nerve-impulse — in a frog's nerve 50 to 60 mm. long, time 0.0014 to 0.0020 seconds. Until then nervous action was supposed to be an instantaneous manifestation of vital force.

CATALYSIS

At a frontier post of the American army, the surgeon, William Beaumont (1785–1853), made good use of an extraordinary opportunity afforded by a persistent opening through the abdominal wall into the stomach of a patient whom he had nursed back to health after a severe gun-shot wound. The full report of his experiments, begun in 1825, is one of the classics of physiological literature (1833). Among many observations, he recorded that the flow of gastric juice is started by the introduction of food, and more freely if by the mouth than if through the opening; that the gastric juice contains, besides free hydrochloric acid, "some other active *chemical* principles"; and "that *no other* fluid produces the same effect on food that the gastric juice does" (*Fulton*, 1931, pp. 79–80).

In Johannes Müller's laboratory, Theodor Schwann (1810–82) first isolated the ferment from the stomach and called it *pepsin* (1835). Two years earlier, Payen and Persoz had isolated from malt a substance called *diastase* that converted starches into sugars. That these organic ferments and certain inorganic substances could produce an effect hitherto unknown, was recognized by the chemist, J. J. Berzelius (1779–1848). He called it *catalysis*, and ascribed it to an ability of these substances "to set into activity affinities which are dormant at this particular temperature, and this not by their own affinity but by their presence alone." Thus began a line of investigation that is still pursued with vigor and astonishing results. The name *enzyme* was applied to organic catalysts by Willy Kühne in 1878.

THE INTERNAL MEDIUM

In Paris, Claude Bernard (1813–1878) assisted and finally succeeded his master, Magendie, at the Collége de France, and also occupied the chair of physiology founded for him in 1854 at the Sorbonne. His great contribution was insistance that an explanation of the external phenomena of life can be found only in the physico-chemical conditions of the inner environment (*milieu intérieur*) and in the reciprocal and simultaneous reactions of this upon the organs, and of the organs upon it, in health and in disease. He was especially interested in the sugar in the blood. In his thesis of 1843 he had shown that cane sugar cannot be utilized in the body until changed to dextrose by the gastric juice. This led to the discovery of the three-fold action of the pancreatic secretion in the intestine: — (1) reduction of starches to dextrose, (2) splitting of fats into glycerine and fatty acids, and (3) completion of the digestion of proteins (1850–56). Bernard discovered dextrose sugar in the blood of starved animals and its origin in a starch-like substance he called *glycogen*, which is formed and stored in the liver. He distinguished two secretions of the liver: (a) its "external secretion," bile poured into the intestine, and (b) its "*internal secretion*," sugar liberated into the blood. This, his most important discovery, suggested researches in an entirely new direction (p. 453). He found the liberation of sugar to be under nervous control; and, while investigating animal heat, he discovered that the caliber of blood-vessels is affected by vaso-dilator nerves, as well as by vaso-constrictors, and thus he opened the way to the whole subject of the nervous regulation of the blood supply. His discoveries, made mainly before 1860, resulted from long series of experiments upon living animals, and caused him to be regarded as the greatest physiologist of his age.

COMPARATIVE ANATOMY IN FRANCE

While the work of Priestley and Lavoisier dominated early nineteenth-century physiology, we see the influence of Bonnet, Goethe, and Buffon among the notable group of comparative anatomists and zoologists gathered in Paris at the beginning of the century. Etienne Geoffroy St. Hilaire (1772–1844) [1] and Lamarck were evolutionists, while Georges Cuvier (1769–1832) upheld the fixity of species. Geoffroy lectured on the vertebrate animals and tried to show that all animal forms are but variations of a uniform plan of structure. Lamarck took the other classes of animals, he called them all Invertebrates, and made a great advance from the classification of Linnaeus. To express the relation of these classes, he adopted, in place of the linear scale of Bonnet, a branching arrangement similar in form, but not in detail, to that used today.

Cuvier was a morphologist in the sense of Goethe, but he would have no scale of nature. Instead, he arranged animals in four Branches — Vertebrates, Molluscs, Articulates, and Radiates — the animals in each branch showing a distinct type of structure in differing degrees of completeness, so that the lowest forms in one branch were simpler than the highest forms in the next branch below. His first important work, lectures on comparative anatomy, 1799–1805, is distinguished by clarity of thought and absence of speculation, and has given him the title of founder of the subject. His best known work is *Le Règne animal*, 1817, and is beautifully illustrated. Cuvier was the first clearly to state the principle that the structures and functions of living things are so correlated that the whole may be inferred from a part, and he applied this successfully in the study of fossils. In 1800 he compared fossil

[1] Not to be confused with Etienne Geoffroy (1725–1810), Parisian physician and entomologist.

elephants with living forms, and his work of 1812 on fossil bones is a classic. To explain the presence of fossils, Cuvier imagined a series of world catastrophies, the last one being the Flood described in the Bible. He was emphatic in denying the existence of fossil man or that man lived with the large fossil mammals found near Paris. His important history of science, covering the years 1789 to 1830, was published after his death.

THE DOCTRINE OF DESCENT

Lamarck was the first to present a comprehensive theory of organic evolution. It was outlined in his *Recherches sur l'organisation des corps vivans*, 1802, more fully developed in his *Philosophie zoologique*, 1809, and worked out in detail in the *Histoire naturelle des animaux sans vertèbres*, 1815–22. Lamarck had seen gradations between species of plants and of animals, individual variations — especially of domesticated forms, and the well-known effects of the use and disuse of parts. His four biological laws may be expressed as follows: —

1. Life naturally tends to increase the size of every body that possesses it up to a certain self determined limit.
2. New organs arise in response to new and reiterated wants and to the changes produced by these wants or by efforts to meet them.
3. The development of organs and their functions is determined by the use of such organs.
4. All changes in organization are conserved by generation and transmitted to offspring.

The great assumption is that *acquired modifications* are inherited. These are induced by changes in environment which produce new wants, and, in animals, new habits to meet them. Plants are without true habits, but produce new races by changes in nutrition and growth. In all there is an inner striving for perfection. Lamarck gives many examples of the supposed effect of habit — as, the snail trying to feel objects in front of it develops tentacles on its head; or, the snake wishing to pass through narrow spaces becomes very elongated and loses its legs because they are not used. These ideas, opposed by the great Cuvier, attracted little attention during Lamarck's life, but later had considerable influence.

Geoffroy accepted Lamarck's theory of evolution, but differed from him in thinking the process to have been a series of jumps. He had done good work in the comparative anatomy of vertebrates, but when he endeavored in his *Philosophie anatomique*, 1818–22, and elsewhere, to stretch the plan of that group over the entire animal kingdom, his ideas became fanciful in the extreme. An attempt to show vertebrate analogies in the anatomy of Sepia (a swimming mollusc) brought on the famous debate with Cuvier at the Academy in 1830. Geoffroy was completely vanquished, and the doctrine of the four types of immutable species remained safe for three decades.

NATURAL THEOLOGY AND THE THEORY OF DESIGN

The victory of Cuvier was all the more decisive because it tended to confirm the popular belief in the theory of design. We have seen how at the end of the seventeenth century John Ray had drawn attention to the remarkable adaptations everywhere discoverable in nature, especially in plants and animals, and had suggested that these adaptations were sufficient to prove the existence of "design" in the universe — a powerful argument in favor of the Mosaic cosmogony. The same idea was urged more at length by others in the eighteenth century as an offset to the growing skepticism of the age, and especially by Joseph Butler (1692–1752), bishop of Durham, in his great work, *The Analogy of Religion, Natural and Revealed, to the Course and Constitution of Nature*, 1736; and by William Paley (1743–1805), archdeacon of Carlisle, in his famous *Natural Theology; or Evidence of the Existence and Attributes of the Diety collected from the Appearances of Nature*, 1802.

The work of Ray was greatly extended through a bequest of £8,000 by the eighth Earl of Bridgewater to the president of the Royal Society for the production of a treatise "On the Power, Wisdom, and Goodness of God as Manifested in the Creation." The result was the series of eight *Bridgewater Treatises*, 1833–37. Some of them were of high rank. The one on *Astronomy and General Physics* was by William Whewell, well known for his history of science. The *Geology and Mineralogy* by William Buckland and the *Habits and Instincts of Animals* by William Kirby are notable contributions to natural history, and *The Hand* by Sir Charles Bell is one of the classics of comparative anatomy.

But at the very time the *Bridgewater Treatises* were being published opposing views were coming to the fore, especially, as we shall see in the field of geology. A notable result was the appearance between 1842 and 1846 of a revolutionary work entitled *Vestiges of the Natural History of Creation*, by an anonymous author, which aroused intense interest in scientific circles and a storm of criticism from those who held to the old cosmogony. It is now known to have been written by Robert Chambers (1802–71), an Edinburgh publisher who preferred to remain unknown from fear of injuring his partners by bringing down upon them the wrath of critics for his heterodoxy. Chambers was an amateur geologist and in his *Vestiges* undertakes to treat the genesis of the earth on more rational and more natural principles than was possible by following the orthodox theory of special creation.

THE FOUNDATION OF MODERN EMBRYOLOGY

Modern embryology begins with the work of Heinrich Christian Pander (1794–1865) of Riga and Karl Ernst von Baer (1792–1876) of Esthonia, fellow students at Würzburg. Pander brought to light the work of Wolff (p. 362) and in a beautifully illustrated monograph on the embryology of the chick, 1817, was the first to distinguish the *germ-layers* from which all organs of the embryo are formed. Von Baer discovered the mammalian ovum in 1827. He found this minute egg in the ovary of a dog, inside the follicle (sac) that de Graaf (p. 295) supposed to be an egg. His *Entwickelungsgeschichte der Thiere*, 1818, 1837, created the science of embryology as a branch of comparative anatomy. Here he first described the notochord, the primitive axis of all vertebrate embryos; developed the germ-layer theory; and stated the important laws that the early stages in different groups of animals are more alike than the developed individuals, and that this similarity is farther back, the greater the difference in the adult forms.

We should mention also the important work in embryology and marine zoology of M. H. Rathke (1793–1860) of Danzig, the successor of von Baer at Königsberg. Among his discoveries are the fish-like gill-slits and their blood-vessels in embryos of birds and mammals. His detailed anatomical description of the Amphioxus is the basis for many more recent studies of this most primitive of vertebrate forms.

THE CELL-THEORY

A new era in the history of biology began with the announcement of the cell-theory in 1838. We have seen that Robert Hooke gave the name "*cell*" to the microscopic cavities in cork (p. 297). The first to base a general theory on minute structure was Xavier Bichat (1771–1802), anatomist, pathologist, and the first modern physiologist in Paris. In his treatise on membranes, 1799–1800, and his *Anatomie générale*, 1801–02, he analyzes the organs into elements of specific texture that he calls *tissues*, 21 kinds, each with its own vital property and liability to disease. Life, health, and disease are the properties of the tissues. He revolutionized pathology and is regarded as the founder of histology, the science of minute anatomy.

Bichat investigated by means of fine dissections, used chemical reagents, boiling, etc. He had no microscope. It was not until about 1835 that improved achromatic objectives with lenses of two kinds of glass cemented together with Canada balsam, and better design of stands, made the compound microscope an instrument of great importance for the advancement of biology.

Robert Brown (1773–1858), an eminent British botanist, made in 1801–05 extensive collections in Australia. He is known to physicists as the discoverer of "the Brownian movement" [1] and to biologists as the discoverer of the *nucleus of the cell*. In a paper on the fecundation of orchids and milkweeds, he reported in 1833 an observation of fundamental importance — that the "nucleus of the cell" is to be found generally in the cells of the various tissues of these and other plants.

The cell-theory was conceived in 1837 at a meeting between Schwann and M. J. Schleiden (1804–81) when they compared plant cells with cells in the embryonic notochord. In his preliminary announcement of 1838 Schwann says, "that all the varied forms in the animal tissues are nothing but transformed cells, that uniformity of structure is found throughout the animal kingdom, and that in consequence a cellular origin is common to all living things. All my work has authorized me to apply to animals as to plants, the doctrine of the individuality of cells."

The cell-theory, that animals and plants are communities of cells, each of which lives its own life and contributes to the life of the whole, was clearly stated by Schleiden in his paper on "Phytogenesis" in *Müller's Archiv* for 1838 and by Schwann in his *Microscopical Researches*, 1839. This great generalization, equaled in importance only by the theory of organic evolution, gave a firm basis for the unification of the biological sciences, and, with some modification, it remains valid today. Schleiden and Schwann regarded the cell-wall as the important part. But gradually it was seen that the living substance is the material within the cell to which Hugo von Mohl (1805–72) in 1846 gave the name *protoplasm* (borrowed from Purkinje), and in 1861 Max Schultze (1825–74) defined the cell as a mass of protoplasm containing a nucleus.

Between 1840 and 1860 much attention was given to the origin of cells. Schleiden thought that the nuclei arose by a sort of crystallization in cell-sap and then expanded into cells. In 1842, the Swiss botanist, Karl Nägeli (1817–91), described the cell-divisions in pollen formation with great care. Schwann had shown the egg to be a cell. Robert Remak (1815–65) proved, 1855, that when the egg divides into new cells, the

[1] A rapid, irregular oscillation of minute particles in liquid suspension.

division starts in the nucleus, and he maintained this to be universally true. Rudolf Virchow (1821–1902) also stated that every cell is derived from a preëxisting cell by division, and he recognized the nucleus as a living part of extraordinary importance for the maintenance and multiplication of the cell. He applied the theory to pathology; and his writings,

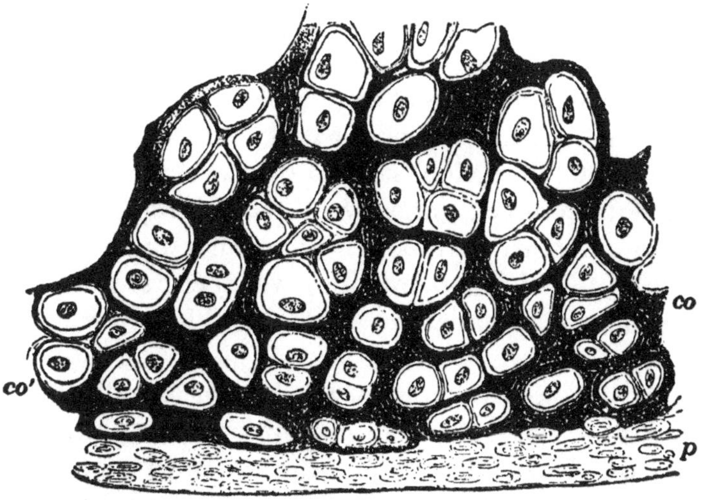

FIG. 59. — A SECTION OF GROWING CARTILAGE SHOWING ITS CELLS AND THEIR NUCLEI. *co*, calcified cartilage; *co'*, calcification beginning; *p*, pericondrium, a covering of fibrous tissue. × 350. After Virchow, *Cellularpathologie*, 1858, p. 370.

summarized in his *Cellularpathologie*, 1858, again revolutionized that subject and made Virchow the leading pathologist of the world.

PROGRESS IN BOTANY

By making morphology the key to classification, A. P. de Candolle (1778–1841) laid the foundation of systematic botany. Meanwhile, Konrad Sprengel (1750–1816) was making brilliant researches, 1812, on the reproduction of plants, showing cross-pollination to be the rule and the importance of insects in this process. At the same time, T. A. Knight (1759–1838) was investigating the movements of plants, and showed, 1806 and 1810, their relation to gravity, or centrifugal force, and to moisture.

With improved microscopes more workers were attracted to the internal structure of plants. Robert Brown in 1831 traced the pollen-tube as far as the embryo-sac; and the Italian mathematician, G. B. Amici, completed the story in 1842 and 1846, by demonstrating in the embryo-sac an ovum that was fertilized by a "fluid" from the tip of the pollen-tube. The sexual organs of ferns were discovered by Suminski in 1848. Of the many investigators of vegetable anatomy and histology between 1830 and 1850, von Mohl and Nägeli stand well above the rest. Von Mohl studied the development of the tissue elements from cells, and *The Vegetable Cell*, 1851, long remained a standard textbook. Nägeli not only studied the histology of the higher plants, but was the first to investigate the morphology and anatomy of Algae with a view to classification.

The most famous botanist of this period was Wilhelm Hofmeister (1824–77). He began in 1840 by tracing the origin of the embryo of flowering plants from the unfertilized ovum in the embryo-sac to the resting embryo in the seed, and confirmed the observations of Robert Brown and of Amici. Then followed a series of monographs, 1849–57, on various mosses, ferns, conifers, and their allies, and in all of them he demonstrated an alternation of a sexual with an asexual generation, each starting from a single cell. He thus established a unity of plan throughout the vegetable kingdom, broke down the barrier between the lower and the higher plants, and prepared the way for Darwin.

Modern palaeobotany was founded by Adolphe Brongniart (1801–76) in his *Histoire des végétaux*, 1828, in which the material was organized in so systematic a manner as to make him the highest authority on the subject. He divided the geological series into four periods showing progress from lower to higher plants. Upon this foundation an imposing superstructure was erected by Robert Goeppert in two great works, 1836 and 1841.

PROGRESS IN GEOLOGY

During the first half of the nineteenth century the work of many investigators built up much of our modern views of the origin of the rocks and of the surface features of the earth. At the beginning, the dispute between Neptunists and Vulcanists still raged, and the catastrophic theory prevailed. While Playfair was elaborating Hutton's views (p. 361) another close friend of Hutton, Sir James Hall (1761–1832), began his experimental researches in geology. He melted diabase and found the result to be glass — as the Neptunists claimed — if cooled quickly; but, if given time to cool very slowly, a crystalline rock was obtained. Lavas gave the same result. He also fused rock material under pressure; and he obtained a substance like marble by heating powdered chalk in sealed tubes. With layers of clay in compression machines, he showed that mountain folds would result from a lateral thrust under a vertical load.

Stratigraphic geology arose naturally in England and France, where strata lie in horizontal succession, are well exposed, and are rich in fossils. The work of Cuvier (p. 424) and his associate, the mineralogist, Alexandre Brongniart (1770–1847) between 1808 and 1811 made the Paris Basin famous. In England William Smith (1769–1839), a land surveyor, while making a series of levellings for a canal, discovered that the strata, seemingly flat, were really inclined at a gentle angle to the east. They terminated abruptly at the west, like so many "slices of bread and butter." In 1799 he published a geological map of the Bath region, the first geological map to show succession of strata (cf. p. 359). In 1815 Smith published his best known work, *A Geological Map of England and Wales* with a quarto memoir of fifty pages. In this he showed that strata could be identified by the fossils found in them and that these fossils occurred in no other strata, either below or above.

There now arose in geology one of those great generalizers so necessary in the advancement of any science. Charles Lyell

(1797–1875), thoroughly familiar with the work of those before him, accepted the uniformitarian point of view — that only those processes which are now active on the earth, and in the same manner and force, have acted throughout all past history. His great work, *Principles of Geology*, appeared in three successive volumes, 1830, 1832, and 1833, and was distinguished especially by the nature of the evidence adduced from fossiliferous deposits. Lyell's writings, logical, clear, and interesting, gave the death blow to the catastrophic doctrine, of which Cuvier was the last great exponent. The erosion of high mountains to sea-level, the deposition of many thousands of feet of rock from sand and mud implied an enormous length of time. This consideration of time as an essential factor in the earth's history prepared the way for Darwin's theory of evolution.

The succession of strata and their fossils was fairly well known down through the Coal Measures into the "Old Red Sandstone" of the Middle Palaeozoic when Roderick Murchison (1792–1871) and Adam Sedgwick (1785–1873) began in 1831 to study the older rocks in Wales and Shropshire. It is pleasant to think of this rugged country, formerly called Cambria, as giving its name to these basal Palaeozoic strata. The detailed study of still older rocks was begun by William Logan (1798–1875), the first director of the Geological Survey of Canada. He noted beneath the oldest Palaeozoic a vast thickness of crystalline rocks in which he found evidence of original bedding, and, convinced that they were highly altered sediments, inferred the former existence of a land from which these sediments could have been eroded.

It had been supposed that scattered superficial boulders and the unconsolidated sediment associated with them, "the drift," had been carried by Noah's flood or by floating ice from their obviously distant origin. It was Louis Agassiz (1807–73), already celebrated for his brilliant researches into the history of fossil fishes, who proved that the drift could not be due to a flood, but instead was due to glaciers. In 1837 he showed by the distribution of boulders from the Alps over the plains to the west and up the southeastern flanks of the Jura Mountains, that Alpine glaciers had extended to these districts. Fine straight lines, or striae, scratched upon solid rock surfaces extended in the same direction as the erratic boulders. Agassiz argued that much of Europe had been covered by glaciers. After coming to America in 1847, he showed that northern North America had also been overrun by glaciers. The surface deposits contain no fossils, and floating icebergs could not have made the long straight striae upon the smoothed rocks beneath, nor could the flood have stopped abruptly at Long Island, piling up a hilly ridge along its northern side. All other drift phenomena — drumlins, kettle holes, moraines, eskers — are explainable only on the glacial theory, which gradually won general acceptance. The theory of the permanency of continents and ocean basins was first definitely announced by J. D. Dana (1813–95) of Yale University in 1847, and this conception is held by the majority of geologists today.

Petrographical geology began with the careful and precise definitions of Werner (p. 360). But the structure of the finer rocks was unknown until Henry Sorby in Edinburgh discovered the inventions of William Nicol, who, utilizing Huygens's discovery of the polarization of light by Iceland spar, had described in 1829 the use of that substance in a prism for examining objects by light vibrating in only one plane; and had devised a method for making slices of hard substances thin enough to be transparent. With these and an improved compound microscope, Sorby was able to discover the minute structure and composition of rocks and to learn much regarding their origin. His *Microscopical Structure of Crystals*, 1858, did for geology what Schleiden and Schwann had done for biology twenty years earlier.

THE ANTIQUITY OF MAN

Stone implements and pottery had been found in several places apparently associated with bones of extinct mammals. But the stratified gravels in the Somme Valley at Abbeville first gave proof of the great antiquity of man, brought to light by the investigations of Casimir Picard (1806–41) and J. Boucher de Perthes (1788–1868). Picard proved that the flaked flints were true implements of a type distinct from the polished tools. His belief that they were contemporary with the strata in which they were found, was accepted by Boucher de Perthes, and he with great tenacity of purpose carried through the investigations that proved their association with the fossil elephant, rhinoceros, etc. His results, published in 1841 and 1847, were not believed, however, until confirmed by English scientists in 1859 and 1860, following the discovery of a primitive human skull in the Neanderthal near Düsseldorf (1857) and of human remains associated with the fossil cave-bear near Torquay in Devonshire (1858).

EVOLUTION BY NATURAL SELECTION

The year 1859 saw the beginning of a new era in biology. In that year Charles Darwin (1809–82) published *The Origin of Species by means of Natural Selection*, which revolutionized biology and furnished ultimately a new point of view for the whole range of science from atoms to galaxies.

When Darwin embarked as naturalist in the surveying expedition of H. M. S. "Beagle," 1831–36, he had the first volume of Lyell's *Principles of Geology*, from which he learned of the orderly sequence of geological events and of Lamarck's theory of evolution. In South America he was "deeply impressed by discovering in the Pampean formation great fossil animals covered with armour like that on existing armadillos; secondly by the manner in which allied animals replace one another in proceeding southward on the continent"; and, thirdly, in the Galapagos Archipelago, 500 miles west of South America, he was impressed by the South American character of the animals "and more especially by the manner in which they differ slightly on each island of the group; none of the islands appearing to be very ancient in a geological sense." These facts seemed to be explainable only on the supposition that species gradually become modified. But he could not accept Lamarck's method. He began to collect evidence in 1837, and he found that selection of variations was the main factor in the development of domestic races. Then in 1838 he read an *Essay on the Principles of Population*, 1803, by T. R. Malthus, who quoted Benjamin Franklin's remark (1751) that, "there is no bound to the prolific nature of plants or animals but what is made by their crowding and interfering with each other's means of subsistance." Malthus applied this to mankind, and Darwin perceived "that under these circumstances favorable variations would tend to be preserved and unfavorable ones destroyed. The result would be the formation of new species." Darwin called this process *natural selection.*

In the meantime Alfred Russel Wallace (1823–1913) in Borneo was making observations that led him to write in 1855, "Every species has come into existence coincident, both in time and space, with a pre-existing closely allied species." Later he also read Malthus, reached the same conclusion, and at once wrote out the theory and sent it to Darwin for publication. On the advice of friends, Darwin published an abstract of his own work with Wallace's paper in 1858. The *Origin of Species* appeared the next year and was sold out on the day of publication.

We cannot honor too highly the unselfish love of truth displayed by both men, and especially the ungrudging loyalty with which Wallace attributed the great discovery to Darwin. Darwin and Wallace were not the first to publish the idea of natural selection, but they were the first to present it as a theory of evolution in a way to carry conviction to scientific minds. Because it did carry conviction, it was attacked with great vigor on all sides, scientific and unscientific alike. It was defended with equal enthusiasm — especially by T. H. Huxley (1825–95) in England, by Ernst Haeckel (1834–1919) in Germany, and by Asa Gray (1810–88) in America.

MAN'S PLACE IN NATURE

The opposition to the theory of organic evolution arose largely from its implications as to human origins. In answer to this, Lyell gathered in his *Antiquity of Man*, 1863, all available evidence of prehistoric man, and in the same year Huxley's *Man's Place in Nature* gave anatomical evidence of the relation of the human species to the higher mammals. Haeckel in his *History of the Creation*, 1868, gave a special account of human evolution, and in his *Anthropogenie*, 1874, presented the evidence from morphology, embryology, and palaeontology. Darwin elaborated his own views in *The Descent of Man*, 1871, and added a new factor to his general theory — *sexual selection*, or success in mating — to account for secondary sexual characters, including ornamentation, so conspicuous in male birds, and horns, spurs, etc. used in fighting rivals.

THE DARWINIAN THEORY MODIFIED

Other factors, besides sexual selection, were added gradually to the theory of organic evolution. One objection to natural selection was that it failed to account for the origin of variations. To meet this, Herbert Spencer (1820–1903), who coined the phrase "survival of the fittest" and adopted evolution as the basis of his *Synthetic Philosophy*, revived Lamarck's theory of *inheritance of acquired characters*. For a time, this met with considerable favor, especially among palaeontologists. Palaeontologists also claimed the discovery of evidence of variation in a definite direction. Whether this "*orthogenic evolution*" is effective is an open question.

Moritz Wagner (1813–87), having studied the geographical distribution of animals, proposed in 1868 the important factor of migration and *isolation* to explain the persistence of new varieties that otherwise might be swamped by intercrossing. He was ably seconded by the Rev. John T. Gulick (1832–1923) as the result of beautiful studies, begun in 1872, on the segregation of terrestrial snails in the Hawaiian Islands. J. G. Romanes hailed the principle of isolation as one of three basic principles of organic evolution, the others being heredity and variation. As a special case of isolation, Romanes suggested in 1887 *physiological selection*, produced by variations in the reproductive apparatus causing sterility between races of the same species.

William Bateson in 1894 gathered a great collection of examples of large variations, which seemed to show that evolution might progress by jumps, instead of gradually by the accumulation of small variations as suggested by Darwin. And at the close of the century Hugo de Vries stated his Mutation Theory, to the effect that ordinary small variations are merely fluctuations in form influenced by external conditions and not inherited, and that the materials for selection are heritable *mutations*, varieties that appear suddenly fully formed, like Minerva from the head of Jove.

NEW DIRECTIONS OF BIOLOGICAL RESEARCH

The doctrine of organic evolution cast a new light on every branch of biology. It stimulated an immense amount of investigation to test the theory in every direction. Classifications were revised to show, so far as possible, community of descent. Phylogeny, the history of the race, became a favorite pursuit. New meaning was given to geographical distribution of animals and plants, centers of distribution were mapped, and it was found that while some species have a cosmopolitan range, most species are more restricted in range than the genera to which they belong, and genera, in turn, have smaller ranges than their families and orders. Barriers were found to be important factors in distribution. In a large way, this was shown by comparing the inhabitants of the northern continents with those of the more isolated land masses of the southern hemisphere.

Special cases of adaptation to conditions of life formed a field of study. Wallace in Brazil and the Malay Archipelago noted many cases of protective coloring. H. W. Bates (1825–92) in 1861 described mimicry in certain butterflies, the edible ones having the appearance of other species that are distasteful to insectivorous animals.

Palaeontology, especially of the higher animals, became a

branch of biology with special emphasis on phylogeny. It included comparative anatomy, as well as the distribution of the fossil remains in time and place. The opponents of evolution had denied the existence of "connecting links." They were supplied by the palaeontologists in considerable numbers. E. D. Cope (1840–97) and O. C. Marsh (1831–99) opened up a vast store of fossils in the "badlands" of the western United States. There Marsh discovered birds with teeth. The finding of the feathered and long-tailed Archaeopteryx in Solenhofen, Bavaria, completed the link between reptiles and birds. The tracing of the series of fossil forms leading up to present-day camels and horses were notable achievements of the nineteenth century.

Embryology and comparative anatomy now became inseparably linked in the search for phylogenies. J. F. Meckel (1781–1833) had stated in 1811 that the embryonic stages of higher animals resemble lower forms. Von Baer had corrected this by pointing out that only the embryonic stages are alike, and, although Agassiz never accepted evolution, he found in embryology the most trustworthy standard of relative rank and affinity among animals. Fritz Müller in Brazil sought proof of the Darwinian theory by comparing the different types of development in closely related forms of Crustacea (*Für Darwin*, 1864). He held the development of the individual to be an "historical document." This view was taken up with enthusiasm by Haeckel and became his *biogenetic law* — that ontogeny (individual development) recapitulates phylogeny. Huxley in 1849 had compared the two cellular layers in the body-wall of jelly-fishes to the two germ-layers, ectoderm and endoderm, in early vertebrate embryos; and the brilliant researches of Alexander Kowalewsky (1840–1901) on the embryology of Amphioxus, 1866–77, in Haeckel's laboratory demonstrated the existence of a cup-shaped embryonic stage with two layers of cells. Haeckel named this stage the *gastrula* (little stomach) and believed that it represents an adult ancestral form, Gastrea, common to all multicellular animals. The germ-layer theory furnished the criteria for homologies, which were accepted as evidence of community of descent.

The realization that primitive forms of life are generally found in the sea, led to great interest in marine biology. Seaside laboratories were established, first at Naples and in Massachusetts; and great explorations, notably the "Challenger" Expedition of 1872–76 added greatly to the knowledge of the variety and distribution of marine life and of the physical geography of the oceans.

PARASITOLOGY

Darwin's studies on evolution were destined to illuminate the specific adaptations and complex life-histories brought to light, chiefly in Germany, by the study of parasites. A *parasite* is any organism that dwells upon or within, and at the expense of, another organism, its *host*. Many parasites have life-histories similar to related free-living forms, as Malpighi demonstrated in the case of the mistletoe. But the origins of many others are so mysterious as to lend color to the belief, held until the middle of the nineteenth century, that they arise by spontaneous generation in the host.

The key to the mystery was supplied in 1841 by J. J. S. Steenstrup (1813–97). In his treatise *On the Alternation of Generations*, he called attention to a remarkable discovery announced in 1819 by Adelbert von Chamisso. On a voyage around the world Chamisso had seen on the surface of the sea a transparent animal, Salpa, in two forms — solitary individuals and individuals united in long chains, and he found that one form is the parent of the other — the daughter is not like its mother, but is like its grandmother. Steenstrup saw that jelly-fishes and certain parasitic flatworms, called flukes, have similar life-histories, a sexual alternating with an asexual generation.

From observations on fish-eating birds C. Th. von Siebold (1804–84) had inferred an *alternation of hosts*, from fish to bird, in certain tapeworms; and in 1848 he suggested that bladderworms are larval tapeworms. This was tested in 1851, when Friedrich Küchenmeister (1821–90) introduced the *experimental method* into parasitology. He obtained intestinal tapeworms in a dog by feeding rabbit's muscle containing bladderworms. Conversely, two years later, he caused bladderworms to develop in a sheep's brain by feeding eggs of another kind of tapeworm from a dog. These experiments, because by a simple method, easy to control and to repeat, created a great sensation.

The study of parasites was then undertaken by many zoologists, the most prominent of whom was Rudolf Leuckart (1822–98). He had in 1847 made clear the distinction between sexual and asexual reproduction (by fission or budding), and in 1851 confirmed and extended Steenstrup's work on alternation of generations. His great treatise *The Parasites of Man* first appeared in 1863 and mainly contained his own work — notably, the life-history of the liver-fluke of the sheep and of the pork tapeworm of man. Students flocked to his laboratory in Leipzig from all parts of the world.

Many parasitic Fungi that infest plants were described by Ferdinand Cohn (1828–98). In 1854 he insisted that the Bacteria are plants related to the lower Fungi, and his classification of Bacteria formed the basis of the system in vogue until well within the twentieth century. Anton de Bary (1831–88) did fundamental work in tracing the life-histories of Algae and, especially, of Fungi from spore to spore. His method of culture and the novelty and importance of his results place him in the first rank among botanists of the nineteenth century. His clear demonstration of the relation of parasite to host furnished conclusive evidence that spontaneous generation of Fungi does not occur. He explained epidemic diseases, and his work on the Potato Blight is a classic (1861). This led to his discovery of alternation of generations in the Wheat Rust, 1863. De Bary's celebrated *Morphology and Physiology of the Fungi*, 1866, revolutionized that branch of botany and entitles him to rank as the founder of modern mycology.

BIOGENESIS: THE GERM-THEORY OF FERMENTATION, PUTREFACTION, AND DISEASE

When the controversy over the origin of species began in 1859, another was raging in France concerning the origin of life. The central figure in this conflict was Louis Pasteur (1822–95), originally a chemist, the founder of modern bacteriology and of the germ-theory of disease.

Pasteur at Lille was continuing in 1854 his microscopical studies of the crystals, especially of the salts of tartaric acid, a by-product of fermentation. A local distillery was having trouble with lactic acid in its fermentations for alcohol. Pasteur found the cause of the trouble to be a living ferment that produced lactic acid, as in sour milk, 1857. This brought on a controversy with Liebig, who had synthesized lactic acid in 1850, and regarded its formation and also alcoholic fermentation as purely chemical processes. Pasteur was called to Paris in 1857. His answer to Liebig was the production of alcohol in a solution of pure sugar and mineral salts by adding a little yeast — a culture, as proved by Schwann in 1837, of microscopic plants.

Pasteur was attacked by others who, in spite of the findings of the parasitologists, were attempting to revive the doctrine of spontaneous generation. In a note to the Academy of Sciences in December, 1858, F. A. Pouchet of Rouen claimed to have obtained micro-organisms by spontaneous generation in solutions exposed only to pure oxygen and nitrogen or to oxygen alone, and denied the validity of Schwann's experiments of 1837 showing the presence of putrefactive organisms in the air. To test this result Pasteur invented the swan-neck flask that would admit air and exclude dust from a solution boiled in the flask. Pasteur proved the presence of organisms in the air; and was led by the criticisms of A. C. Bastian to the discovery of other sources of bacteria and of their spores that resist boiling. The conclusion that spontaneous generation does not occur, called by Huxley the doctrine of *Biogenesis*, was confirmed by John Tyndall (1820–93) with his dust-free box.

The manufacture of vinegar in Orleans was reformed through Pasteur's study of the process. He found acetic acid to be produced by specific bacteria, contrary to the opinion of Liebig, who held the process to be simple oxidation. Pasteur's investigations of the "diseases" of wine and of beer were equally fruitful of practical results, and led to the important law of *specific ferments*, foreshadowed in his earlier work — that each kind of fermentation is due to a particular kind of organism. His method for preserving bottled wine by heating to 55° C is the "pasteurization" now used for the protection of milk.

While engaged in these investigations, Pasteur was sent to Alais in the south of France to study the disease then devastating the silk industry. Much of the years 1865 to 1870 was spent on this problem. Pasteur found the silkworms afflicted not by one disease, but by two caused by different organisms, and thus he laid the foundation of the *germ-theory of disease*.

The theory was firmly established by his work on anthrax, a disastrous disease of cattle communicable to man. Several observers had seen the rod-shaped organism (bacillus) in the blood of animals dead of the disease, and C. J. Davaine (1812–82), also in Paris, had shown these organisms to be the sole cause of the disease. But his experiments (1863–73) were not convincing. In 1876 Robert Koch (1843–1910) at Posen demonstrated the formation of resistant spores, and that they form when an animal dead of anthrax is buried. He reared several generations of the bacilli from spores in drops of serum, and proved this bacillus capable of inducing anthrax in small animals when injected or when mixed with their food. Pasteur's method was to find a medium in which the bacilli would thrive. Urine, made neutral, provided a growth so abundant as to permit experiments on a scale never before attempted. With pure cultures he had undoubted proof that these bacilli are the sole cause of anthrax, and he demonstrated a source of the contamination of pastures when he collected earthworm castings over sheeps' graves and with them produced anthrax by inoculation into guinea pigs.

But that was not all. In the course of studies of chicken cholera begun in 1879, Pasteur found by chance that inoculation with an old culture made chickens *immune* to a fresh inoculation that promptly would kill other chickens. This led to a method of attenuating anthrax cultures for use as a vaccine, and to the famous experiment of May, 1881, when fifty sheep were inoculated with a virulent culture. Half of them had been vaccinated. They survived, the others died. Vaccination against anthrax quickly became a custom.

Pasteur's greatest triumph was the conquest of rabies. He did not find the organism. But he did find that a portion of a the spinal cord from a rabid animal, if exposed to pure dry air, gradually loses its virulence and can be used to immunize another animal. Then came the crucial test on a boy bitten by a mad dog, July 6–16, 1885. The boy survived, as did many others later, and a grateful public dedicated the Institut Pasteur in 1888.

Koch was called to Berlin in 1880. His greatest contribution to bacteriology was a brilliantly simple and effective technique. In 1881 he introduced the use of a solid medium (gelatin) for making pure cultures, he fixed bacteria to a glass slide by heat, and made them visible under the microscope by staining with the aniline dyes first used by Carl Weigert in 1875. His discovery of the tubercle bacillus, demonstrated in London in 1881, with the aid of the differential stain prepared for the purpose by Paul Ehrlich, created a sensation. His laboratory became the Mecca of bacteriologists, and discoveries of many other disease germs followed rapidly. The discovery of the cause of tuberculosis soon led to the demonstration by Dr. E. L. Trudeau (1848–1915) at Saranac Lake, N. Y., that the disease is curable.

THE REVOLUTION IN SURGERY

The practice of surgery was revolutionized during the nineteenth century by three discoveries — anaesthesia, asepsis, and X-rays. The second of these resulted directly from the work of Pasteur.

The first surgical operation under ether was performed at Athens, Ga., by Dr. C. W. Long in 1842, but was not published until much later. In 1844 Horace Wells at Hartford, Conn., began to use nitrous oxide to render patients unconscious while extracting teeth, but he failed in a surgical operation. Painless surgery was given to the world through the enterprise of T. G. Morton (1819–68), one-time partner of Wells, and through the courage of Dr. John Collins Warren (1778–1856), a surgeon in Boston. Morton, then a student at the Harvard Medical School, by experiments on animals had convinced himself that ether could safely be used, and Warren permitted him to etherize a patient during an operation at the Massachusetts General Hospital, October 16, 1846. It was a complete success. The beneficent effect of ether was named *anaesthesia* by Dr. Oliver Wendell Holmes. Its use spread rapidly. In 1847 Sir J. Y. Simpson adopted chloroform, first prepared by Liebig in 1831. Anaesthesia abolished the pain of operation and gave time for procedures hitherto impossible.

Puerperal fever killed many women in childbirth; a broken skin usually resulted in suppuration, fever, and frequently in death. Dr. O. W. Holmes in Boston in 1843 and Ignaz Semmelweis in Vienna a little later pointed out that puerperal fever was due to a contagion on the hands of the operators. Effective measures to prevent these accidents of obstetrics and surgery came from the insight and resourcefulness of Joseph Lister (later Lord Lister, 1827–1912). At Glasgow he began to study inflammation (1853) and by its odor detected putrefaction (*sepsis*) in suppurating wounds. Nothing could be done about it according to Liebig's theory of oxidation, but Pasteur's studies on spontaneous generation about 1860 pointed the way. When Lister heard that carbonic acid had a deodorizing effect upon sewage, he adopted it for use on dressings or by spraying the site of operation to kill or exclude germs from the air. Later experiments by himself and others showed that air-borne germs and putrefaction were not the chief causes of surgical troubles, and attention was directed toward other sources of infection. Steam-sterilization of dressings was introduced by E. von Bergmann of Berlin in 1886, and in 1890 W. S. Halstead, then in New York, adopted rubber gloves. Lister in 1890 demonstrated that surgical technique could be made simpler and more effective by disregarding air-borne germs and relying upon sterilization of hands and apparatus. *Asepsis* (freedom from germs) was recognized as indispensable for success. Hospitals ceased being unhealthy, and operations which had been regarded from time immemorial as unjustifiable were adopted with complete safety.[1]

As a result of purely physical inquiry, W. C. Röntgen at Würzburg announced in January, 1896, the discovery of X-rays, an event of prime importance in twentieth century physics (p. 405). The astonishing ability of these rays to penetrate some substances opaque to ordinary light was recognized at once as capable of furnishing a means of diagnosis of the highest value in medicine and surgery. The X-rays make visible internal conditions otherwise difficult or impossible to ascertain. Among the first to investigate this application were Oliver Lodge and Herbert Jackson in England and F. H. Williams in Boston, Mass.

CARRIERS AND VECTORS OF DISEASE

Davaine made experiments to show that flies may distribute germs of anthrax; and in 1891 M. B. Waite proved that bacteria causing "fire blight" of pears are carried by bees from diseased to healthy trees. Such *carriers* of disease play an important rôle, but a simple one, compared to the *vectors*, which are parasites in which other parasites of the same host must dwell to complete their life-cycle. This mode of transmission was discovered in China by Patrick Manson (1844–1922), who in 1884 showed that a minute worm in the blood of persons suffering from elephantiasis is swallowed by a mos-

[1] For a fuller and very interesting account, read "Address by the President," Sir Joseph Lister (Liverpool Meeting B.A.A.S.) *Science* **4**, 409–429, Sept. 25, 1896.

quito and undergoes development in its body. He did not follow it further.

The first account of the transmission of a parasitic protozoan (unicellular animal) by a blood sucking parasite appeared as the first *Bulletin* of the U. S. Bureau of Animal Industry, in 1893. This great classic of parasitology describes a detailed investigation of the Texas fever of cattle by Theobald Smith (1858–1934). He found the organism causing the disease to be a protozoan living in the red blood-cells, found the vector to be a tick, and demonstrated the extraordinary life-cycles of both organisms.

These results were unknown, however, to Ronald Ross (1857–1932) in India when in 1895 he began an investigation of malaria that was to promote human welfare throughout the world. The disease derived its name from the idea that it was caused by bad air from swamps. This *miasma theory* was dispelled by Alphonse Laveran (1845–1922) when in 1880 he discovered an organism in the red blood-cells of patients and proved it to be the cause of one type of the disease. Italian investigators soon found the organisms of the two other types of human malaria and recognized them as Protozoa. How these parasites are transmitted from man to man was the mystery ultimately solved by Ross.

The mosquito theory was suggested by Manson. But Ross encountered unexpected difficulties owing to the presence of several kinds of bird malaria and the inability of all except one kind of mosquito to transmit human malaria. An amazing phase of the life history of malarial parasites occurs in the stomach of the mosquito. It is a process of fertilization like the union of egg and sperm in the higher animals. The discovery of this process by W. G. MacCallum of Baltimore (1897) furnished the clue that enabled Ross to announce, July 25, 1898, the discovery of the sequence of stages in the transmission of malaria from bird to bird through the mosquito. With this knowledge the Italians quickly solved the problem of human malaria. A sequel was the discovery by an American commission in Cuba in 1900 under the direction of Walter Reed, that a certain common tropical mosquito transmits the unknown virus of yellow fever — a discovery that has saved countless lives.

CYTOLOGY AND GENETICS

The statistical method was introduced by Francis Galton (1822–1911) for the study of heredity, a prime factor in organic evolution. With large numbers of observations, he applied the theory of probability, as Quetelet had done (1835) for human populations, and adduced his *Law of Ancestral Heredity* — that, on the average, of the total characteristics of the child, one half comes from the parents, one quarter from the grandparents, one eighth from the great-grandparents, and so on. Karl Pearson (1857–1936) continued this line of investigation with mathematical precision, and founded the important science of Biometry. It should be added that his journal, *Biometrika*, has had a very profound influence, not only in genetics, but upon all modern statistics and its applications in other fields of research.

In 1875 Darwin offered "pangenesis" as a possible explanation of inheritance of acquired characters. This was a revival of the oldest theory of heredity, the theory of Democritus that the germ of a new individual receives organized contributions from all parts of the parent.

The year 1875, however, was one of remarkable progress in biology. Improvements in microscopes and in methods of cutting and staining sections were attracting many workers to the study of cells, *cytology*, especially the germ-cells; and in that year they began a series of epoch-marking discoveries that proved fatal to Darwin's hypothesis. This series of discoveries, mostly made in Germany, proved that two fundamental processes of reproduction are essentially alike in animals and plants. These processes are (a) cell-division, and (b) the fertilization of the egg. It was found that the division of a cell is preceded by the division of its nucleus, and this usually is accomplished by a process called karyokinesis or mitosis. In *mitosis* a part of the nucleus that is easily colored by certain dyes becomes divided into bodies later called *chromosomes* (Waldeyer, 1888), and Walther Flemming discovered (1882) that each chromosome splits lengthwise and the halves separate to each daughter-cell (Fig. 60). Thus the chromosomes, the nucleus, and the cell each divide in turn.

It had been known since first seen by Oskar Hertwig and by E. van Beneden in 1875, that fertilization essentially consists in the fusion of the egg nucleus with the nucleus of the

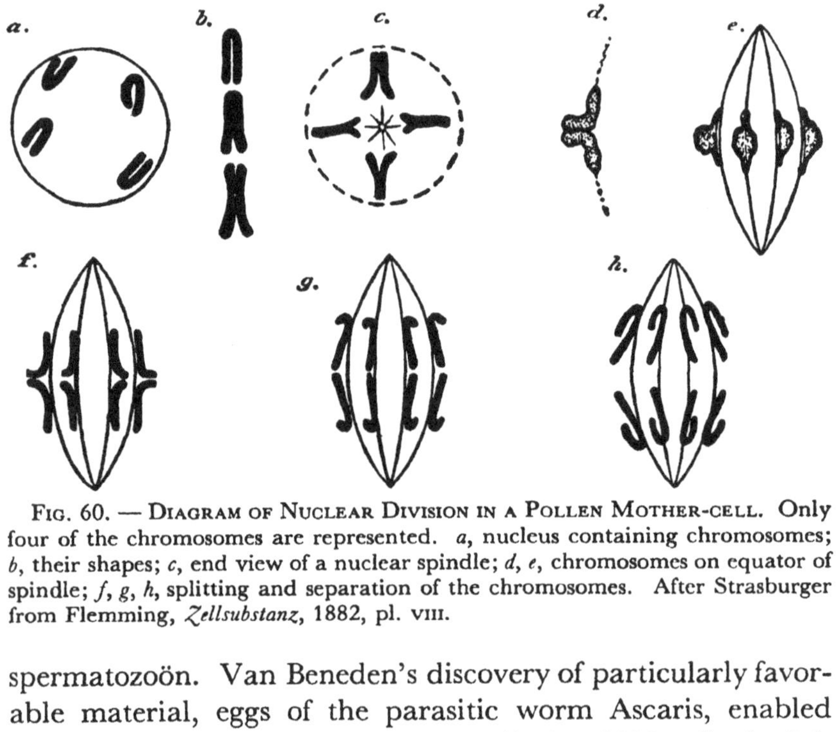

FIG. 60. — DIAGRAM OF NUCLEAR DIVISION IN A POLLEN MOTHER-CELL. Only four of the chromosomes are represented. *a*, nucleus containing chromosomes; *b*, their shapes; *c*, end view of a nuclear spindle; *d*, *e*, chromosomes on equator of spindle; *f*, *g*, *h*, splitting and separation of the chromosomes. After Strasburger from Flemming, *Zellsubstanz*, 1882, pl. VIII.

spermatozoön. Van Beneden's discovery of particularly favorable material, eggs of the parasitic worm Ascaris, enabled him to announce the *law of equal contribution*, 1883. In the fertilized egg of this worm, as in unripe eggs and sperm mother-cells (1884), van Beneden found four chromosomes. He showed how by cell-division the number is *reduced* before fertilization to two in the ripe egg; and to two in the spermatozoön, which on entering the egg forms a "male pronucleus." When this fuses with the "female pronucleus" the four chromosomes remain distinct and split during cell-division, the new embryonic cells thus receiving their chromosomes equally from each parent. This epoch-marking discovery laid the foundation for a scientific theory of heredity.

Likewise, in 1883 August Weismann (1834–1914) began the series of essays in which he formulated his theory of heredity, the *continuity of the germ-plasm.* Beginning with a total denial of any inheritance of acquired characters, he located the physical basis of heredity in the chromosomes, and postulated their individuality and continuity throughout the life-cycle. In 1887 Weismann emphasized the theoretical importance of the "reducing division" and defined it as a kind

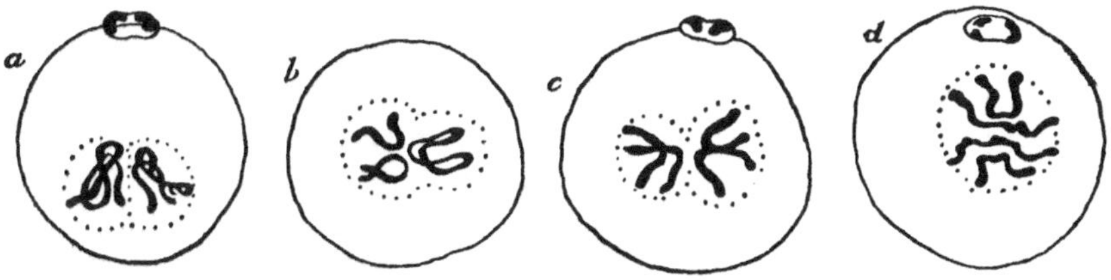

FIG. 61. — THE LAW OF EQUAL CONTRIBUTION. *a*, fertilized egg of Ascaris, male and female pronuclei in contact, two chromosomes in each; *b*, *c*, fusion of the pronuclei; *d*, four chromosomes together on the spindle for the first cleavage of the egg (cf. Fig. 60, *c* and *d*). × *c*. 500. After van Beneden, *Archives de Biologie*, 4, pl. 19 *bis*, 1883.

of karyokinesis in which undivided chromosomes are separated into two groups, each of which forms one of the two daughter nuclei. Many investigators took up the problem, but it required the discovery of *synapsis* by J. E. S. Moore in 1895 to explain the mechanism. This became clear in 1900 when the brilliant work of W. S. Sutton, at McClung's laboratory in Kansas, disclosed germ-cells containing *pairs* of chromosomes distinguishable by size and shape. The members of each pair, one apparently derived from each parent, join together at synapsis so that at reduction they can be drawn apart, one to each daughter-cell, in such a way that each ripe germ-cell, or *gamete*, receives one set of single chromosomes. The paired condition will be restored by the union of two gametes in the fertilized egg, which by ordinary cell-division will form the body and germ-cells of the offspring.

Thus the stage was set for a great event of the closing year of the century — the discovery of Mendel's theory of heredity. This was announced first by de Vries and soon after by two other botanists. From independent experiments in plant breeding they had reached identical conclusions; and then found their results anticipated in a memoir, a model of orderly and clear exposition, hidden away in libraries since its publication in 1866.

This memoir was the outcome of some eight years of experimental plant breeding in the monastery garden at Brünn in Moravia (now *Brno,* Czechoslovakia) by Gregor Mendel (1822–84). Superior to his predecessors in originality and insight — he selected varieties of the *same* species differing only in one character (as color or shape of seed) and thus capable of giving a definite result. When he crossed garden peas having a pair of contrasted characters, the first hybrid generation were all alike, showing the "dominant" character, e.g., yellow or round. The opposite character, green or wrinkled, Mendel called "recessive" because it reappeared in offspring produced by self-pollination of the hybrids — on the average in the proportion, 3D : 1R, the fundamental *Mendelian ratio.* Continuing his careful experimental and statistical work through succeeding generations, Mendel showed that dominant and recessive characters reappear according to simple numerical laws, and that these hold true in the presence of several pairs of contrasting characters. His greatest contribution was the theory of *segregation* — that the factors for any pair of contrasting characters are always in separate gametes, of approximately equal number, which unite at random in fertilization. If this be granted, as Mendel pointed out, the facts of Mendelian heredity, as then known, accord with the theory of probability.

Any gamete, then, may be supposed to carry a single set of the Mendelian factors, *and* a single set of the chromosomes, characteristic of its variety or species. Observation and theory harmonize; and the brilliant researches that followed the republication of Mendel's paper by William Bateson in 1902 brought to light many new aspects of cytology and genetics during the first quarter of the twentieth century.

VITAMINS AND HORMONES

Science is a community of more or less related disciplines differing in content and age. A new discipline is born when a gradually developing body of fact is fertilized by a new idea or a new method. We have just witnessed the birth of twentieth century genetics from the union of Mendelism with nineteenth century cytology. A similar union of pathology and physiology led to the appearance of "vitaminics" and of endocrinology in the first quarter of the twentieth century. Both of these branches of physiological medicine relate to the internal medium of the body as affected by materials mainly derived from the food or by substances produced within the body.

Scurvy, long known as a cause of disability and death among sailors and soldiers, was found in the eighteenth century to be curable by adding fruits, etc., to the diet; and in 1804 lime juice was made compulsory in the rations of the British navy. In the second half of the nineteenth century it became known that rickets, which was deforming so many children, could be cured by giving cod-liver oil. In 1882 Admiral Takaki banished beri-beri from the Japanese navy by adding vegetables and meat to the usual diet of polished rice. The first suggestion that these diseases are due to a lack of substances *accessory* to metabolism came from investigations made by C. Eijkman in Java during the years 1890 to 1906. Eijkman discovered beri-beri in hens fed on milled rice; and he cured it with an *extract* of rice bran.

In the meantime the study of nutrition begun by Magendie was carried on by many physiologists. They classified chemically the constituents of a complete diet and calculated their values in calories. But in 1888 it was found that on a complete diet, when *highly purified*, the animals died; and it was inferred that natural foods contain very small quantities of some unknown substance essential to life.

In 1912 F. G. Hopkins by exact quantitative methods proved this to be true, and that it was the want of the unknown substance, not loss of appetite, that resulted in the death of experimental animals. In the same year Casimir Funk, a Polish biochemist working in London, tried to isolate Eijkman's anti-beri-beri substance, obtained a crystalline material that would cure the disease in birds, and named it "anti-beri-beri vitamine." He predicted the existence of three other vitamines preventive of scurvy, rickets, and pellagra, respectively, and founded the *vitamin*[1] *theory*. The great interest aroused by the papers of Hopkins and of Funk culminated in the discovery by E. V. McCollum and M. Davis that the unknown food factors and vitamins were the same (1913 and 1915). They showed that there were at least two accessory food factors, called A and B, both necessary for growth; and that B would prevent beri-beri. Soon the number of known vitamins was increased to five, others have been added, and this new knowledge has been applied in preventive medicine, agriculture, and food technology with increasing benefit.

Johannes Müller's distinction between secretion and excretion left out of account the ductless glands, long known anatomically, but regarded as of little use. Now they are the chief object of study in the science of Endocrinology.

For centuries there had been a vague idea that each organ might contribute something to the blood. But it was Claude Bernard's work on glycogen (p. 423) that focused attention on this subject. At the same time Thomas Addison (1793–1860) was preparing his famous monograph (1855) on a disease he found associated with a pathological condition of the suprarenal bodies (ductless glands attached to the kidneys). This was followed up by C.-E. Brown-Séquard (1817–94), who showed by experiments on animals, 1856, that Addison's disease is due to the want of something normally supplied to the blood by the suprarenals. Others took up the

[1] Note the change in spelling after it was found that Funk had erred in classifying vitamins as amines.

study of thyroid deficiency, and the curative methods adopted about 1892 were dramatic in their success. Brown-Séquard's epoch-marking papers of 1889 and 1891 on testicular and pituitary extracts stimulated enormous activity along these lines.

The result was much confusion of theory until two physiologists, W. M. Bayliss and E. H. Starling announced the doctrine of the hormones in 1902. The hormones, chemical messengers, were revealed as specific chemical substances contributed to the internal medium, blood and lymph, by definite organs, and controlling the growth or activities of other parts of the body. This humoral system of coördination, supplementing the nervous system, at once became the subject of intensive study. Many hormones were found, and a considerable number have been isolated. Not all are products of ductless glands, so "endocrin gland" was adopted to designate any structure producing an internal secretion.

The interest in endocrinology has spread during the twentieth century through all fields of biology and to all branches of clinical medicine, vastly enlarging biological horizons and conferring untold benefits upon mankind.

SCIENCE IN THE DAWN OF THE TWENTIETH CENTURY

At the beginning of the nineteenth century, science as such had no existence either as a branch of learning or as a special discipline — still less as a preparation for practical life. Mathematics, highly esteemed largely because of its ancient origin and associations, and natural philosophy, had a limited recognition; but the term science meant as yet hardly more than knowledge or learning. The eighteenth century had, however, sowed broadcast the seeds of science and the nineteenth soon began to reap the harvest. Before 1850 scientific schools as distinct from others had been founded both within and without the older colleges and universities. New associations and academies for the advancement or promotion of science soon sprang up; science courses appeared in some of the public schools; funds for scientific research began to be provided; and thousands of eager and enthusiastic students began to prefer science, and especially applied science, to the older "classical" learning. Meantime, the marvellous achievements of invention and of industry had caught and fixed public interest and attention, so that by the opening of the twentieth century, no branch of learning stood in higher favor than science, either for its own sake or as a preparation for useful service in contemporary life.

As we look back over the nineteenth century, we see a vast growth in the body of science. With each new discovery a host of students arose to test its implications. As new methods and better apparatus were made available, more workers entered the field and specialization increased. Progress became more and more a gradual growth of ideas fed by small increments of knowledge, until, toward the close of the century, there was hardly any major discovery that could be attributed to any one man; advance in theory more frequently depended on critical insight and ability to generalize from the works of other investigators.

The master keys of science, now everywhere employed for unlocking the problems of the cosmos, are: first, *the principles of mathematics*, which admit mankind into the mysteries of the relations of number and space — the abstract skeleton of science — and second, *the principles of evolution and of energy*, which reveal some, at least, of the secrets of form and of function, not only of the earth and of plants and animals, but of the heavens; something of the prodigious forces of the universe and their orderly behavior; something of that apparently infinite and eternal energy which, while forever changing, is never lost; something, though as yet but little, of the nature and the processes of life.

REFERENCES FOR READING

Compton, Piers, *The Genius of Louis Pasteur*, 1932.
Darwin, Charles, *Voyage of H. M. S. "Beagle,"* 1860.
——, *Life and Letters*, ed. F. Darwin, 1893.
Drachman, J. M., *Studies in the Literature of Science*, 1930.
Duclaux, Emile, *Pasteur, the History of a Mind*, 1920.

FRANKLIN, K. J., *Short History of Physiology*, 1933, Chapter IX.
FULTON, J. F., *Physiology* (Clio Medica V.), 1931, pp. 60–115.
GREEN, J. R., *History of Botany 1860–1900*, 1909.
GUMPERT, MARTIN, *Trail-Blazers of Science*, 1936.
HARRIS, L. J., *Vitamins in Theory and Practice*, 1935.
HARVEY-GIBSON, R. J., *Outlines of the History of Botany*, 1919.
HERDMAN, SIR W. A., *Founders of Oceanography*, 1923.
HYLANDER, C. J., *American Scientists*, 1935.
LOCY, W. A., *Biology and its Makers*, 1908, Chapters VII–XX.
MERRILL, G. P., *First Hundred Years of American Geology*, 1924.
NEWSHOLME, SIR ARTHUR, *Story of Modern Preventive Medicine*, 1929.
OLMSTED, J. M. D., *Claude Bernard, Physiologist*, 1938.
POULTON, E. B., *Charles Darwin and the Theory of Natural Selection*, 1896.
RUSSELL, E. S., *Form and Function*, 1916.
SIGERIST, H. E., *The Great Doctors*, 1933.
YOUMANS, W. J., *Pioneers of Science in America*, 1896.
WALKER, M. E. M., *Pioneers of Public Health*, 1930.

From previous lists: GEIKIE, *Founders*, pp. 317–473; LOCY, *Growth Biol.*, Ch. XV–XX; LONG, *Path.*, Ch. VI–XII; MERZ, *Hist.*, II, pp. 200–464; SINGER, *Hist. Med.*, pp. 186–362; *Living Things*, pp. 192–568.

Appendix A

Some Inventions of the Eighteenth and Nineteenth Centuries. Applied Science and Engineering

He who seeks for immediate practical use in the pursuit of science, may be reasonably sure that he will seek in vain. Complete knowledge and complete understanding of the action of the forces of nature and of the mind, is the only thing that science can aim at. The individual investigator must find his reward in the joy of new discoveries . . . in the consciousness of having contributed to the growing capital of knowledge. . . . Who could have imagined, when Galvani observed the twitching of the frog muscles as he brought various metals in contact with them, that eighty years later Europe would be overspun with wires which transmit messages from Madrid to St. Petersburg with the rapidity of lightning, by means of the same principle whose first manifestations this anatomist then observed. — HELMHOLTZ.

The place of inventions in the history of science is hard to define. Conditioned as they doubtless are by a favorable environment — at least for survival — they do not always obviously arise as a direct or logical consequence of preceding discoveries, or even of known principles, but seem sometimes to spring almost *de novo* from the brain of the inventor. And yet such an origin is probably more apparent than real. The steam-engine could hardly have come from Watt without Newcomen and Black as his predeçessors, the telegraph from Morse or the telephone from Bell except after Franklin, Oersted, and Faraday. Probably the truth is that if we knew all the facts, instead of only some of them, we should find every invention the natural descendant, near or remote, of science already existing. And as inheritance often seems to skip a generation or two and children sometimes show no discoverable resemblance to their immediate forbears, so inventions may come without disclosing any resemblance to parent inventions or ideas, while yet really intimately related to knowledge that has gone before.

Nor is it easy to estimate the reciprocal debt of science to inventions and the arts. That this debt is large there can be no doubt. To illustrate this fact it is hardly necessary to do more than mention examples, such as the service of the compass to the sciences of geography, navigation, and surveying; of the telescope and the chronometer to astronomy; of the microscope to biology; of the air pump to natural philosophy; or of the abacus or the Arabic numerals to arithmetic.

Among the more notable of the inventions of the nineteenth century were the locomotive, the steamboat, the friction match, the sewing-machine, the steel pen, the telegraph, the telephone, and the phonograph; labor-saving machinery; explosives; and the internal combustion engine, with its numerous offspring (motor vehicles, airplanes, motor boats, etc.).

POWER: ITS SOURCES AND SIGNIFICANCE

The recent progress of science and of civilization has been accompanied by a remarkable extension of man's control over his environment, which has come largely with his ability to develop, transmit, and utilize chemical, gravitational, and electrical energy or *power.* The ancients and the men of the Middle Ages used chiefly the power of man and other animals and of winds (windmills) and to some extent water (i.e., gravitation), as in water-wheels, but knew little of heat power or chemical power and nothing of electrical power, or of power transmission of any kind — except in moving herds, treadmills, or marching armies. The first step in the modern direction was apparently toward chemical power, in the invention of gunpowder.

GUNPOWDER, NITROGLYCERINE, DYNAMITE

Gunpowder is believed to have been known to the Chinese long before it appeared in Europe. An explosive mixture of charcoal, sulphur, and nitre was apparently also known to the Arabians, but the first important appearance of gunpowder in Europe was about the fourteenth century, and since the sixteenth it has played an all-important part in war and in peace. Its effects upon society and civilization have been profound, and with society and civilization the progress of science is always closely bound up.

The manufacture of gunpowder marks the beginning of the manufacture of power, if we may describe the controlled accumulation, storage, and liberation of energy by that convenient term. In 1845 gun-cotton was invented by Schönbein, and in 1847 nitroglycerine by Sobrero, and both explosives were found to be far more copious and powerful sources of energy than gunpowder. It was Alfred Nobel, however, a Swedish engineer, who after mixing nitroglycerine with gunpowder first made practical use of this for blasting. It was also Nobel who in 1867 made nitroglycerine less dangerous by diluting it with inert substances such as silicious earth — mixtures to which he gave the name dynamite.[1]

The production and utilization of power from gravitational sources, such as water-power and wind power, goes back to the earliest times — sails, windmills, and water-wheels being of very ancient origin. Power from fuel begins with Newcomen, Watt, and the steam-engine. Electrical power is at present chiefly derived indirectly from gravitational (hydraulic) or from chemical (fuel) sources.

THE STEAM-ENGINE

The last half of the eighteenth century was not merely an era of great revolutions: it was also an age of great inventions and among these, first in importance as well as first to arise, was the steam-engine.

Various and more or less successful attempts to utilize heat or steam as a source of power had been made before Watt's time, such, for example, as those of Hero in Alexandria (*c.* 75 B.C.) the Marquis of Worcester (1663) and Newcomen (1705). Of these only Newcomen's need be dwelt upon here. In Newcomen's engine a vertical cylinder with piston was used, the piston-rod, also vertical, being fixed above to one end of a walking-beam of which the other end carried a parallel rod. Thus the rise and fall of the piston caused a corresponding fall and rise of a parallel rod, which could be attached to anything, e.g., to a pump. The cylinder was connected with a steam boiler by a pipe fitted with a stopcock, and was filled with steam below the piston by opening the stopcock. The steam pressing upon the boiler raised the piston and depressed the parallel (pump)

[1] Nobel died in 1896, bequeathing his fortune, estimated at $9,000,000, to the founding of a fund which supports the international "prizes" — usually $40,000 each — which bear his name and are annually awarded to those who have most contributed to "the good of humanity." Five prizes have been usually given: viz., one in physics, one in chemistry, one in medicine or physiology, one in literature, and one for the promotion of peace.

rod. The stopcock was then closed, a "vent" in the cylinder was opened, cold water was introduced from another pipe to condense the steam, whereupon a vacuum formed, and the atmospheric pressure depressed the piston and lifted the pump rod. By having the various stopcocks carefully worked by hand a certain regularity of operation could be obtained, but before long improvements were made and the stopcocks were caused to work automatically. But since the cold (condensing) water chilled the cylinder, much heat was necessarily wasted.

Watt began by inventing (in 1765) a separate condenser, for cooling the steam without cooling the cylinder — thus saving a vast amount of heat. He next abandoned altogether the use of atmospheric pressure for depressing the piston, employing steam above as well as below the piston, to lower as well as to lift it: and with these improvements, to which he added many others, he soon had in his possession a serviceable and automatic steam-engine, rudimentary in many respects, but not essentially unlike that of today.

THE SPINNING JENNY, THE WATER-FRAME, AND THE MULE

In 1770 James Hargreaves patented the spinning jenny, a frame with a number of spindles side by side, by which many threads could be spun at once instead of only one, as in the old, one-thread, distaff or the spinning wheel. In 1771 Arkwright operated successfully in a mill a patent spinning machine which, because actuated by water power, was known as the "water-frame." In 1779 Crompton combined the principles involved in Hargreaves's and Arkwright's machines into one, which, because of this hybrid origin, became known as the spinning "mule." This proved so successful that by 1811 more than four and a half million spindles worked as "mules" were in operation in England.

A similar machine for weaving was soon urgently needed, and in 1785 the "power loom" of Cartwright appeared, although it required much improvement and was not widely used before 1813.

THE COTTON GIN (ENGINE)

With the inventions just described facilities arose for the manufacture of cotton as well as woollen, but the supply of raw cotton was limited, chiefly because of the difficulty of separating the staple (fibers) from the seeds upon which they are borne. Cotton had for centuries been grown and manufactured in India, the fibers being separated from the seeds by a rude hand machine known as a *churka*, used by the Chinese and Hindus. By this it was impossible to clean cotton rapidly. The invention therefore in 1793 by Eli Whitney of Connecticut of the saw cotton gin which enormously facilitated this separation was one of the most important inventions ever made. This consisted in a series of saws revolving between the interstices of an iron bed upon which the cotton was so placed as to be drawn through while the seeds were left behind. The value of the saw gin was instantly recognized, and the output of cotton in America was rapidly and immensely increased by its use.

STEAM TRANSPORTATION

Boats and ships propelled by man power or by the wind have been used from time immemorial, and parallel rails for wheeled conveyors moved by animal power or by gravity preceded the steam locomotive. The steamboat and the steam vehicle appeared at (or in the case of the latter even before) the opening of the nineteenth century.

The first practically successful steamboat was a tug, the *Charlotte Dundas*, built and operated in Scotland for the towing of canal boats by Symmington in 1802. The first commercially successful steamboat was Fulton's *Clermont*, on the Hudson, in 1807. The first steam-engine to run on roads appears to have been Cugnot's in France in 1769. The first to run on rails was Trevithick's, in 1804, built to fit the rails of a horse railway. This engine also discharged its exhaust steam into the funnel to aid the draught of the furnace — a device of fundamental importance to the further development of the locomotive. The first practically successful locomotive was Stephenson's *Rocket* (1829).

The development of the locomotive into a more efficient tractor gradually permitted greater speed and longer and heavier trains of cars, so that eventually it became difficult to control their momentum by the hand brakes then in use. This difficulty was met by George Westinghouse, Jr., with the invention in 1869 of a simple form of air brake controlled from the locomotive. His invention in 1872 of the automatic air brake with its triple valve situated on each car, but also controlled from the locomotive, is an outstanding event in the history of railway engineering. He made further improvements in 1887 and 1897 in response to the demand arising from the increasing speed of trains.

The compound (double or triple expansion) engine, which dates from 1781 (Hornblower), 1804 (Woolf), and 1845 (McNaughton), embodies what is perhaps the greatest single improvement in the steam-engine in the nineteenth century. The steam-turbine had begun toward the end of the century to replace the reciprocating engine for certain purposes, particularly for the generation of electric power.

ILLUMINATING GAS

Illuminating gas, made by the destructive distillation of coal, was invented and introduced in 1792 by William Murdock, who in 1802 had so far perfected the process that even the exterior of his factory in Birmingham was illuminated with gas in celebration of the peace of Amiens.

FRICTION MATCHES

Friction matches were preceded early in the nineteenth century by splinters of wood coated with sulphur and tipped with a mixture of chlorate of potash and sugar. These when touched with sulphuric acid ignited. It was not, however, until 1827 that practical friction matches were made and sold. These were known, after their inventor, as "Congreves" and consisted of wooden splints coated with sulphur and tipped with a mixture of sulphide of antimony, chlorate of potash, and gum. When subjected to severe friction, specially arranged for, these took fire. The phosphorus friction match was introduced commercially in 1833.

THE SEWING-MACHINE

Very few labor-saving inventions surpass in efficiency sewing-machines. These also were invented in the nineteenth century and had a gradual development, in which various inventors participated. The first which need be mentioned was that of a French tailor, named Thimonier, patented in 1830. It is said that although made of wood and clumsy, eighty of these machines were in use in Paris in 1841, when an ignorant mob wrecked the establishment in which they were located and nearly murdered the inventor. The most important ideas embodied in modern machines are, however, of strictly American origin, the work of Walter Hunt of New York, and of Elias Howe of Spencer, Massachusetts, being of principal importance (1846). Other Americans, especially Singer, Grover, Wilson, and Gibbs, afterwards contributed to the present excellence and variety of the sewing-machine.

PHOTOGRAPHY

Scheele, the Swedish chemist, appears to have been the first to study the effect of sunlight on silver chloride. Others, including Rumford and Davy, observed the chemical properties of light, but it was Wedgwood who, in 1802, made the first photograph by throwing shadows upon white paper moistened with nitrate of silver. Wedgwood was unable, however, to fix his prints.

Daguerreotypes, taken on silver plated copper, date from 1839, and were made by covering the copper with a thin film of silver iodide — a compound sensitive to light. The image was developed by mercury vapor and fixed by sodium hyposulphite. The discovery of the fixing power of hyposulphite was in itself alone of immense importance. With the name of Daguerre, who began experimenting in 1826, that of a fellow countryman and partner, J. N. Niepce, is intimately associated.

The subsequent development of photography is due to a host of workers. The collodion film which underlies all modern work was first introduced in 1850. It is said to be a practically perfect medium because totally unaffected by silver nitrate. Modern photography began when in 1851 Scott Archer coated glass plates with collodion containing bromides and iodides and sensitized them just before exposure by dipping them into the solution of a silver salt. This inconvenient and cumbersome wet method was finally superseded by the use of dry plates, introduced by Maddox in 1871. An improved method of making dry plates was patented in England in 1879 by George Eastman (1834–1932) of Rochester, N. Y. In 1889 he invented the "Kodak" and the first rollable nitrocellulose photographic film to be used instead of glass plates. The latter invention made it possible in 1893 for T. A. Edison (1847–1931) to develop the "kinetoscope," the first projection apparatus for showing pictures of moving objects. It consisted of a projection lantern with a transparent, flexible film carrying consecutive pictures in front of the lens. The coöperation of Eastman and Edison in devising methods for taking and projecting pictures of objects in motion laid the foundations of the great moving picture industry of today.

ANAESTHESIA. THE OPHTHALMOSCOPE

Anaesthesia, or insensibility to pain, during dental surgical operations was introduced, if not discovered, by Wells, a dentist of Hartford, Connecticut, who himself took nitrous oxide gas for anaesthesia in 1844. The first public demonstration of surgical anaesthesia under ether was made by a dentist, Morton, and a surgeon, Warren, at the Massachusetts General Hospital in Boston in 1846. Anaesthesia by chloroform was introduced by Simpson of Edinburgh, in 1847.

The ophthalmoscope, an instrument for examination of the interior of the eye, of inestimable value to medicine, was invented by Helmholtz in 1851. It is said that when von Graefe, an eminent ophthalmologist, first saw with it the interior of the eye he cried out, "Helmholtz has unfolded to us a new world."

INDIA-RUBBER

India-rubber, the coagulated and dried juice of the rubber tree, first reported by Herrera, "who in the second voyage of Columbus observed that the inhabitants of Hayti played a game with balls made 'of the gum of a tree' and that the balls although large were lighter, and bounced better, than the windballs of Castile," was at the end of the eighteenth century still a curiosity, employed by Priestley, among others, as an eraser or "rubber."

Rubber is a hydrocarbon, soft when pure but readily hardened by "vulcanization," i.e., treatment with sulphur or certain sulphur compounds (chloride, carbon bisulphide), a process introduced by Goodyear in 1839.

ELECTRICAL APPARATUS; TELEGRAPH, TELEPHONE, ELECTRIC LIGHTING, ELECTRIC MACHINERY

The first important application of electricity to the service of man was the telegraph. This is too well known to require more than the briefest description. An electric circuit in a wire "made" or "broken" at one point is likewise made or broken at all other points. Hence, it is only necessary to employ a preconcerted system of make-and-break signals to dispatch messages. This plan was first employed in 1836 by S. F. B. Morse, a native of Charlestown, Massachusetts, and the first telegraph line between two cities was installed between Baltimore and Washington in 1844. The first transatlantic cable was laid in 1858.

The telephone, invented by Alexander Graham Bell in 1876, is even more familiar. This, also, depends on the making and breaking of an electric circuit, not (as is usual in the telegraph) by a key manipulated by the finger, but by sound waves of the human voice impinging upon a thin metallic diaphragm (the transmitter) and reproduced at a distance by corresponding vibrations induced by an electro-magnet in another diaphragm (the receiver).

The so-called wireless telegraphy and wireless telephony, now called radio, differ from ordinary telegraphy and telephony in the use of signal waves set up in the ether instead of in the current carried by a wire. Radio is one of the marvels of the twentieth century and therefore the history of the series of inventions that have brought it to its present perfection lies mainly beyond the scope of this book. These inventions began, however, in the first half of the nineteenth century. Soon after he had invented the electric telegraph Morse, in 1842, made the first successful attempt at wireless telegraphy by sending signals through the water of a canal 80 feet wide in the city of Washington. He was followed by other inventors with various methods. The demonstration by Hertz (1888) of electro-magnetic waves in air that had been predicted by Maxwell, aroused great interest among physicists. (See page 404.) In 1890 E. Branly in Paris observed that metallic filings introduced in an electric circuit have high resistance. But the discharge of an electric spark near by reduces the resistance so that the current can pass. If this powder is then disturbed by a mechanical jar the resistance is restored. Oliver Lodge made an improved Branly coherer with a mechanical tapper as a detector for Hertzian waves, and in 1892 Sir William Crookes suggested the use of Hertzian waves in wireless telegraphy.

This suggestion was followed by Guglielmo Marconi, who received in 1896 a British patent for his system. This consisted essentially of a Hertzian primary circuit, or oscillator, and a tuned secondary circuit, or detector, each connected with telegraphic instruments. But Marconi introduced many improvements, including in the receiver, in place of the spark-gap, a specially designed coherer with an automatic tapper to break the current almost as soon as formed. He added also an insulated conductor attached to one pole of the oscillator and ending high in the air and at the other pole a connection to ground. The receiver was similarly equipped.

Finally, on the 27th of March, 1899, with antennae (as we now call them) 150 feet high Marconi demonstrated the success of his system by sending and receiving telegraphic messages through the air across the English Channel to and from Wimreux near Boulogne, a distance of thirty-two miles.

As early as 1883 Edison had noticed that the effect of placing a cold plate in an incandescent lamp was a flow of negative electricity from the filament to the cold plate. J. A. Fleming applied this "Edison effect" to wireless telegraphy in 1904, and in 1906 Lee De Forest of New York used this principle in his great invention, the "Audion," the thermionic valve or vacuum tube, that has made radio telephony so immensely successful.

The electric light, which had long been known as a laboratory experiment, became of practical utility about 1880, with the invention of the incandescent lamp, first the carbon arc and then the carbon filament, the former by Brush (1878), the latter by Edison (1879).

But before the electric light could come into general use it was necessary to find a mechanical source for large electric currents. This has been supplied by the electric generator, or dynamo. The basic principle of the dynamo was discovered independently in 1831 by Faraday at the Royal Institution in London and by Joseph Henry at Albany, N. Y. They found that electricity can be generated from magnetism. Arago had noticed that when a plate of copper and a magnet are free to rotate parallel to each other, if one be revolved the other will follow its motion. In an experimental attempt to explain this observation Faraday arranged a copper disk to revolve between the poles of a large steel magnet and obtained from the disk a direct current, and thus made the first dynamo ever invented. This discovery immediately interested inventors. In 1831, H. Pixii in Paris produced alternating currents by rotating a permanent horseshoe magnet beneath a pair of fixed coils, and the next year, following a suggestion of Ampère, he added a commutator and produced the first direct current magnetic machine after Faraday's experimental model.

Soon other inventors took up the problem, among them Soren Hjorth of Copenhagen, who in 1854 described the "reaction principle" of the self-exciting dynamo, in which the effect of the field magnets is enhanced by current from the armature, or revolving part of the machine. But it was some eight years later when the first dynamo was constructed on this principle. In 1860 Antonio Pacinotti of Florence made an armature of a ring of soft iron with a series of coils of insulated copper wire wound upon it and connections from between the coils radiating to the commutator at the center. His dynamo was the first to give a true continuous current like a battery current. The Pacinotti ring in modified form was combined with Hjorth's principle by Z. T. Gramme of Paris to produce in 1870 the first highly efficient and durable direct-current dynamo. Later it was discovered that this device would serve equally well as an excellent electric motor. Like Belle's telephone, it was shown as a novelty in 1876 at the Centennial Exhibition in Philadelphia. The first three-phase dynamo, the modern type, was made at Philadelphia in 1879 by Elihu Thomson and patented in 1880. The first commercial application of the step-down transformer occurred in 1887.

The dynamo not only has furnished current for extensive systems of lighting but also for electric motors supplying mechanical power for many purposes, and in addition has been used as a source of heat. The electric furnace has many applications, especially in metallurgy. An outstanding example is the reduction of aluminum. The earliest electric furnaces were of the arc type and were used for chemical experiments. Sir William Siemens was the first to apply the electric arc furnace to commercial operations. In 1885 Ferranti described the fundamental principles of the modern induction furnace. At the age of twenty-two C. M. Hall, a student at Oberlin, Ohio, invented the only commercially successful process for producing aluminum. It was known that by electrolysis aluminum could be separated from cryolite (sodium-aluminum-fluoride). But the process was impracticable. Hall had the original idea that aluminum oxide could be electrolyzed when dissolved in melted cryolite. He applied for a patent in 1886 and production began on a commercial scale in 1888. About the same time a similar process was patented in France by Paul Héroult.

THE PHONOGRAPH

The phonograph was invented by Edison in 1877, and was the culmination of attempts extending over many years to record and reproduce sound waves. In these attempts Young, König, Fleeming, Jenkin, and many others participated.

THE LINOTYPE

This is one of the outstanding American inventions. For centuries type-setting, or composition, had been done by hand. The types were picked out of compartments in a wooden case and returned there after use. One of the earliest attempts at composition by mechanical means was a devise patented by William Church in England in 1822. This was a machine for casting types and leaving them in reservoirs from which they were set by pressing keys, like a typewriter, and after use they were melted. Between 1840 and 1880 many inventors attempted to solve the problem and some of their machines were actually used for a time. In all of these inventions composition was by types previously cast.

Another method occurred to Ottmar Mergenthaler, an instrument maker in Baltimore. After several years of experimentation, he secured in 1884 a patent for a machine to make the molds and in 1885 for his linotype machine. This sets up a series of molds and casts a complete line of type in one piece. Part of the issue of July 3, 1886, of the *New York Tribune* was composed on the first of twelve machines ordered for that journal. A similar device, the monotype, was invented by Tolbert Lanston of Washington, D. C., in 1885. It casts each letter separately. The linotype and monotype are used extensively for printing newspapers and books where speed of production is important.

FOOD PRESERVING BY CANNING AND REFRIGERATION

In 1810 Appert of France succeeded in preserving foods in closed vessels by heating and sealing while hot. In 1816 a small amount of food preserved in this way found its way into the British Navy, where its value was recognized to some extent as a preventive of scurvy. It was not, however, until after the American Civil War that the industry began to assume anything like the vast extent and importance it has since reached.

Refrigeration in various forms has been used for food preserving probably from the earliest times, but the present enormous industry of cold storage has all grown up since the middle of the nineteenth century with the invention and development of refrigerators (domestic and commercial) and especially of machines for producing and distributing compressed air or other vapors or brine, ammonia, and other liquids at very low temperatures. These have been perfected rather rapidly since 1860, but did not become common before 1880. The first cargo of fresh meat successfully exported from America to Europe was shipped in March, 1879, and from New Zealand to Europe in February, 1880, arriving after a passage of 98 days in excellent condition.

THE INTERNAL-COMBUSTION ENGINE

For a century or thereabouts the steam-engine stood without a rival as a thermodynamic machine and prime mover. Innumerable attempts had been made meantime to construct other kinds of engines to convert heat more directly into power for mechanical work; but it was not until 1876 that the internal-combustion engine as improved by Otto became a practical success.

In the steam-engine, the furnace in which the heat is generated is external to the cylinder in which that heat does its work, the steam being merely an intermediary. It is therefore an external-combustion engine. Obviously, if the fuel burned is made to liberate its heat in the cylinder instead of the furnace, the steam can be dispensed with. This is what actually happens in the internal-combustion engine. The present enormous extent of the use of such engines for motors of all kinds, testifies to the importance of this invention.

ANILINE DYES

Aniline was first obtained from indigo in 1826 by O. Unveerdorben and named by him *crystallin.* In 1834, F. Runge prepared a similar substance from coal tar, and in 1840 C. J. Fritzsche obtained from indigo an oil which he called *anilin* — a word derived from the Sanskrit *Nila*, the indigo plant. A. W. Hofmann, at the Royal College of Chemistry in London, proved these substances to be identical. The commercial importance of aniline in the dyestuffs industry dates from the discovery of mauve, aniline purple, by W. H. Perkin in 1856, while an assistant in Hofmann's research laboratory. This was the first of the notable series of aniline dyes now so well known, and the forerunner of the immense color industry of today.

THE MANUFACTURE OF STEEL; BESSEMER

The making of steel by the decarbonization of cast-iron, a process which initiated what has been called the "age of steel," was introduced by Henry Bessemer (1813–98) in 1856. Bessemer's attention was drawn to the subject by his recognition of the necessity of improving gun-metal. Bessemer's process was at first only partially successful, but since others have shown how to improve it (by the addition of Spiegeleisen, etc.) it has reached enormous proportions.

AGRICULTURAL APPARATUS AND INVENTIONS

Beginning about 1850 an era of improved agricultural apparatus began, of which one result has been the opening of vast tracts of farm lands which might otherwise have remained unproductive. Steel plows, better harrows, mowing-machines, horse-power rakes, haymaking machinery, and especially harvesters of ingenious design for cereal crops (first introduced by McCormick in 1834), threshing-machines and spraying-machines are today common, where these were almost unknown before 1875. Machinery has also been applied to dairying, first to the making of butter and cheese, and more recently even to the milking of cows. Progress has also been made in the preservation of milk and of eggs by condensing, drying, freezing, etc. by new and economical processes invented and applied since that time.

APPLIED SCIENCE. ENGINEERING

Very much as discoveries and inventions blend together and as both spring from a common source, manifested as curiosity, inquiry, experimentation, and correlation (i.e., from science), so applied science, including engineering, comes from a common ancestry, i.e., from correlated knowledge — which is science. Both terms are loosely used and both cover today a multitude of diversified human activities.

With the progress of science, arts, and invention, engineering and other forms of applied science have developed so that these frequently have their own schools, either with or apart from universities and colleges; the school for miners at Freiberg, in Saxony, begun in 1765, being now only one of hundreds of technological and scientific schools for the training of engineers and others. Up to 1850 most engineers in America were trained in military schools and were primarily military engineers. But from that time forward the civil, as opposed to the military, engineer began to appear, and from the parent stem of civil engineering we now have mechanical, mining, electrical, sanitary, chemical, marine, and other branches of engineering, often highly specialized. The term "engineer" is now very widely employed, with more or less appropriateness, to occupations remote from those of the military or civil engineer, as for example, the "illuminating engineer," the "efficiency engineer," the "public health engineer," etc. We may soon expect to have added to these many others, such as the agricultural engineer, the forest engineer and even the fishery engineer.

An historical sketch of applied science and engineering would obviously include the work of Archimedes, Vitruvius, Frontinus, and Leonardo, and proceed with the applications made of the discoveries and inventions of the Renaissance and modern times. Some of this ground is covered in the present volume, and more of it in the series of books by Smiles entitled *Lives of the Engineers.*

The great inventions of former ages were made in counties where practical life, industry, and commerce were most advanced; but the great inventions of the last fifty years in chemistry and electricity and the science of heat have been made in the scientific laboratory: the former were stimulated by practical wants; the latter themselves produced new practical requirements, and created new spheres of labor, industry, and commerce. Science and knowledge have in the course of this century overtaken the march of practical life in many directions. — MERZ, I, 92.

REFERENCES FOR READING

ABBOT, C. G., *Great Inventions*, 1934 (Smithsonian Scientific Series, Vol. 12).

BYRN, E. W., *Progress of Invention in the Nineteenth Century*, 1900.

CRESSY, EDWARD, *Discoveries and Inventions of the Twentieth Century*, Ed. 3, 1930.

HART, I. B., *Great Engineers*, 1928.

HYLANDER, C. J., *American Inventors*, 1934.

LEONARD, J. M., *Tools of Tomorrow*, 1935.

MUMFORD, LEWIS, *Technics and Civilization*, 1934.

ROUTLEDGE, ROBERT, *Discoveries and Inventions of the Nineteenth Century*, 1876.

SMILES, SAMUEL, *Lives of the Engineers*, 1862.

THOMPSON, HOLLAND, *The Age of Invention*, 1921 (Chronicles of America Series, Vol. 37).

Appendix B

Some Important Names, Dates, and Events in the History of Science and Civilization

NOTE: Most of the dates before 600 B.C. are only approximate. Authorities differ on many.

c. = *circa*, about, *fl.* = flourished.

Before the Sixth Century B.C.

SCIENCE		GENERAL HISTORY, LITERATURE, ART, ETC.	
4236.	Egyptian civil calendar		
c. 3500.	Pictographic writing in Sumer.	*c.* 3360.	Egypt united by Menes.
c. 3000.	Cuneiform writing fully developed.	*c.* 2950.	First dynasty of Ur.
	Hieroglyphics in final form.		
	Imhotep, physician and architect.	2980–2475.	Pyramid Age.
		2630–2575.	Sargon of Akkad.
c. 2500.	Bronze becomes common in Egypt.		
c. 2300.	Sumerian tablet on pollination of the date palm.	*c.* 2205–2180.	Ibi-Sin last king of Ur.
2000/1700.	Ahmes (Rhind) Papyrus.	2160–1788.	Feudal Age of Egypt.
		1948–1905.	Hammurabi king of Babylon.
1920–1901.	Venus Tablets, astronomical data.		Abraham.
1700/1600.	Smith Surgical Papyrus.	1580–1150.	The Empire of Egypt.
c. 1550.	Earliest known water-clock, Egypt.		
c. 1500.	Ebers Medical Papyrus.		

	Science		General History, Literature, Art, etc.	
Before the Sixth Century B.C.	*c.* 1200.	Egyptians learn heat treatment of iron.	*c.* 1200.	Cnossus, Crete, sacked by Greeks.
	1150/1130.	Star tables at Luxor.	*c.* 1100.	Gades (Cadez) founded by Phoenicians.
	1000/900.	Ionian Greeks receive the Phoenician alphabet.	*c.* 1000–960.	David king of Israel.
			c. 850.	Homer. Carthage founded.
	776.	Era of Olympiads begins.	*c.* 800.	Hesiod.
			c. 753.	Rome founded (legendary).
			732.	Damascus falls to Assyria.
			705–681.	Nineveh flourishes under Sennacherib.
		Libraries at Nineveh.	668–626.	Assurbanipal.
		Botanical and medical tablets.	*c.* 660.	Byzantium founded.
			612.	Nineveh destroyed.
			c. 610.	Sappho and other Greek poets.
Sixth Century B.C.	*c.* 600.	Reform of calendar and of weights and measures attributed to Solon.	*c.* 638–558.	Solon.
			c. 600.	Marseilles founded.
		Nicho's sailors circumnavigate Africa.	609–593.	Nicho, king of Egypt, attempts to connect the Nile and the Red Sea.
		Astronomy in Chaldea.	604–561.	Nebuchadnezzar.
	c. 624–548 +.	Thales of Miletus.	586.	Fall of Jerusalem. Exile of the Jews.
	c. 610–545.	Anaximander.	*c.* 560.	Croesus.
			c. 560–483/77.	Buddha.
	fl. c. 546.	Anaximenes.	551–479.	Confucius.
	fl. c. 540.	Xenophanes.	538.	Babylon taken by Cyrus.
	fl. c. 532.	Pythagoras.		
		Nabu-ri-mannu.	525.	Persian conquest of Egypt.
		Alcmaeon.		
			521–485.	Darius emperor of Persia.

	Science		General History, Literature, Art, etc.	
Fifth Century B.C.	*fl. c.* 500.	Heraclitus.	526–456.	Aeschylus.
	fl. c. 465.	Parmenides.	*c.* 500.	Carthagenians explore west coast of Africa.
	fl. c. 462.	Zeno of Elea.		
	c. 499–428.	Anaxagoras.	490–429.	Pericles.
	fl. 450–430.	Hippocrates of Chios (Mathematician).	490.	Marathon, Battle of.
			c. 484–425.	Herodotus.
			480.	Thermopylae, Battle of.
	469–399.	Socrates.		
	c. 490–*c.* 435.	Empedocles.	480.	Salamis, Battle of.
	c. 460.	Leucippus.	*c.* 484–407.	Euripides.
	c. 460–370.	Democritus.	*fl. c.* 443–429.	Pheidias.
	c. 460–?	Hippocrates of Cos (Physician).	450–385.	Aristophanes.
			460–400.	Thucydides.
	fl. c. 450.	Philolaus (Motion of Earth).	*c.* 430.	The plague at Athens.
			404.	End of Peloponnesian war.
	fl. 432.	Meton (Calendar).		
	c. 428–347.	Archytas.	399.	Trial and death of Socrates.
	fl. c. 420.	Hippias of Elis.		
Fourth Century B.C.	428–347.	Plato.	*c.* 430–355 +.	Xenophon.
			384–322.	Demosthenes.
	c. 408–355.	Eudoxus.	*c.* 370.	Diogenes. Scopas. Praxiteles.
	384–322.	Aristotle.	356–323.	Alexander the Great.
	c. 375–325.	Menaecchmus.	338.	Chaeronea, Battle of.
	c. 388–312.	Heraclides of Pontus.		
	c. 372–287.	Theophrastus.		
	c. 350–260.	Zeno (Stoic).	332.	Alexandria founded.
	341–270.	Epicurus.	323.	Death of Alexander.
	c. 325.	Eudemus.	312.	Seleucid era began.
	fl. c. 323–285.	Euclid.		
Third Century B.C.	*c.* 300.	Herophilus.	306–285.	Ptolemy I Soter.
	c. 304–258.	Erasistratus.	294.	Museum and Library of Alexandria.
			285–247.	Ptolemy II Philadelphus.
	287–212.	Archimedes.	*c.* 285.	Theocritus.
	c. 273–192.	Erastosthenes.	283.	The Pharus at Alexandria.
	fl. c. 280.	Aristarchus.	280.	The Colossus of Rhodes.
	c. 260–200.	Apollonius.	269.	Silver money first coined in Rome.

Century	Date	Science	Date	General History, Literature, Art, etc.
Third Century B.C.	238.	Decree of Canopus (Leap Year).	247–222.	Ptolemy III Euergestes.
			212.	Fall of Syracuse.
			c. 210.	The Great Chinese Wall begun.
Second Century B.C.			234–149.	Cato the Censor.
				Paper made in China.
			c. 207–125.	Polybius.
			c. 166.	Terence.
			161.	Philosophers and rhetoricians banished from Rome.
			155.	Athenian philosophers welcomed in Rome.
	c. 135.	Ctesibius.	146.	Carthage destroyed (rebuilt in 123).
	fl. c. 161(?)–127.	Hipparchus.		Greece made a Roman province.
First Century B.C.	116–27.	Varro.	106–43.	Cicero.
	c. 98–55.	Lucretius.	102–44.	Julius Caesar.
	c. 75.	Hero of Alexandria.	70–19.	Virgil.
	fl. c. 70–?	Geminus.	59 B.C.–A.D. 17.	Livy.
	c. 63 B.C.–A.D. 24.	Strabo.	37.	Pollio founds First Public Library.
	46.	Julian Calendar.	30.	Death of Cleopatra, Egypt becomes a Roman province.
			27.	Octavius Caesar becomes Emperor Augustus.
				Beginning of Roman Empire.
	14.	Vitruvius, *De Architectura.*		Golden Age of Roman Literature.
			c. 8 B.C.?	Birth of Jesus Christ.[1]

[1] Almost certainly the computations of the Christian era by Dionysius (*c.* A.D. 525) were wrong. It is highly probable that Christ was born before A.D. 1. (Sarton, *Introd.*, I, 236.)

	Science		General History, Literature, Art, etc.	
First Century A.D.	4 B.C.–A.D. 65.	Seneca.	28 or 29.	The Crucifixion.
	23–79.	Pliny.	30 or 31.	Conversion of St. Paul
	c. 40–103.	Frontinus.		
	fl. c. 50.	Dioscorides, *Materia medica.*	54–68.	Nero emperor. Upper Nile explored.
		Nicomachus.		
Second Century A.D.	129–199.	Galen.	98–117.	Trajan, Roman emperor.
	140.	Ptolemy, *Almagest.*	161–180.	Marcus Aurelius emperor.
	fl. c. 127–132.	Theon of Smyrna.		
Third Century A.D.	*c.* 250.	Diophantus.	*c.* 160–230.	Tertullian.
	c. 300.	Pappus.	203–244.	Plotinus.
		Zosimos.		
Fourth Century A.D.	*fl.* 365–395.	Theon of Alexandria.	272–337.	Constantine (First Christian emperor)
			324.	Byzantium becomes Constantinople.
	c. 325–400.	Oribasius.	325.	Council of Nicaea.
			330.	Constantinople capital of Roman Empire.
			337.	Empire divided.
			331–363.	Julian the Apostate.
			340/50–420.	St. Jerome.
			354–430.	St. Augustine.
Fifth Century A.D.	410–485.	Proclus.	431.	Nestorius expelled from Constantinople.
	fl. c. 470.	Martianus Capella (*Liberal Arts*).		
	499.	Aryabhata.	*c.* 389–461.	St. Patrick. Latin culture in Ireland.
			476.	Last emperor of the West deposed by Odoacer the German.

	Science		General History, Literature, Art, etc.	
Sixth Century A.D.			493–526.	Theodoric king of the Ostrogoths.
			c. 480–544.	St. Benedict.
		Philoponus.		
	c. 480–524.	Boëthius.	527–565.	Justinian emperor of the East.
	c. 490–580.	Cassiodorus.	529.	Justinian closes the Academy at Athens. Issues *Code of Roman Law*.
	502–575.	Aeteus of Amida.		
	c. 525–605.	Alexander of Tralles.		Benedict founds Monte Cassino.
	552.	Silk worm eggs brought to Constantinople.	540–604.	Gregory the Great.
Seventh Century A.D.	560–636.	Isadore of Seville.	569–632.	Mohammed.
	c. 628.	Brahmagupta.	622.	The Hegira to Medina.
			624.	Bedr, Battle of.
			634–644.	Omar caliph.
			641.	Arabs take Alexandria.
			661–749.	Umayyad Dynasty.
			673–735.	Bede.
Eighth Century A.D.			704.	Samarkand taken by Arabs.
			711.	Moorish conquest of Spain.
			732.	Tours, Battle of.
	735–804.	Alcuin of York.	*c.* 742–814.	Charlemagne.
			750–900 +.	Abbasid Dynasty.
			756.	Caliphate of Cordova.
			762.	Baghdad founded.
			787.	Abbey schools in France.
	794.	Paper made in Baghdad.	795.	Iceland discovered.

	Science		General History, Literature, Art, etc.	
Ninth Century A.D.	*c.* 809–877.	Hunain, translator.	813–833.	Al-Mamun, Abbasid caliph.
	c. 825.	Al-Khwarizmi, *Algebra.*	*c.* 856.	Baghdad library and school of translation.
Tenth Century A.D.	*c.* 850–923/4.	Al-Razi, or Rhazes.	910.	Abbey of Cluny founded.
	c. 930–1003.	Gerbert (pope Sylvester II).		
	?–*c.* 932.	Ishaq al-Israili.	962.	Otto crowned emperor of the Holy Roman Empire.
	995.	"House of Science" at Cairo.		
Eleventh Century A.D.	965–1020.	Alhazen.		
	980–1037.	Ibn-Sina, or Avicenna.		
	973–1048.	Al-Biruni.		
	fl. c. 1056–1087.	Constantine the African.	*c.* 1050.	Turkish invasion of Islam.
			1066.	Battle of Hastings.
	1077–1268.	Salerno medical school.	1071.	Turks take Jerusalem.
			1071–1077.	Norman conquest of Sicily.
		Anatomia porci.	1085.	Toledo taken by Alfonso VII.
	c. 1043–1131/2.	Omar Khayyam.	1097.	First Crusade.
			1100.	Bologna, students of Roman law.
Twelfth Century A.D.	1091/4–1161/2.	Ibn Zuhr, or Avenzoar.		Three great schools in Paris.
	c. 1090–1165.	John of Seville.	1098–1179.	Hildegard.
	fl. 1141–1150.	Robert of Chester.		Troubadours.
			1152–1190.	Frederick I (Barbarossa) emperor.
	1100–1166.	Al-Idrisi.		
	1114–1178+.	Bhaskara.	1156.	Medical faculty at Bologna.

	Science		General History, Literature, Art, etc.	
Twelfth Century A.D.	1126–1198.	Ibn Rushd, or Averroës.	1160+.	Roman law studied at Montpellier.
	1130–1160.	Salernus.	1167.	Students settle in Oxford.
	c. 1114–1187.	Gerard of Cremona.	c. 1189.	Paper made in Spain.
	1135–1204.	Maimonides.	1189–1199.	Richard I king of England.
	1157–1217.	Alexander Neckam.	1200.	University of Paris chartered by the king.
Thirteenth Century A.D.	c. 1170–1240+.	Leonardo Fibonacci of Pisa.		
	1193–1280.	Albertus Magnus.	1194–1250.	Frederick II emperor.
	1214–1294?	Roger Bacon.	1204.	Constantinople pillaged by army of Fourth Crusade.
			1210.	Aristotle's *Physics* proscribed in Paris.
			1215.	*Magna Charta.*
	1225–1274.	St. Thomas Aquinas.	1223–1284.	Alfonso X of Castile.
			1231.	University of Paris completely organized.
			1245.	Knowledge of Aristotle required for M.A. degree in Paris.
			1249.	University College, Oxford.
	fl. 1269.	Peter the Stranger, *On the magnet.*	c. 1254–1324.	Marco Polo.
			1258.	Fall of Baghdad to Hulaku the Mongol.
			1265–1321.	Dante Alighieri.
			1268.	End of Norman rule in Sicily. Decline of medicine at Salerno.
	c. 1289.	Spectacles invented in Italy.	1284.	Peterhouse College, Cambridge.

	Science		General History, Literature, Art, etc.	
Fourteenth Century A.D.	c. 1275–1326.	Mondino de'Luzzi, anatomy.		
	1300.	Portolani (charts).	1304–1374.	Petrarch.
	1300–1368.	Guy de Chauliac, *Great Surgery.*	1313–1375.	Boccaccio.
			1337–1453.	The Hundred Years' War.
			c. 1340–1400.	Chaucer.
			1347–1348.	The Black Death.
	1364.	University of Vienna.	1379–1440.	Brunelleschi.
Fifteenth Century A.D.	1401–1464.	Nicholas of Cusa.		
	1423–1461.	Purbach.		
	1436–1476.	Regiomontanus.	1444–1511.	Bramante.
		Tartar Observatory (Samarkand).	c. 1450.	*Invention of Printing.*
			1453.	*Fall of Constantinople to the Turks.*
	1452–1519.	Leonardo da Vinci.		
	1473–1543.	Copernicus.	1471–1528.	Dürer.
			1475–1564.	Michelangelo.
	1486–1567.	Stifel.	1492.	*Discovery of America.* Moors expelled from Spain.
	1490–1555.	Agricola.	1497.	*Vasco da Gama rounds Cape of Good Hope.*
	1493–1541.	Paracelsus.		
Sixteenth Century A.D.	1501–1576.	Cardan.	1467–1536.	Erasmus.
	1503.	*Margarita philosophica.*	1483–1546.	Martin Luther.
			1509–1547.	Henry VIII.
			1509–1564.	Calvin.
	1510–1558.	Recorde.	1513.	*Balboa reaches Pacific Ocean.*
	c. 1500–1557.	Tartaglia.		
	1510–1589.	Palissy.	1517.	*Protestant Reformation.*
	1512–1594.	Mercator.		
	1514–1564.	Vesalius.	1519–1522.	*First Circumnavigation of the Globe by Magellan.*
	1514–1576.	Rheticus.		
	1516–1565.	Gesner.		
	1522–1565.	Ferrari.	1524–1580.	Camoens.
	1540–1603.	Vieta.	1530.	*Spinning wheel.*
	1543.	Copernicus, *De Revolutionibus.*	1542–1614.	El Greco.
			1547–1616.	Cervantes.
		Vesalius, *Fabrica.*	c. 1552–1599.	Spenser.
	1543–1615.	Giambattista della Porta.	1558–1603.	Queen Elizabeth.
			1564–1616.	Shakespeare.
	1544–1603.	Gilbert.	1573–1637.	Ben Jonson.

	Science		General History, Literature, Art, etc.	
Sixteenth Century A.D.	1546–1601.	Tycho Brahe.		
	1548–1600.	Bruno.		
	1548–1620.	Stevin.		
	1550–1617.	Napier.		
	1560–1621.	Harriott.		
	1560–1626.	Francis Bacon.		
	1564–1642.	Galileo.		
	1571–1630.	Kepler.		
	1575–1660.	Oughtred.		
	1577–1644.	Van Helmont.	1577–1640.	Rubens.
	1578–1657.	W. Harvey.		
	1582.	*Gregorian Calendar.*	1588.	*Defeat of the Spanish Armada.*
	1591–1626.	Snellius.		
	1593–1662.	Desargues.	1598.	*Edict of Nantes.*
	1596–1650.	Descartes.	1599–1660.	Velazquez.
	1598–1647.	Cavalieri.	1600–1681.	Calderon.
Seventeenth Century A.D.	1601–1665.	Fermat.	1603.	Accademia dei Lincei.
	1602–1686.	Von Guericke.	1605.	*Don Quixote.*
	1608–1647.	Torricelli.	1607.	*First Permanent English Colony in America.*
	1608–1679.	Borelli.		
	1616–1703.	Wallis.		
	1623–1662.	Pascal.	1607–1690.	Rembrandt.
	1624–1689.	Sydenham.	1608–1674.	Milton.
	1627–1691.	Boyle.	1618–1648.	*Thirty Years' War.*
	1627–1705.	John Ray.		
	1628.	Harvey, *De motu cordis.*		
	1628–1694.	Malpighi.		
	1629–1695.	Huygens.	1622–1673.	Molière.
	1630–1677.	Barrow.	1631–1700.	Dryden.
	1632–1723.	Leeuwenhoek.		
	1635–1703.	Hooke.	1636.	*Harvard College founded.*
	1635–1672.	Willughby.		
	1641–1712.	Grew.		
	1643–1727.	Newton.	1638–1715.	Louis XIV.
	1646–1716.	Leibniz.	1639–1699.	Racine.
	1637.	*Discours sur la Méthode. (Analytic Geometry.)*	1649–1660.	*English Commonwealth.*
	1644–1710.	Roemer.	1657.	Accademia del Cimento.
	1654–1705.	Bernoulli, James.		
	1656–1742.	Halley.		
	1660–1734.	Stahl.	1660–1731.	De Foe.

	Science		General History, Literature, Art, etc.	
Seventeenth Century A.D.	1668–1738.	Boerhaave.	1662.	Royal Society of London.
	1677–1761.	Hales.	1672–1725.	Peter the Great.
	1682–1771.	Morgagni.	1666.	Academie des Sciences charted by Louis XIV.
	1687.	Newton, *Principia*.		
	1698–1746.	Maclaurin.	1683.	*Siege of Vienna by Turks.*
	1699–1739.	Dufay.	1688–1744.	Pope.
	1699–1777.	Jussieu.	1694–1778.	Voltaire.
Eighteenth Century A.D.	1700–1782.	Bernouilli, D.		
	1705.	*Newcomen's Engine.*		
	1706–1790.	Franklin.		
	1707–1778.	Linnæus.		
	1707–1783.	Euler.		
	1707–1788.	Buffon.		
	1708–1777.	Haller.		
	1715–1786.	Guettard.	1709–1784.	Samuel Johnson.
	1717–1783.	d'Alembert.		
	1725–1815.	Desmarest.	1711–1776.	Hume.
	1726–1797.	Hutton.	1712–1778.	Rousseau.
	1728–1793.	Hunter.	1724–1804.	Kant.
	1728–1799.	Black.	1728–1774.	Goldsmith.
	1731–1810.	Cavendish.		
	1733–1804.	Priestley.	1732–1790.	Washington.
	1736–1813.	Lagrange.		
	1736–1806.	Coulomb.		
	1736–1819.	Watt.		
	1737–1798.	Galvani.	1737–1794.	Gibbon.
	1738–1822.	Herschel, F. W.		
	1742–1786.	Scheele.		
	1743–1794.	Lavoisier.		
	1744–1829.	Lamarck.	1743.	American Philosophical Society.
	1745–1827.	Volta.		
	1746–1818.	Monge.		
	1749–1827.	Laplace.		
	1750–1817.	Werner.	1749–1832.	Goethe.
	1752–1833.	Legendre.		
	1753–1814.	Rumford.	1759–1796.	Burns.
	1765.	*Watt's Steam-Engine.*	1759–1805.	Schiller.
	1766–1844.	Dalton.		
	1767.	*Spinning Jenny.*	1765.	Freiberg School of Mines.
	1769–1832.	Cuvier.		

	Science		General History, Literature, Art, etc.	
Eighteenth Century A.D. (*Continued*)	1769–1859.	Humboldt.		
	1769.	*Spinning Frame.*	1769–1821.	Napoleon I.
	1769–1839.	Smith, William.	1770–1850.	Wordsworth.
	1771–1802.	Bichat.		
	1772–1844.	Geoffroy St.-Hilaire.		
	1773–1829.	Young.		
	1773–1858.	Brown, Robert.	1771–1832.	Scott.
	1774.	*Discovery of Oxygen.*	1772–1834.	Coleridge.
	1775–1836.	Ampère.	1773–1859.	Metternich.
	1776–1837.	Treviranus.	1775–1781.	*American Revolution.*
	1776–1847.	Dutrochet.		
	1777–1855.	Gauss.		
	1778–1829.	Davy.		
	1778–1841.	De Candolle.		
	1779–1848.	Berzelius.		
	1781–1848.	Stephenson.		
	1781.	*Discovery of Uranus.*	1780.	American Academy of Arts and Sciences.
	1783.	*Air Balloon.*		
	1783–1855.	Magendie.		
	1784–1846.	Bessel. *Parallax of stars.*	1789–1794.	*French Revolution.*
	1787–1854.	Ohm.		
	1789.	Lavoisier, *Traité de Chimie.*		
	1791–1867.	Faraday.		
	1791–1872.	Morse.		
	1792.	*Cotton Gin.*		
	1792–1876.	von Baer.	1792–1822.	Shelley.
	1793–1856.	Lobachevski.	1794.	*École Polytechnique.*
			1795–1821.	Keats.
	1796.	*Vaccination.*	1795–1881.	Carlyle.
	1796–1832.	Carnot.		
	1797–1875.	Lyell.		
	1800.	*Continuous electric current.*	1800.	Royal Institution.

	Science		General History, Literature, Art, etc.	
Nineteenth Century A.D.	1801–1858.	Müller.		
	1803–1873.	Liebig.	1802–1885.	Victor Hugo.
	1804–1881.	Schleiden.		
	1807–1873.	Agassiz, L.	1802–1894.	Kossuth.
	1809–1882.	Darwin, Charles.	1803–1882.	Emerson.
	1810–1882.	Schwann.		
	1810–1888.	Gray, Asa.	1804–1865.	Cobden.
	1811–1877.	Leverrier.	1805.	*Battle of Trafalgar.*
	1811–1899.	Bunsen.	1805–1872.	Mazzini.
	1813–1878.	Bernard.	1807–1882.	Longfellow.
	1813–1898.	Bessemer.	1807–1882.	Garibaldi.
	1817–1911.	Hooker.	1807–1892.	Whittier.
	1818–1889.	Joule.	1809–1865.	Lincoln.
	1819–1892.	Adams, J. C.	1809–1892.	Tennyson.
	1820–1903.	Spencer, Herbert.	1810–1861.	Cavour.
	1820–1910.	Florence Nightingale.	1811–1863.	Thackeray.
			1812–1870.	Dickens.
	1821–1894.	Helmholtz.	1815.	*Battle of Waterloo.*
	1821–1902.	Virchow.		
	1822–1884.	Mendel.	1815–1898.	Bismarck.
	1822–1898.	Leuchart.	1817–1862.	Thoreau.
	1822–1888.	Clausius.	1834.	*Poor Law Reform in England.*
	1822–1895.	Pasteur.		
	1822–1908.	Gibbs, Wolcott.		
	1823–1913.	Wallace.		
	1824–1877.	Hofmeister.		
	1824–1887.	Kirchhoff.		
	1824–1907.	Kelvin.		
	1825–1895.	Huxley.		
	1827–1912.	Lister.		
	1828.	*Synthesis of Urea.*		
	1828.	*Stephenson's "Rocket."*		
	1829–1896.	Kekulé.		
	1830.	Lyell, *Principles of Geology.*		
	1831–1888.	DeBary.		
	1831–1879.	Maxwell.		
	1833–1837.	*Bridgewater Treatises.*	1837.	*Accession of Queen Victoria.*
	1834–1906.	Langley.		
	1834–1907.	Mendeléjeff.		
	1834–1914.	Weismann.		
	1835–1909.	Newcomb.		
	1836.	*The Telegraph.*		

	Science		General History, Literature, Art, etc.	
Nineteenth Century A.D. (*Continued*)	1838–1839.	*The Cell Theory.*		
	1838–1907.	Perkin.		
	1839.	*The Daguerreotype.*		
	1839–1903.	Gibbs, J. W.	1840–1893.	Symonds, J. A.
	1842–1919.	Rayleigh.		
	1843–1910.	Koch.		
	1845–1923.	Röntgen.	1842–1910.	James, William.
	1846.	*Anæsthesia.*	1846–1848.	*War between United States and Mexico.*
		Discovery of Neptune.	1846.	*Repeal of the Corn Laws.*
	1847–1922.	Graham Bell.		
	1847.	*Die Erhaltung der Kraft.*	1848.	*Abdication of Louis Philippe of France.*
	1847–1931.	Edison		Société de Biologie formed in Paris.
	1848–1935.	De Vries.		
	1852–1908.	Becquerel.	1852.	*Accession of Napoleon III.*
	1852–1911.	Van't Hoff.	1853–1856.	*Crimean War.*
			1854.	*First (Perry) Treaty between United States and Japan.*
	1857–1894.	Hertz.		
	1857–1932.	Ross, Ronald.		
	1858.	Atlantic cable.		
		Cellularpathologie.		
	1858–1934.	Smith, Theobald.		
	1859.	*The Origin of Species.*	1859.	*Peace of Villafranca.*
			1861.	*First Italian Parliament.*
	1859–1860.	Spectroscope.	1861–1865.	*Civil War in the United States.*
	1859–1906.	Curie, Pierre.		
	1859–1927.	Arrhenius.	1861–1888.	William I of Germany.
			1863.	National Academy of Sciences.
			1866.	*War between Prussia and Austria.*
			1867.	*End of the Shogunate of Japan.*
	1867–1934.	Curie, Marie.	1867–1933.	Galsworthy.
	1868.	Antiseptic surgery.	1869.	First American transcontinental railway.
	1875.	Fertilization of the egg.		Suez Canal.
	1876.	Koch, *Anthrax.*	1870–1871.	War between Prussia and France.

	Science		General History, Literature, Art, etc.	
Nineteenth Century A.D. (*Continued*)	1876.	Telephone.	1888.	Pasteur Institute in Paris.
		Electric motor.	1888–1918.	William II of Germany.
	1877.	Phonograph.	1890.	*Promulgation of Constitution of Japan.*
			1894.	*War between China and Japan.*
			1898.	*War between Spain and the United States.*
			1899.	*Boer War in South Africa.*
	1881.	Pasteur, *Immunity.*		
	1896.	Röntgen, *X-rays.*		First radio message across the English Channel.
		Langley, First flight.		
		Marconi, Radio telegraph.		
	1898.	P. and M. Curie. *Radium.*		
	1900.	Sutton, Paired chromosomes.	1900.	*Boxer Uprising in China.*
		De Vries and others, *Mendel's theory of heredity.*		
	1900–1901.	Planck, *Quantum theory.*	1901.	Edward VII king of Great Britain.
	1902.	*Hormones.*	1914.	Panama Canal opened.
	1912.	*Vitamin theory.*		

Appendix C

A Short List of Reference Books

Particular attention may be called to the important and valuable publications of the John Crerar Library, Chicago, prepared by Aksel G. S. Josephson: viz., *A List of Books on the History of Science (1911); Supplement to the Same (1916)*; and *A List of Books on the History of Industry and Industrial Arts (1915)*.

A suggestive list of references is given in *The Development of the Scientific Point of View, a Syllabus*, by Dorothy Stimson (Baltimore, 1932). But above all, a student should have access to the "Critical Bibliography" which appears in each issue of *Isis*.

* indicates Everyman's Library; ** indicates Home University Library; *** indicates Loeb Classical Library. A collection of essays is entered under the title.

GENERAL

Abbot, C. G., *Great Inventions*, Washington, 1932.
Adams, F. D., *Birth and Development of the Geological Sciences*, Baltimore, 1939.
Allbutt, Sir T. C., *Greek Medicine in Rome*, London, New York, 1921.
Allman, G. J., *Greek Geometry from Thales to Euclid*, Dublin, 1889.
Andrade, E. N. daC., *Mechanism of Nature*, Rev. Ed. London, 1938.
Andrae, Tor, *Mohammed, the Man and His Faith*, tr. T. Menzel, New York, 1936.
Arber, Agnes, *Herbels, Their Origin and Evolution; a chapter in the History of Botany, 1470–1670*. New Ed. Cambridge, New York, 1938.
Archimedes, *Works*, tr. T. L. Heath, Cambridge, 1897.
Aristotle, *Works Translated into English*, ed. W. D. Ross, Oxford, 1908–31, 11 vols.
Arnold, Sir T. W., "Muslim Civilization during the Abbasid Period," *Cambridge Medieval History*, Vol. 4, Ch. 4, 1928.
Avebury, Lord, *see* Lubbock.
Aykroyd, W. R., *Three Philosophers (Lavoisier, Priestley, Cavendish)*, London, 1935.

Ball, W. W. R., *Short Account of the History of Mathematics*, Ed. 4, London, 1908.

Baynes, N. H., *The Byzantine Empire*, London, 1925. **
Bell, E. T., *Men of Mathematics*, New York, 1937.
Berry, Arthur, *Short History of Astronomy*, London, 1898.
Bligh, N. H., *Evolution and Development of the Quantum Theory*, London, 1926.
Boas, Franz, *Mind of Primitive Man*, Rev. Ed., New York, 1938.
Breasted, J. H., *Ancient Times*, Ed. 2, Boston, 1935.
——, *The Conquest of Civilization*, Ed. 2, New York, 1938.
——, *The Edwin Smith Surgical Papyrus*, Chicago, 1930, 2 vols.
Brewster, David, *Martyrs of Science*, London, 1880.
Budge, Sir E. A. W., *Egypt*, London, 1925. **
——, *The Divine Origin of the Craft of the Herbalist*, London, 1928.
——, *The Rosetta Stone* (Decrees of Memphis and Canopus, Vol. 1), London, 1904.
Bullock, William, "History of Bacteriology," Great Britain, Medical Research Council, *System of Bacteriology*, Vol. 1, pp. 15–103, London, 1930.
Burnet, John, *Early Greek Philosophy*, Ed. 4, London, New York, 1930.
Butler, Samuel, *Evolution, New and Old*, New Ed., New York, 1911.
Byron, Robert, *The Byzantine Achievement, A.D. 330–1453*, London, New York, 1929.

Cajori, Florian, *History of Mathematics*, Ed. 2, New York, 1919.
——, *History of Physics*, Rev. Ed., New York, 1929.
Cambridge Ancient History, ed. J. B. Bury, *et al.*, New York, 1923–36, 11 vols.
Carus, J. V., *Geschichte der Zoologie*, München, 1872.
Cary, Max, *The Legacy of Alexander*, New York, 1932.
——, and E. H. Warmington, *Ancient Explorers*, New York, 1929.
Castiglioni, A., *Italian Medicine* (Clio Medica, VI), New York, 1932.
——, *The Renaissance of Medicine in Italy*, Baltimore, 1934.
Celsus, *De Medicina*, tr. W. G. Spencer, Cambridge, Mass., 1936, 2 vols. ***
Chace, A. B., *The Rhind Mathematical Papyrus*, Oberlin, 1927–29, 2 vols.
Chikashige, M., *Alchemy and Other Chemical Achievements of the Ancient Orient*, Tokyo, 1936.
Childe, V. G., *New Light on the Most Ancient East*, London, New York, 1934.
Clark, G. N., *Science and Social Welfare in the Age of Newton*, Oxford, 1937.
Clerke, Agnes M., *Popular History of Astronomy during the 19th Century*, London, 1902.
——, *The Herschels and Modern Astronomy*, New York, 1901.

Clerke, Agnes M., *Modern Cosmogonies*, London, 1905.
Clodd, Edward, *Pioneers of Evolution from Thales to Huxley*, London, 1903.
——, *Story of the Alphabet*, New York, 1900 (re-issued 1938).
Cole, F. J., *Early Theories of Sexual Generation*, Oxford, 1930.
——, *History of Protozoology*, London, 1926.
Cole, Fay-Cooper, *The Long Road from Savagery to Civilization*, Baltimore, 1933.
Contenau, G., *Manuel d'archéologie orientale*, Paris, 1927, 1931, 3 vols.
Corner, G. W., *Anatomy* (Clio Medica, III), New York, 1930.
Coulton, G. G., *The Medieval Scene*, Cambridge, 1930.
Cressy, Edward, *Discoveries and Inventions of the Twentieth Century*, London, 1914.
Crew, Henry, *The Rise of Modern Physics*, Ed. 2, Baltimore, 1935.
Crowther, J. G., *British Scientists of the Nineteenth Century*, London (*Men of Science*, New York), 1936.
——, *Famous American Men of Science*, London, 1937.
——, *Progress of Science*, London, 1934.
Cumston, C. G., *Introduction to the History of Medicine from the Time of the Pharaohs to the End of the XVIIIth Century*, London, New York, 1926, 1927.
Curtis, J. G., *Harvey's Views on the Circulation of the Blood*, New York, 1915.

Dalton, J. C., *The Experimental Method in Medical Science*, New York, 1882.
Dampier, Sir W., *History of Science*, Cambridge, New York, 1930, 1932.
Dannemann, F., *Die Naturwissenschaften in ihrer Entwicklung und in ihrem Zusammenhange*, Leipzig, 1910–23, 4 vols.
Darrow, Floyd, L., *Masters of Science and Invention*, New York, 1923.
Darwin, C. G., *The New Conception of Matter*, London, 1931.
Darwin, Charles, *Descent of Man, and Selection in Relation to Sex*, London, 1871.
——, *Journal of Researches . . . during the Voyage of H. M. S. Beagle* (1860), London, 1906. *
——, *Life and Letters*, ed. F. Darwin, London, 1887.
——, *Origin of Species by means of Natural Selection*, London, 1859, Ed. 6, 1866.
——, *Variation of Animals and Plants under Domestication*, London, 1868, Ed. 2, 1875.
Davis, W. S., *A Short History of the Near East (A.D. 330–1922)*, New York, 1922.

Dawson, W. R., *The Beginnings, Egypt and Assyria* (Clio Medica, I), New York, 1930.
Delage, Y., and M. Goldsmith, *Theories of Evolution*, New York, 1912.
Delaporte, L. J., *Mesopotamia*, tr. V. G. Childe, London, New York, 1925.
Descartes, René, *Philosophical Works*, tr. E. S. Haldane and G. R. T. Ross, Cambridge, 1911, 1912.
De Sitter, William, *Kosmos*, Cambridge, 1932.
Dobell, C., *Antony van Leeuwenhoek and His "Little Animals,"* New York, 1932.
Drachman, J. M., *Studies in the Literature of Natural Science*, New York, 1930.
Draper, J. W., *History of the Conflict between Religion and Science*, New York, 1875, 1903.
Dreyer, J. L. E., *History of the Planetary Systems from Thales to Kepler*, Cambridge, 1906.

Ebbell, B., *The Papyrus Ebers*, Copenhagen, 1937.
Eddington, Arthur, *Space, Time and Gravitation*, Cambridge, New York, 1920, 1923.
Einstein, Albert, *The World as I See It*, tr. A. Harris, New York, 1934.
Elgood, Cyril, *Medicine in Persia* (Clio Medica, XIV), New York, 1934.
Essig, E. O., *History of Entomology*, New York, 1931.
Euclid, *The Elements*, ed. Isaac Todhunter, New York. *
——, *The Thirteen Books of Euclid's Elements*, tr. T. L. Heath, Cambridge, 1908.
Evans, Sir A. J., *The Palace of Minos*, London, 1921–1935, 4 vols.

Fabricius of Aquapendente, H., *De venarum ostiolis, 1603*, tr. J. Franklin, Springfield, Ill., 1933.
Fahie, J. J., *History of Wireless Telegraphy, 1838–1899*, Edinburgh, 1899.
Fairbanks, Arthur, *First Philosophers of Greece*, London, 1898.
Farrington, B., *Science in Antiquity*, London, 1936.
Findlay, A., *A Hundred Years of Chemistry*, London, New York, 1937.
Fisher, H. A. L., *History of Europe*, Boston, 1935.
Fleming, A. P. M., and H. J. Brocklehurst, *History of Engineering*, London, New York, 1925.
Foster, Sir M., *History of Physiology during the XVI, XVII and XVIII Centuries*, Cambridge, 1901.
Franklin, K. J., *Short History of Physiology*, London, 1933.
Fulton, J. F., *Physiology* (Clio Medica, V), New York, 1931.

Galen, *On the Natural Faculties*, tr. A. J. Brock, London, Cambridge, Mass., 1916. * * *
Galileo Galilei, *Dialogues concerning Two New Sciences*, tr. Henry Crew and Alfonso de Salvio, New York, 1914.
Garrison, F. H., "History of Endocrine Doctrine." *Endocrinology and Metabolism*, ed. L. F. Barker, New York, 1922, Vol. I, pp. 45–78.
——, *Introduction to the History of Medicine*, Ed. 4, Philadelphia, 1929.
Geikie, Sir A., *Founders of Geology*, Ed. 2, London, 1905.
Glotz, Gustave, *The Aegean Civilization*, tr. M. R. Dobie and E. M. Riley, New York, 1925.
Gow, James, *Short History of Greek Mathematics*, Cambridge, 1884.
Grant, Sir A., *Aristotle*, Edinburg, 1910.
Grant, Robert, *History of Physical Astronomy*, London, 1852.
Green, J. R., *History of Botany 1860–1900*, Oxford, 1909.
——, *History of Botany in the United Kingdom*, London, 1914.
Greene, E. L., *Landmarks of Botanical History*, Washington, 1909.
Greene, W. C., *The Achievement of Greece*, Cambridge, Mass., 1923.
——, *The Achievement of Rome*, Cambridge, Mass., 1933.
Gregory, Sir R., *Discovery*, New York, 1925.
Gumpert, Martin, *Trail Blazers of Science*, New York, 1936.
Gunther, R. T., *Early Science in Oxford*, Oxford, 1920–37, 11 vols.

Haddon, A. C., *History of Anthropology*, London, 1910.
Harris, L. J., *Vitamins in Theory and Practice*, Cambridge, 1935.
Hart, I. B., *Makers of Science: Mathematics, Physics, Astronomy*, Oxford, 1924.
Harvey, William, *Exercitatio anatomica de motu cordis*, tr. C. D. Leake, Springfield, Ill., 1928.
——, *On the Movement of the Heart and Blood*, London, New York, 1908. *
——, *Work and Life by R. Willis*, London, 1847.
Harvey-Gibson, R. J., *Outlines of the History of Botany*, London, New York, 1919.
——, *Two Thousand Years of Science*, New York, 1929.
Haskins, C. H., *The Rise of Universities*, New York, 1923.
——, *Studies in the History of Mediaeval Science*, Cambridge, Mass., 1924.
Hawes, C. H., and Harriet B. Hawes, *Crete the Forerunner of Greece*, Ed. 2, New York, 1911.
Hawks, E., *Pioneers of Plant Study*, London, New York, 1928.
Heath, Sir T. L., *Appollonius*, Cambridge, 1896.
——, *Archimedes*, New York, 1920.
——, *Diophantus of Alexandria*, Cambridge, 1885, 1910.

Heath, Sir T. L., *The Copernicus of Antiquity, Aristarchus of Samos*, Oxford, 1913, New York, 1920.
——, *Greek Astronomy*, New York, 1932.
——, *History of Greek Mathematics*, Oxford, 1921, 2 vols.
——, *Manual of Greek Mathematics*, Oxford, 1931.
Heiberg, J. L., *Mathematics and Physical Science in Classical Antiquity*, tr. D. C. Macgregor, Oxford, 1922.
Heidel, W. A., *The Heroic Age of Science*, Washington, 1933.
Helmholtz, H., *Popular Lectures on Scientific Subjects*, tr. E. Atkinson, New York, 1873.
Herdman, Sir W. W., *Founders of Oceanography*, London, 1923.
Herodotus, (*History*), tr. A. D. Godley, London, Cambridge, Mass., 1920, 1922. ***
Hertz, H., *Electric Waves*, tr. D. E. Jones, London, 1893.
Hill, G. F., *Development of Arabic Numerals in Europe*, Oxford, 1915.
Hippocrates, (*Works*), tr. W. H. S. Jones and E. T. Withington, London, Cambridge, Mass., 1923–31. ***
History of Science Society, *Johann Kepler, 1571–1630*, Baltimore, 1928.
——, *Sir Isaac Newton, 1727–1927*, Baltimore, 1928.
Hitti, P. K., *History of the Arabs*, London, 1937.
Hjort, Johan, *The Human Value of Biology*, Cambridge, Mass., 1938.
Hobson, E. W., *The Domain of Natural Science*, Cambridge, 1923.
Holland, R. S., *Historic Inventions*, Philadelphia, 1911.
Holmyard, E. J., *The Great Chemists*, London, 1928.
Hopkins, A. J., *Alchemy, Child of Greek Philosophy*, New York, 1934.
Hull, G. F., *Elementary Survey of Modern Physics*, New York, 1936.
Hume, Martin, *The Spanish People*, New York, 1901.
Hylander, C. J., *American Inventors*, New York, 1934.
——, *American Scientists*, New York, 1935.

Jastrow, Morris, *The Civilization of Babylonia and Assyria*, Philadelphia, 1915.
Jeans, Sir J. H., *The Mysterious Universe*, New York, 1930.
——, *New Backgrounds of Science*, New York, 1933.
——, *Through Space and Time*, New York, 1934.
Judd, J. W., *The Coming of Evolution*, Cambridge, 1911.

Kaempffert, W. B., *Popular History of American Inventions*, New York, 1924.
——, *Inventions and Society*, Chicago, 1930.
Keith, Sir Arthur, *Antiquity of Man*, New Ed., London, Philadelphia, 1927, 2 vols.
Krumbhaar, E. B., *Pathology* (Clio Medica, XIX); New York, 1937.

Lamarck, J. B., *Zoological philosophy*, tr. Hugh Elliot, London, 1914.
Lambert, S. W., and G. M. Goodwin, *Medical Leaders from Hippocrates to Osler*, Indianapolis, 1929.
Larmor, Sir J., *Origins of Clerk Maxwell's electric ideas as described in familiar letters to William Thomson*, Cambridge, New York, 1937.
Langdon, S., and J. K. Fotheringham, *The Venus Tablets of Ammizaduga*, Oxford, 1928.
Leeuwenhoek, A. van, *Select Works*, tr. S. Hoole, London, 1798, 1807, 2 vols.
Legacy of Greece, ed. R. W. Livingstone, Oxford, 1922.
Legacy of Islam, ed. Sir Thomas Arnold and Alfred Guillaume, Oxford, 1931.
Legacy of Israel, ed. C. R. Bevan and Charles Singer, Oxford, 1927.
Legacy of Rome, ed. Cyril Bailey, Oxford, 1923.
Leonard, J. N., *Tools of Tomorrow*, New York, 1935.
Leonardo da Vinci, *Notebooks of Leonardo da Vinci*, tr. Edward MacCurdy, New York, 1938, 2 vols.
Levy, Hyman, *The Universe of Science*, New York, 1933.
Little, A. G., *see Roger Bacon.*
Locy, W. A., *Biology and Its Makers*, New York, 1908 (*Story of Biology*, 1934).
——, *The Growth of Biology*, New York, 1925.
Lodge, O. J., *Pioneers of Science*, New York, 1904.
Lones, T. E., *Aristotle's Researches in Natural Science*, London, 1912.
Long, E. R., *History of Pathology*, Baltimore, 1928.
Lubbock, J., Lord Avebury, *Prehistoric Times*, London, New York, 1913.
Lucas, A., *Ancient Egyptian Materials and Industries*, Ed. 2, London, 1934.
Lucretius, *De rerum natura*, tr. W. H. D. Rouse, London, Cambridge, Mass., 1924. ***
Lund, F. B., *Greek Medicine* (Clio Medica, XVIII), New York, 1936.

MacCullum, T. W., and S. Taylor, *The Nobel Prize Winners and the Nobel Foundation 1901–1937*, Zurich, 1938.
Mack, Ernst, *Science of Mechanics*, tr. T. J. McCormack, Ed. 4, Chicago, 1919.
Mackay, Ernest, *The Indus Civilization*, London, 1935.
Makers of British Botony, ed. F. W. Oliver, Cambridge, 1913.
McKie, D., and N. H. de V. Heathcote, *The Discovery of Specific and Latent Heats*, London, 1935.
McMurrich, J. P., *Leonardo da Vinci, the Anatomist (1452–1519)*, Washington, 1930.

Merrill, G. P., *The First Hundred Years of American Geology*, New Haven, 1924.
Merz, J. T., *History of European Thought in the Nineteenth Century*, Edinburgh, 1896–1914, 4 vols.
Miall, L. C., *Early Naturalists*, London, 1912.
Millikan, R. A., *Electrons (+ and −), Protons, Photons, Neutrons, and Cosmic Rays*, Chicago, 1935.
——, *Science and the New Civilization*, New York, 1930.
Moore, F. J., *History of Chemistry*, Ed. 2, New York, 1931.
Moret, Alexander, *The Nile and Egyptian Civilization*, London, New York, 1928.
Morgan, Jacques de, *Prehistoric Man*, tr. J. H. Paxton and V. C. C. Cullum, New York, 1925.
Moritz, R. E., *Memorabilia Mathematica*, New York, 1914.
Mottelay, P. F., *Bibliographical History of Electricity and Magnetism*, London, 1922.
Muir, M. M. P., *The Story of Alchemy and the Beginnings of Chemistry*, London, 1902.
Mumford, Lewis, *Technics and Civilization*, New York, 1934.

Needham, Joseph, *History of Embryology*, Cambridge, 1934.
Newberry, J. S., "The Prehistory of the Alphabet," *Harvard Studies in Classical Philology*, **45**, 105–156, 1934.
Newsholme, Sir A., *Evolution of Preventive Medicine*, Baltimore, 1927.
——, *Story of Modern Preventive Medicine*, Baltimore, 1929.
Newton, Sir Isaac, *Mathematical Principles of Natural Philosophy and System of the World*, tr. M. Motte, ed. F. Cajori, Berkeley, Calif., 1934.
——, *Opticks*, Ed. 4 reprinted, London, New York, 1934.
Nordenskiöld, Erik, *History of Biology*, New York, 1928.

Osborn, H. F., *From the Greeks to Darwin*, Ed. 2, New York, 1899.
——, *Men of the Old Stone Age*, Ed. 3, New York, 1921.
Osler, Sir William, *The Evolution of Modern Medicine*, New Haven, 1921.

Partington, J. R., *The Composition of Water*, London, 1928.
——, *Origins and Development of Applied Chemistry*, London, 1935.
——, *Short History of Chemistry*, London, New York, 1937.
Peake, H. J., *Early Steps in Human Progress*, London, Philadelphia, 1933.
Pearson, Karl, *Grammar of Science*, Ed. 2, London, 1900, Ed. 3, Part 1, Physical, 1911.
Peattie, D. C., *Green Laurels, The Lives and Achievements of the Great Naturalists*, New York, 1936.

Petrie, Sir F., *Measures and Weights*, London, 1934.
Planck, Max, *The Philosophy of Physics*, tr. W. H. Johnston, New York, 1936.
Pliny, *Natural History*, tr. H. Rackham, Cambridge, Mass., 1938+, 10 vols. ***
Poggendorff, J. C., *Biographisch-literarisches Handwörterbuch zur Geschichte der exacten Wissenschaften*, Leipzig, 1863–1931.
Poincaré, Henri, *Foundations of Science*, tr. G. B. Halsted, New York, 1921.

Rádl, E., *History of Biological Theories*, tr. E. J. Hatfield, Oxford, 1930.
Randale, J. H., *Making of the Modern Mind*, Boston, 1926.
Rashdall, Hastings, *The Universities of Europe in the Middle Ages*, New Ed., Oxford, 1936.
Read, John, *Prelude to Chemistry, an Outline of Alchemy*, London, 1936, New York, 1937.
Redi, Francesco, *Experiments on the Generation of Insects* (1688), tr. M. Bigelow, Chicago, 1909.
Riesman, David, *The Story of Medicine in the Middle Ages*, New York, 1935.
Roger Bacon, Essays by Various Writers, ed. A. G. Little, Oxford, 1914.
Rolleston, Sir H., *Internal Medicine* (Clio Medica, IV), New York, 1930.
Rolt-Wheeler, F. W., *Monster-hunters*, Boston, 1916.
Ross, W. D., *Aristotle*, Ed. 2, London, 1930.
Routledge, Robert, *Discoveries and Inventions of the Nineteenth Century*, London, 1876.
Royal Society, *Catalogue of Scientific Papers 1800 to 1900*, London, 1867–1925, 19 vols.
Russell, Bertrand, *The Scientific Outlook*, London, 1931.
Russell, E. S., *Form and Function: . . . History of Animal Morphology*, London, 1916.
Russell, H. N., *The Solar System and Its Origin*, New York, 1935.
Rutherford, Ernest, Lord, *The Newer Alchemy*, New York, 1936.

Sachs, Julius von, *History of Botany*, tr. H. E. F. Garnsey, Oxford, 1890.
Sandys, Sir J. E., *Harvard Lectures on the Revival of Learning*, Cambridge, Mass., 1905.
Sarton, George, *The History of Science and the New Humanism*, New York, 1931.
——, *Introduction to the History of Science*, Washington, 1927–31, 2 vols in 3.
——, *The Study of the History of Mathematics*, Cambridge, Mass., 1936.

Sarton, George, *The Study of the History of Science*, Cambridge, Mass., 1936.

Schmidt, R. R., *The Dawn of the Human Mind, a Study of Palaeolithic Man*, tr. R. A. S. Macalister, London, 1936.

Science and Civilization, Essays, ed. F. S. Marvin, Oxford, 1923.

Seneca, *Physical Science in the Time of Nero*, tr. John Clarke, London, 1910.

Shipley, A. E., *The Revival of Science in the Seventeenth Century*, Princeton, 1913.

Shryock, R. H., *The Development of Modern Medicine*, Philadelphia, 1936.

Sigerist, H. E., *The Great Doctors*, tr. E. and C. Paul, New York, 1933.

Singer, Charles, *The Discovery of the Circulation of the Blood*, London, 1922.

——, *The Evolution of Anatomy*, London, 1925.

——, *From Magic to Science*, London, 1928.

——, *Greek Biology & Greek Medicine*, Oxford, 1922.

——, *Short History of Biology*, Oxford (*Story of Living Things*, New York), 1931.

——, *Short History of Medicine*, Oxford, 1928.

Smiles, Samuel, *Lives of the Engineers*, London, 1857–65, 5 vols., New Ed., 1874–79.

Smith, D. E., *History of Mathematics*, Boston, 1923–25, 2 vols.

——, *Rara Arithmetica*, Boston, 1908.

——, and J. Ginsburg, *Numbers and Numerals*, New York, 1937.

——, and L. E. Karpinski, *The Hindu-Arabic Numerals*, Boston, 1911.

Smith, Preserved, *History of Modern Culture*, New York, 1930, 2 vols.

Soddy, Frederick, *Matter and Energy*, New York, 1912.

Stillman, J. M., *The Story of Early Chemistry*, New York, 1924.

Strabo, *The Geography of Strabo*, tr. H. L. Jones, London, Cambridge, Mass., 1917–32, 8 vols. * * *

Stubbs, S. G. B., and E. W. Bligh, *Sixty Centuries of Health and Physick*, London, 1931.

Studies in the History and Method of Science, ed. Charles Singer; Oxford, Vol. 1, 1917; 2, 1921.

Sullivan, J. W. N., *Science, a New Outlook*, New York, 1935.

Swann, W. F. G., *Archeology of the Universe*, New York, 1934.

Taylor, Clara M., *The Discovery of the Nature of the Air, and of Its Changes During Breathing*, London, 1934.

Taylor, H. O., *Deliverance: The Freeing of the Spirit in the Ancient World*, New York, 1915.

Taylor, H. O., *The Mediaeval Mind*, New York, 1930, 2 vols.
——, *Thought and Expression in the 16th Century*, New York, 1930.
Theophrastus, *Enquiry into Plants*, tr. Sir Arthur Hort, London, Cambridge, Mass., 1916. ***
Thompson, R. C., *The Assyrian Herbal*, London, 1924.
——, *A Dictionary of Assyrian Chemistry and Geology*, Oxford, 1936.
Thomson, Sir J. A., *The Great Biologists*, London, 1932.
Thomson, Sir J. J., *Recollections and Reflections*, New York, 1937.
Thorndike, Lynn, *History of Magic and Experimental Science During the First Thirteen Centuries of Our Era*, New York, 1923–34, 4 vols.
——, *Science and Thought in the Fifteenth Century*, New York, 1929.
Thureau-Dangin, F., *Textes Mathématiques Babyloniens*, Leiden, 1938.
Tilden, Sir W. A., *Famous Chemists*, London, New York, 1921.
Todhunter, Isaac, *History of the Mathematical Theory of Probability from the Time of Pascal to that of Laplace*, Cambridge, 1865.
Trattner, E. R., *Architects of Ideas*, New York, 1938.
Tropfke, Johannes, *Geschichte der Elementar-Mathematik in systematischer Darstellung*, Berlin, 1921–24, 7 vols.
Turnbull, H. W., *The Great Mathematicians*, London, 1929.
Tyndall, John, *Faraday as a Discoverer*, New York, 1901.

Ullman, B. L., *Ancient Writing and Its Influence*, New York, 1932.
Usher, A. P., *History of Mechanical Inventions*, New York, 1929.

Van Loon, Hendrik, *The Story of Inventions*, Garden City, 1934.
Vitruvius, *On Architecture*, tr. Frank Granger, London, Cambridge, Mass., 1931. ***

Walden, J. W. H., *Universities of Ancient Greece*, New York, 1909.
Walker, Helen M., *Studies in the History of Statistical Method*, Baltimore, 1929.
Walker, M. E. M., *Pioneers of Public Health*, Edinburgh, 1930.
Waterfield, R. L., *A Hundred Years of Astronomy*, New York, 1938.
Weidlein, E. R., and W. A. Hamor, *Science in Action*, New York, 1931.
Westway, F. W., *The Endless Quest, Three Thousand Years of Science*, London, 1934.
Wethered, H. N., *The Mind of the Ancient World: A Consideration of Pliny's Natural History*, London, 1937.
Whewell, William, *History of the Inductive Sciences from the Earliest to the Present Time*, Ed. 3, London, 1857, 3 vols.
White, A. D., *History of the Warfare of Science and Theology*, New York, 1910.

Whitehead, A. N., *Introduction to Mathematics*, New York, 1911.
——, *Science and the Modern World*, New York, 1925.
Wolf, A., *History of Science, Technology, and Philosophy in the 16th and 17th Centuries*, London, 1935.
——, *History of Science, Technology, and Philosophy in the Eighteenth Century*, New York, 1939.
Woolley, C. L., *The Sumerians*, Oxford, 1928.
The World and Man as Science Sees Them, ed. F. R. Moulton, Garden City, 1937.

Zeno of Elea, *A Text*, tr. H. D. P. Lee, Cambridge, New York, 1936.
Zittel, K. von, *History of Geology and Palaeontology to the End of the Nineteenth Century*, tr. (Mrs.) M. M. Olgivie-Gordon, London, 1901.

BIOGRAPHY

Bradford, Gamaliel, *Darwin*, Boston, 1926.
Brewster, David, *Memoirs of Sir Isaac Newton*, Edinburgh, 1855.
Cardan, Jerome, *The Book of My Life*, tr. J. Stoner, New York, 1930.
Cheyne, Sir W. W., *Lister and His Achievement*, London, 1925.
Compton, Piers, *The Genius of Louis Pasteur*, London, New York, 1932.
Dickinson, H. W., *James Watt, Craftsman and Engineer*, Cambridge, New York, 1936.
Dreyer, J. L. E., *Tycho Brahe*, Edinburgh, 1890.
Duclaux, Émile, *Pasteur, The History of a Mind* (1896), tr. E. F. Smith and F. Hedges, Philadelphia, 1920.
Fahie, J. J., *Galileo, His Life and Work*, London, 1903.
Foster, Mary L., *Life of Lavoisier*, Northampton, 1926.
Goebel, K. von, *Wilhelm Hofmeister, Work and Life*, tr. H. M. Bower, London, 1926.
Green, E. L., *Carolus Linnaeus*, Philadelphia, 1912.
Gorgas, Marie D., and B. J. Hendrick, *William Crawford Gorgas*, Garden City, 1924.
Iltis, H., *Life of Mendel*, tr. E. and C. Paul, New York, 1932.
Jackson, B. D., *Linnaeus*, London, 1923.
Jordan, E. O., G. C. Whipple, and C-E. A. Winslow, *A Pioneer in Public Health, William Thompson Sedgwick*, New Haven, 1924.
Koenigsberger, L., *Hermann von Helmholtz*, tr. F. A. Welby, Oxford, 1906.
McKie, Douglas, *Antoine Lavoisier, the Father of Modern Chemistry*, London, 1935.

Malloch, A., *William Harvey*, New York, 1929.
Mégroz, R. L., *Ronald Ross*, London, 1931.
More, L. T., *Isaac Newton*, New York, 1934.
Morley, Henry, *Jerome Cardan*, London, 1854, 2 vols.
Olmstead, J. M. D., *Claude Bernard, Physiologist*, New York, 1938.
Pearson, Karl, *The Life, Letters and Labours of Francis Galton*, Cambridge, 1924.
Peterson, H., *Huxley, Prophet of Science*, London, 1932.
Poulton, E. B., *Charles Darwin and the Theory of Natural Selection*, New York, 1896.
Power, D'Arcy, *William Harvey*, London, 1897.
Prowe, L. F., *Nicolaus Coppernicus*, Berlin, 1883–84.
Sullivan, J. W. N., *Isaac Newton*, New York, 1938.
Vallery-Radot, R., *Life of Pasteur*, tr. Mrs. R. L. Devonshire, London, New York, 1902, 2 vols.
West, Geoffrey, *Charles Darwin, a Portrait*, New Haven, 1938.
Wyatt, R. B. H., *William Harvey (1578–1657)*, London, 1924.

SOURCE BOOKS

Alembic Club Reprints, Edinburgh, 1–17, 1893–1907.
Brunet, Pierre, *Histoire des Sciences, Antiquité*, Paris, 1935.
Cambridge Readings in the Literature of Science, arranged by W. C. D. Whetham and Margaret D. Whetham, Cambridge, 1924.
Fulton, J. F., *Selected Readings in the History of Physiology*, Springfield, Ill., 1930.
Harvard Classics, ed. C. W. Eliot, Vol. 38, *Scientific Papers;* New York, 1897.
Knickerbocker, W. S., *Classics of Modern Science (Copernicus to Pasteur)*; New York, 1927.
Long, E. R., *Selected Reading in Pathology from Hippocrates to Virchow*, Springfield, Ill., 1929.
Magie, W. F., *Source Book in Physics*, New York, 1935.
Mather, K. F., and S. L. Mason, *Source Book in Geology*, New York, 1939.
Neudrucke von Schriften und Karten über Meteorologie und Erdmagnetismus, ed. G. Hellmann, Berlin, Heft 1–15, 1893–1904.
Ostwald's Klassiker der exakten Wissenschaften, Leipzig, 1–196, 1889–1922.
Scientific Memoirs, ed. J. S. Ames, New York, 1898–1902, 15 vols.
Shapley, Harlow, and Helen E. Howarth, *Source Book in Astronomy*, New York, 1929.
Smith, D. E., *Source Book in Mathematics*, New York, 1929.

JOURNALS

Ambix, Journal of the Society for the Study of Alchemy and Early Chemistry, ed. F. S. Taylor, London (quarterly), 1, 1937 +.

Annals of Medical History, ed. F. R. Packard, New York (bimonthly), 1, 1917 +.

Annals of Science, ed. D. McKie, London (quarterly), 1, 1936 +.

Archeion, ed. Aldo Mieli, Rome, Paris (quarterly), 1, 1919 +.

Bibliotheca Mathematica, ed. Gustaf Eneström, Leipzig, 1884–1915.

Bulletin of the Institute of the History of Medicine, *see* Johns Hopkins University.

Isis, *Organ of the History of Science Society*, ed. G. Sarton, Bruges (quarterly), 1, 1913 +.

Janus, *Archives internationales pour l'histoire de la Medicine*, ed. H. F. A. Reypers, Haarlem, 1, 1896 +.

Johns Hopkins University, Institute of the History of Medicine, *Bulletin*, ed. H. E. Sigerist, Baltimore (quarterly), 1, 1913 +.

Medical Classics, ed. E. C. Kelly, Baltimore (monthly), 1, 1936 +.

Osiris, ed. G. Sarton, Bruges (irregular), 1, 1936 +.

Quellen und Studien zur Geschichte der Mathematik, Astronomie und Physik, ed. O. Neugebaur, Berlin (irregular), 1, 1929 +.

Royal Society of Medicine, *Proceedings*, *Section of the History of Medicine*, London (monthly) 1, 1907 +.

Scripta Mathematica, New York (quarterly), 1, 1932 +.

Society of Medical History of Chicago, *Bulletin*, Chicago (irregular), 1, 1911 +.

Index

This book has been produced in line with the EU GPSR guidelines about the safety of products.

The General Product Safety Regulation is the European Union's updated framework for ensuring that all consumer products, including books, are safe for consumers.

This book has been printed by CPI books GmbH. The printer has issued safety certificates for the materials - like ink, paper and glue - being used.

The product identifier is: 9786256629301

The author is responsible for the content of the book, is the publisher of the works and bears full responsibility for it.

The book has been published via Bookmundo. Bookmundo enables any author to share their stories with the rest of the world via printed books and ebooks and a broad distribution network.

Bookmundo will act as an intermediary in regard to questions about safety and will address them to the printer / author. Should there be any question in regard to the safety of the product, please contact us.

Bookmundo
Delftsestraat 33
3013AE Rotterdam
The Netherlands
info@bookmundo.com

Kolibri 360 GmbH
Ferdinand-Jühlke-Straße 7
99095 Erfurt, Deutschland
produktsicherheit@kolibri360.de